2nd International Conference on Comprehensive Art and Cultural Communication (CACC 2023)

SHS Web of Conferences Volume 167 (2023)

Wuhan, China
14 – 16 April 2023

Editors:

Luqman Lee
Zhihuang Yan
Afizan Bin Azman

ISBN: 978-1-7138-7346-4

Printed from e-media with permission by:

Curran Associates, Inc.
57 Morehouse Lane
Red Hook, NY 12571

Some format issues inherent in the e-media version may also appear in this print version.

For additional information, please contact EDP Sciences – Web of Conferences
at the address below.

EDP Sciences – Web of Conferences
17, Avenue du Hoggar
Parc d'Activité de Courtabœuf
BP 112
F-91944 Les Ulis Cedex A
France

Phone: +33 (0) 1 69 18 75 75
Fax: +33 (0) 1 69 28 84 91

contact-edps@webofconferences.org

Additional copies of this publication are available from:

Curran Associates, Inc.
57 Morehouse Lane
Red Hook, NY 12571 USA
Phone: 845-758-0400
Fax: 845-758-2633
Email: curran@proceedings.com
Web: www.proceedings.com

TABLE OF CONTENTS

The Feminist View on Maud Watts' Awakening in the Film Suffragette 1
Yang Binzhe

An Analysis of the Three Realms of Musical Performing Arts in Bel Canto 5
He Hao, Ng Elena

Discussion on the Cultural Differences of Love and Marriage Between China and Japan from Their Television Dramas—— Taking the First Half of My Life and We Married as a Job as Examples........ 10
Su Xueying

Overview of Artificial Intelligence Painting Development and Some Related Model Application........ 15
Liu Mingyang

Content Research Strategy of "Three Agricultural Videos" in the Internet Era – Take the Example of the Three Farmers Food Skills Category........ 20
Zhang Yimeng

An Analyze on Intersection of Two Female Figures in the Film in the Mood for Love and an Interpretation of Its Metaphor in Cultural Identity 24
Li Qi

The Value and Character of Music as a Tool for Political Mobilization........ 28
He Yulin

Background of Western Modern Art in the 20th Century 32
Liao Lei

A Comparative Study of Two Architectural Reconstruction Design of Casa Batlló........ 36
Lu Yang

The Inner Dynamics of Kidult Culture: Trending Toys and Games, Theme Parks, and Animated Movies 40
Gu Peiying, Liang Wenrui

A Preliminary Exploration of the Cross-Cultural Production, Performance and Dissemination Models of Contemporary Chinese Adaptations of Shakespeare's Plays (1990-2020)........ 45
Shanghua Huangfu

Leaving and Death: The Awakening of Individual Modernity in Ethnic Minority films——Based on the Reading of "A Mongolian Tale" and "Jinpa"........ 49
Cui Kaixiang

The Return of Actors' Body Essence in Theaters 53
Zhou Shiran

Film Literary Criticism and Image Aesthetics: Around Film Criticism........ 57
Wang Yunning

A Narratological Study of Streaming Science Fiction Films: The Case of "Dune" 61
Zhou Wei

Research on Field Investigation Methods of Folk Drama 66
Qiao Huhe

A Brief Analysis of the Evolution of Audio Language Communication in the Internet Age 70
Zhang Huiwen

A Short Film Interpreting the Chinese Characters: 'Trace Back' 75
Yu Zhe, Chung Szu-Ming

Research and Analysis of Audiovisual Art Based on the Background of New Media 79
Diao Guanwen, Zhao Huiwen

Research and Analysis of Interaction in Virtual Reality Films in the Context of New Media 84
Zhang Long, Li Zecheng, Gahb Jahng Surng

Research and Dynamic Analysis of Interactive Methods of Film in the Context of New Media 89
Li Jinhao, Liu Jie, Gahb Jahng Surng

Current Situation of TPACK for Art Vocational Teachers: Problem Analysis and Countermeasures 94
Zhao Tao, Li Shijian, Zhang Xiujuan

Analyse the Influence of "cultural and Creative + Blind Box" Mode on the Commercialization of Chinese Traditional Culture IP 100
Luo Yining

Research on the External Dissemination of Yao Pangwang Festival Culture Based on the Belt and Road Initiative 104
Li Zhangyan

Retro Futurism: The Resurgence of Y2K Style in the Fashion Field 108
Yang Xiaochun

Cultural Heritage Game Design Based on the Collective Memory Reconstruction Model 113
Liu Wenqian, Nie Xiaomei

A Study on Visual Innovation of Macau Souvenirs Packaging in the Context of Multicultural Communication 121
Zheng Yue, Li Xianlu, Zhang Caiyu, Wang Xuan

Speech Signal Processing and Analysis Based on Mongolian Long-Tune Folk Songs 126
Hu Sile, Liu Shu, Ma Shuzhen, Hu Axu

The Contemporary Application of Confucius' Being Carefree 131
Xing Enhui

A Comparative Study of Chinese and Japanese Folk Culture 135
Li Ling, Na Cui

Application of Green Plant Elements in Office Space 139
Wang Wenchao, Chi Jiaqi, Niu Siqi

Research on Smart Home Interior Design in Youth Living Space 144
Chi Jiaqi, Zhao Mingcen

Analysis of the Application of Light-Transmitting Concrete in Interior Design 148
Wang Wenchao, Chi Jiaqi, Niu Siqi

Analysis the Character of J.S.Bach's Sinfonia in a Major (BWV 798) – Based on the Dynamic and Research on Related Acoustic Parameters 152
He Zixuan

A New Path of Visual Communication from the Perspective of the Metaverse —— Take Liu Yexi's Internet Communication Spectacle as an Example 158
Li Dingtai, Liang Tingwei

Research and Analysis of the Development of News Film and Television Based on the Integrated Media Environment 163
Tian Yi Nong, Diao Guanwen

Optimization Study of Media Creation Ideas Based on the New Media Environment 167
Zhao Hui Wen, Tian Yi Nong

Heterogeneous Isomorphism - Spatial Gene Transcreation Design for Traditional Villages in Hunan 171
Lin Jin, He Chuan, Shi Shuang, Song Jiongfeng

Food as a Medium: The Framework and Cross-Culture Implications of Short Videos of Overseas Chinese Sharing Chinese Food in Foreign Countries 178
Zhang Keke

The Construction of Cross-Cultural Communication Strategies Amid the Interaction of Narrative Subject——the Identification of the Other and the Telling of the Self in a Long Cherished Dream 182
Xing Fenglin, Li Yi

A Review and Prospects of Chinese Character Research in the Last Thirty Years 187
Niu Yunbo

"Large Fat Sighings" of a "large Bulky Figure": Identity-Building and Formation of New Order in Persuasion 196
Zhang Yukang, Wei Mengyang

An Analysis of the Protection and Inheritance of Intangible Cultural Heritage from the Perspective of Cultural and Creative Industries: A Case Study on the "Huojing" in Bashu 202
Gan Xili

Author Index

The Feminist View on Maud Watts' Awakening in the Film Suffragette

Binzhe Yang [1,*]

[1]Lanzhou University

Abstract. Beginning in the 20th century, the philosophy of the British women's suffrage movement evolved in a manner that encouraged the awakening of more women. From the middle-class women of the 19th century, the group of participants in the movement steadily shifted below, and the action for women's suffrage grew in popularity. Suffragette uses the feminist awakening of the fictional protagonist, Maud Watts, to illustrate the immense fight and sacrifice women made for the right to vote at the time. The usual character of Maud Watts is quite valuable for research purposes. The great battle of women suffragists for the right to vote might be better comprehended by examining her transformation from apathy to spontaneous pursuit of women's rights.

1 Introduction

Britain was the first country to embark on the path of industrialization and urbaniza-tion which divide women into three classes and each with its own task in the service of men. Upper-class women were the least numerous and least affected by the indus-trial revolution of British women's groups. They did not have to care for the house-hold; all they had to do was dress themselves every day for men to enjoy and play with. Lower middle-class women who needed to earn a living on their own usually worked as governesses for wealthy families, a unique occupation between house-wives and servants in British society. However, society required a large number of cheap women from the lower classes as laborers[1]. As a result of the industrial revolution, there were more factories, which required more labor to keep up with demand. Women made up a sizable share of the workforce in factories because, in a patriarchal society, female labor was inexpensive and easy to manipulate. Simultaneously, industrial change precipitated a crisis in class and gender relations. The contradictory and reconciled interests of gender differences in labor division evolved alongside the industrialization process. Women's salaries for the same number of hours are much lower than men's, and the age of entry is younger.

Suffragette is based on Maud Watts, a laundrywoman, and WSPU marches. It tells the story of the great battle and sacrifices made by women suffragists for the right to vote in Britain at the beginning of the 20th century. Maud Watts, the 'angel in the house'[2] who saw nothing wrong with patriarchal power, is shaken out of her lethargy. When her numb existence is broken and she is denied her rights, including custody of her child, she speaks up for her rights and the rights of women in society.

The film itself is a microcosm of the living conditions of women factory workers in early 20th-century Britain. Society is overwhelmed by the idea of the superiority of men over women and the most concrete manifestation of which is the legal nullity of the wife in married life. The English common law practice of coverture meant that a woman's possessions and herself passed to her husband at marriage[3]. "Husband and wife are the same person." Maud Watts' awakening path shows how a lower-class woman gains self-awareness and fights an unjust society. Maud Watts, the most symbolic traditional and virtuous female icon from the outset, is constantly given new meanings to help us understand the suffragists' brave campaign for women's right to vote.

2 Early twentieth-century research on women's suffrage in the United Kingdom

The history of the women's suffrage movement in Britain was developed almost simultaneously with the movement, largely through propaganda and participant memoirs. Ray Strachey's The Cause 1928 and Sylvia Pankhurst's The Suffragette Campaign 1931 are two key historical examinations of the women's suffrage movement.

Andrew Rosen's 1974 book The Militant Campaign of the Women's Social and Political Union 1903-1914 praised the Pankhursts' role in the women's movement. In her book on the politics of the women's movement, she used archive information to illustrate the importance of constitutionalism.

Since the 1980s, women's suffrage scholarship has flourished. Both Marion Ramelson and Sheila

* Corresponding author: binzhesyydnr@outlook.com

Rowbotham describe the women's movement's class distribution and tensions. Hannah Mitchell's autobiography, The Hard Way Up, addresses socialists and middle-class suffragettes' solidarity and collaboration, as well as their internal tensions and the formation of the Women's Freedom League.

3 Flim Suffragette

The Making of a Feature Film Suffragette was commissioned by June Purvis (Editor of Women's History Review) and written by the film's director, Sarah Gavron, to ad-dress the film's choice of WSPU women's groups, as well as the creative process and creative thinking.

Brain Mcfarlane compares historical film and television studies on the topic of women's suffrage in No Surrender, demonstrating that Suffragette was the first film to realistically confront the struggle for women's suffrage, discussing the film's significance in terms of both content and subject matter.

Deeds, Words, and Drama: A Review of the Film Suffragette by Gwen Seabourne, published in 2016, discusses the reasons for the film Suffragette's choice to depict working-class women, as well as the reasons and significance of the selection of historical footage from the WSPU, primarily through oral research with director Sarah Gavron.

Ana Stevenson's Screening Women's History in the Film Suffragette: Between Intersectional Feminist Activism and Historical Memory, published in 2022, employs Jemima Repo's con-cept of feminist commodity activism to locate t-shirts in the film's marketing process.

The author believes that examining the journey of the heroine Maud Watts' awakening will enrich the understanding of the film and related historical facts, as well as better appreciate the struggles and great sacrifices of the suffragists.

4 Maud Watts, a suffragette pushed by the times

The story of the film Suffragette takes place in 1913 England. The dominant British norms of women's roles and responsibilities in the mid to late nineteenth century were primarily characterized by traditional customs and tradition-al gender perceptions of men's superiority over women, men as the lords and rulers of women, and women as subordinate to men, with no separate identity or status. In social connections, women are often submissive to males, and in economic ties, mar-ried women have no property rights and may solely rely on men[3]. Women have no spiritual autonomy, and their dads or husbands govern their marriage, job, and edu-cation. It was a situation in which the male was in command and the woman was in charge of the allocation of responsibilities.

Mary Wollstonecraft and John Stuart Mill influenced 19th-century British women's ideas. She critiques mainstream culture, focusing on the oppression of women by a patriarchal society. Male dominance and restraints have left women dumb and foolish, compelling them to use masculine speech. Men and women have equal rationality, according to Mill. Second, the 19th-century division of labor and gender injustice resulted from men maximizing their own interests. At the beginning of patriarchal society, physical distinctions were viewed as the rule of men's dominance over women, therefore the legal system was based on their relationships[4]. Mary and Mill's principles empow-ered weak women and inspired the women's suffrage movement in 19th-century Britain.

At the beginning of the 20th century, the women's suffrage movement in Britain transitioned from being led by middle-class women to a more popular group, and its motto changed from 'Women's Suffrage' to 'Votes for Women.'[5]. It brought more women into the voting rights fight. Maud's development and progress during this time period are influenced by three factors: the direction of the women around her, those with a vested stake in male dominance, and her self-reflection and internal analysis.

4.1 Women's Guidance

Without aroused women around her, the heroine couldn't become a suffragette. Maud helps Miller who was late to work because of involving in sabotage the day before when the boss censure her, so Miller informs her about suffragist meetings. When Miller convinced Maud to join, she responded, "That's unethical." "Morality strangles," Miller said. "Do you want me to obey the law? Then make the law honor-able and respectable." Miller invites Maud to hearings on the campaign for women's right to vote. Miller couldn't speak since her husband beat her the day before so Maud had to represent Miller. "I began at 7. I started working at 12. 17-year-old washer, 20-year-old foreman, now 24." When a politician commended her early suc-cess, she responded, "Women don't get to be laundresses for long. You suffer chest pains and coughing, crushed fingers, ulcers, burns, and migraine-inducing exhaust fumes. A girl couldn't work last year due to poisoned lungs." Capitalism and patriar-chy influence the social gendering of the division of labor, where men and women are unequal in the type of labor they perform, the level of treatment they receive, and their participation in labor-related activities, positioning women as a lower paid, less skilled, and less skilled labor force. This prompts Maud to reflect, and as her testi-mony demonstrates, she is progressively becoming conscious of societal injustice, as opposed to initially defending the pro-male bosses. Miller is the mentor who gradually leads Maud to the road of the women's suffrage movement.

Edith Ellyn, a female doctor at the clinic where Women's Suffrage campaigners met for weekly discussions and gatherings, was the one who organized many of the groups Maud eventually joined. She explained to Maud t her husband had never studied chemistry and that his father wanted him to be a pharmacist, but she was highly interested in medicine. The unfairness against women is seen throughout society, where dads and spouses have the power of speech, and education for women is considered a luxury. Edith told Maud that women would not be able to attain equal rights if they did not fight at every arrest and jail. She is a doctor with a caring spouse, yet she still strives for equal rights for women across society.

Emmeline Pankhurst, the WSPU's founder, retains a special place in the hearts of all women's suffrage supporters. The film performs the most crucial role in its three-minute presence. Mrs. Pankhurst's words reach home in the film, and Maud's eyes light up as she listens to her as if for the first time, as if she had embraced hope. As she states later in the game with the Inspector, "If males have the right to fight for freedom, why not women?" She is working hard to become a legislator.

4.2 Oppression from the patriarchy

The term 'patriarchy' stems from the ancient notion and has two meanings: first, the rule of women by men in a social system; second, the domination of the family by husbands and fathers provides the political system its character. The political signifi-cance of patriarchy evolved in the 17th century, and the family questioned the mon-archy's authority, which Parliament had entrusted to the patriarchal line. Society believed men and fathers had God-given power over wives, children, and servants[6].

Taylor, Maud Watts' boss, is a typical guy with a vested interest in society. He sneers at Miller who wants to fight for women's voting rights and molests Miller's 12-year-old daughter Madge. He sits on a pedestal of male power and mocks the women. Taylor tells her that Madge reminds him of herself when she was younger. 12-year-old female workers are sexually abused by their bosses with little recourse, and it makes her realize how accustomed and numb she was to it all before. When Maud's employer ambiguously touches her waist again after her release from prison, she finally fights back, stops repressing it, takes an iron to her boss's hand, and quits her job.

Maud's husband is Sonny Watts. "He rejects his wife's activity and is outwardly cruel but inwardly tortured by the repercussions," director Gavron says[7]. He's stuck in his era's customs and afraid of societal consequences. Sonny Watts is also victim-ized, but he oppresses women. After his wife was imprisoned, he worried for himself and agreed with the stories that drove Maud Watts out of the house. Maud, anxious about her husband and neighbors' weird looks, removed her medals as soon as the WSPU presented them. Maud couldn't care for her son when she left but her husband couldn't even change his pajamas for school, so he sold him to an affluent family. Maud began WSPU activism after selling his son. Maud understood she had to break the law since women had no custody rights and children were considered husbands' property in early 20th-century England.

Even the women mock at the suffragists. On Maud's release night, neighbors shut their doors and windows. A woman approaches her with the newspaper and says, "Shame on you." According to John Stuart Mill, men's privilege has made women weak and subservient at home and in society by limiting their ability to dispose of property, raise children, choose a job, etc[8]. Women have absorbed this as their moral code, therefore they don't protest it because laws don't support them. As a result, this group of women feels embarrassed of Maud.

Whether they come from the working class or the aristocracy, men dominate society. Regardless of whether they experience oppression in class society, men hold the most authority in matters of gender. Women were also expected to educate them-selves on how to be good moms and wives in addition to listening to what males had to say. Maud is a brand-new person who is struggling to come to. She is brave even though she transitions from numbness to resistance. She decides to oppose the vested interests of a patriarchal society in response to the unfairness and lack of women's rights. Such an image is also profoundly instructive for the reflection of modern women.

4.3 Maud's Internal examination in a patriarchal culture

Maud evolves from opposition to hesitancy to suffragist passion. Initially, she thought Miller's and others' actions were unlawful and unethical so she was hostile to WSPU activities. When her husband asked her whether she was sure to enter the ac-tivity, she answered, "Only to look." Maud denied being a suffragist when arrested and jailed. After witnessing abuses, she began to contemplate her own situation and the predicament of women. Maud asked her husband early one morning about how their daughter's life would be if they had one. "Same as you," her spouse replied. This made her realize that the situation of women's affairs affects millions of women's destinies. The next generation of women will suffer if they don't resist.

Maud was arrested and imprisoned for an act of activity after becoming a deter-mined suffragette, and while imprisoned, she went on a hunger strike, as had her pre-decessors, and was force-fed by the jail officials. Many women were brutalized dur-ing the history of force-feeding: "Whether given from a cup, via the nostrils with a tube (the most popular way) or down the throat into the stomach (the most excruci-ating method), suffragists toiled alone and often with dread of mental or bodily injury"[9]. Maud overcame such harsh experiences, and her inner will experience a full transformation.

5 Conclusion

As one of the main sources of labor in a divided domain without active status, women are submissive to the "angel in the house" who sacrifices her status and rights for the family and country. In patriarchal societies, women have been crowned 'angels of the house,' so much so that we cannot deny their presence. However, in British culture and law, women were 'non-existent' beings with no rank or place. The husband was the lone 'owner' of his wife and children as the head of a flourishing household[10]. Maud Watts ponders what is happening to women, spurred by the women around her who are also suffering. She is furious by the unrelenting pressure imposed on women by people with a vested interest in male authority. As she thought, she became more eager to join the suffragettes.

Maud Watts, a fictional suffragette, epitomizes early 20th-century Britain's femininity. Maud Watts' walk to enlightenment shows the sacrifices suffragists took to win the vote. Although women ultimately got the vote, we still have a long way to go in terms of economic inequality and societal perspectives, things Maud Watts of today must fight for.

6 Author's contribution

A better understanding of the film and the great struggle and sacrifice made by suffragists for the right to vote by examining the awakening of feminist consciousness of Maud Watts, the virtual protagonist in the film Suffragette.

Acknowledgements

I'd like to thank my thesis supervisor, Hongmei Pang of Tsinghua University's Language Teaching Centre. Mrs. Pang gave me assistance and recommendations on the research direction of my thesis and provided timely counsel on the difficulties and doubts I had while writing it, as well as numerous excellent suggestions for improvement. I also thank Mr. Zhou Hongxu for his encouragement and help. Last but not least, I'd want to thank my parents for their emotional and material support during the pandemic and for spending a stressful and joyful period with me.

References

1. Berg, M. (1993). Women's property and the industrial revolution. The Journal of Interdisciplinary History, 24(2): 233-250.
2. Coventry, P. (1866) The angel in the house. Macmillan, London.
3. Tabili, L. (2002) Poor Women's Lives: Gender, Work, and Poverty in Late-Victorian London. Victorian Studies. Indiana University Press, Bloomington. 44(2): 339-340.
4. Wollstonecraft, M. (2014) A Vindication of the Rights of Woman. Yale University Press, London.
5. Rosen, A. (2013) Rise Up, Women!: The Militant Campaign of the Women's Social and Political Union. Routledge, London.
6. Dahlerup, D. (2018) Confusing concepts–confusing reality: a theoretical discussion of the patriarchal state. In Women and the State. Routledge, London. 93-127.
7. Gavron, S. (2015) The Making of the Feature Film Suffragette. Women's History Review, 24(6), 985–995.
8. Smith, E. S. (2001) John Stuart Mill's" The Subjection of Women": A Re-Examination. Polity, 34(2): 181-203.
9. Purvis, J. (1995) The Prison Experiences of the Suffragettes in Edwardian Britain. Women's History Review, 4(1), 103-133.
10. Caine, B. (1982). Feminism, suffrage and the nineteenth-century English women's movement. Women's Studies International Forum, 5(6), 537–550.

https://doi.org/10.1051/shsconf/202316701002

An Analysis of the Three Realms of Musical Performing Arts in Bel Canto

Hao He[1], Elena Ng [2*]

[1]Singapore Raffles Music College pte ltd. 456 Alexandra Road, Fragrance Empire Building #09-02 Singapore 119962

[2]Singapore Raffles Music College pte ltd. 456 Alexandra Road, Fragrance Empire Building #09-02 Singapore 119962

ABSTRACT: The three realms of music performance are intuition, interest, and style. Intuition is a feeling in music performance; interest is a choice that one persists in, and style is the result of effort and perception based on it. As the highest artistic achievement in the world of singing today, Bel canto (Italian: Bel canto) refers to the singing method in which the larynx blows the vocal cords with exhaled air while maintaining the inhalation position so that the open resonance cavity can resonate completely and evenly. This article, based on the research of the three realms of musical performing art, analyses the application of the three realms of musical performing art in Bel canto, aiming to let Bel canto performance better grasp the application of this art.

1 Introduction

The three realms in the art of musical performance are intuition, interest, and style. Intuition is a feeling in musical performance; interest is a choice that one persists in, and style is the result of effort and perception. As the highest artistic achievement in the world of singing today, we analyzed how the realms of the art of musical performance can be applied to the enhancement of Bel canto.

2 Review of the literature

2.1. A review of research on the state of the art of musical performance

Music is a flowing melody that brings out people's joy, anger, sadness, and happiness. Jiaxiao WU (2011)[1] believes that there are three realms in the art of music performance, of which intuition, that is, talent, is a prerequisite, strong interest is a condition, and style is the final result. Each of the three realms is indispensable. Chunyu MA(2017)[2] divides music performance into three realms, and he believes that the three become interdependent and closely related to each other, of which the first realm is the human perception of the world. Xiaomang DENG (2021)[3] proposes that life is like a performance, and performance is also an experience of life. Only when it is based on the state of human existence and inner experience can we realize the musical triad of performing art.

In conclusion, there is almost universal consensus on the state of the art of music performance, which is intuition, interest and style. Intuition is the most basic talent, the most fundamental of the three realms; interest is the choice to be able to persevere, to have a strong interest in order to have a lasting musical performance; style is the final result of continuous exploration and research. All three are interlinked and indispensable.

2.2. A review of research on Bel canto

Most of the literature by Chinese scholars on Bel canto has focused on the origins, development, and analysis of characteristics and integration of Bel canto with other singing styles. For the origin of Bel canto, a review of the literature reveals that Yirong HE (2016)[4] argues that after the introduction of Bel canto to China by castrati singers, it has now developed into a very accessible and grounded art for everyone. Shan JIANG (2017)[5] discusses the origin and development of Bel canto in the same chronological order as the former in his article The Origin and Development of Bel canto. In a superficial analysis of the origins and development of Bel canto in the first part of the 19th world, the historical role of the castrati singers in the origins and development of Bel canto is analyzed by Xiaobo MA (2017)[6]. In terms of the analysis of the characteristics and fusion of Bel canto with other singing styles, Yan LIU (2011)[7] briefly describes the comparison between Bel canto and folk singing in terms of their origins and performance styles, and he mentions the development of the fusion of Bel canto and folk singing in China. Jian ZHANG (2013)[8] compares the similarities and differences between Bel canto and folk singing. Later Jingting LI(2022)[9] describes the characteristics of the Bel canto style and explains in detail how to use the

*Corresponding author. Email: 1344159622@qq.com

vocal cords and the breath in two ways. In the literature, the three realms of music performing arts are undoubtedly intuition, interest, and style, and the study of Bel canto, in addition to the study of its origins and characteristics, is now tending towards a study of its integration with other singing styles. On the one hand, the research is repetitive, with most of the literature focusing on the three realms of English performing arts and the origins and characteristics of Bel canto and lacking in the integration of the two aspects.

3 Analysis of the three realms of musical performing arts

Musical arts performance is a complex process of applying systematic theory to cognition. Theories of musical performance have been explored in many ways in ancient and modern times, including thematic studies of performance styles, specific instruments, and vocal techniques from each period, from different performance styles, performance genres, and the improvement of performance skills, as well as the appreciation and understanding of music. According to the definition of music performance major in general Chinese colleges and universities, music performance mainly studies the basic knowledge and skills of music, instruments, dancing, performance, conducting, masters the analytical methods and deductive skills of musical works of different styles and genres, and performs music performances in musical theater (troupe), cultural and educational units, etc. Music not only cultivates emotions but also enhances temperament, improves the ability to express oneself in speech and speak confidently and boldly, improves the ability to coordinate the body and perform comfortably on stage, improves various qualities, including imagination, creativity, and imitation, improves interpersonal skills, social skills, leadership skills, organizational skills, improves artistic qualities and cultivates the emotions, and improves reading and retelling ability. In the performing arts, intuition is a prerequisite and having good talent and intuition is paramount to making art. Developing an interest in music deserves our attention and needs to be addressed. In exploring the art of musical performance, we need to look at style horizontally and vertically, a topic worthy of long-term consideration, and one of these three realms is indispensable in the art of musical performance.

3.1. Intuition in musical performance empowers the creative power of artistic thinking

Intuition is very important to the artist. The possession of intuition is a prerequisite for creation. Baidu explains that intuition is an intuitive feeling, a point of view that has not been analyzed and reasoned. Intuition refers to a particular way of thinking that is not controlled by human will. It is a form of thinking based on human occupation, experience, knowledge, and instinctive existence. It seems that intuition is common to all. In the case of the artist, however, this intuitive approach develops into an intuition of the language of art, or artistic intuition. Artistic intuition contains a deep inner psychological element, and intuition is a fundamental rational activity. It is a process of perception and profound psychological experience of the objective world. Intuition is by no means a mere perception, but a synthesis of multiple psychological functions interwoven with emotional imagination and perception.

Intuition is the intuitive feeling that a person has about something or someone; it is an opinion that can be reached for no reason at all, without any rigorous logical analysis or rigorous reasoning. Intuition is usually a special way of thinking that is not controlled by human will; it is a form of thinking that exists based on human instinct. Intuition is characterized by its rapidity, immediacy, and instinctive awareness, and deliberate development often achieves little. In the art of music performance, the musician's artistic intuition plays a huge role in his or her perception of the objective world and his or her grasp of artistic forms. It enables musicians to constantly discover and inspire their artistic thinking and constantly improve their ability to create music artistically. Intuition is thus the extraordinary ability to recognize the truth by direct and sensitive speculation, a gift that is free, flexible, contingent, and unreliable, also highly susceptible to disappearing with time.

3.2. Interest in music performance as a fundamental motivation for artistic creation

Interest is another realm of musical art performance, which, unlike intuition, is neither an uncaused feeling nor can it be swayed or dictated by the will of others. Everyone gives priority to and actively explores what interests them and shows a fascination for it. Interest is based on spiritual needs and is more than a superficial interest in things; any interest arises from the emotional satisfaction one experiences as a result of acquiring knowledge of or participating in such activities. If a person feels a strong spiritual need for something or an activity, he or she will be keen to touch it, observe it, actively engage in it and pay attention to exploring its mysteries. Interest is also linked to awareness and emotion; if one has not known something or an activity for a long time, one will not be emotionally attached to it, and therefore will not be interested in it. Conversely, the deeper the knowledge, the more intense the emotion, and the stronger the interest.

Interests are positive and tendentious, selective attitudes and emotions that an individual develops in response to particular things, activities, and people. Each individual gives priority attention to and actively explores things that interest him or her. Interest is a prerequisite for hobby, and hobby is the expansion and action of interest. Hobby is not only a mood of priority attention and desire for something but also some kind of practical action. Interest is particularly important in the art of musical performance, where it is necessary not only to master the technique but also to grasp the spiritual connotations of the art, and therefore interest plays a large role in musical performance. The musician Chopin's keen interest in Polish music and culture shows up significantly in his works for the rest of his life, and he composed such beautiful melodies for us that we still recall them to this day. Interest is therefore the basic motivation for learning music, the prerequisite for maintaining a close connection with music, enjoying it, and beautifying our lives with it, and the mental strength to take the initiative in our studies and research.

3.3 Different styles in musical performance are what express the richness of the music

It is because of interest that the art of musical performance can be expressed more richly, and a variety of styles can be developed. A style is a representative and unique appearance of a work of art as a whole. The formation of style has its own subjective and objective reasons. Subjectively, due to their different life experiences, ideological concepts, artistic qualities, emotional tendencies, personality traits, and aesthetic ideals, artists are bound to form a creative personality consciously or unconsciously in their artistic creation that is different from other artists, with relative stability and significant characteristics. Style is the natural expression and concrete manifestation of a creative personality. Objectively, the formation of an artist's creative personality is inevitably influenced by the social and historical conditions of the time, society, nationality, class, and family environment to which he or she belongs. The object of the artwork, the subject chosen, and the genre and art discipline to which the work is subordinated also have an intrinsic role in the formation of style.

The formation of musical performance art style is different from the general artistic characteristics or creative personality, it is expressed through artworks, and therefore relatively stable, and can also reflect more intrinsically and profoundly the ideology, aesthetic ideals, and spiritual temperament of the times, the nation, or the artist. This is why we often say that style is the external imprint of an artist's inner identity. The formation of a style is a sign that an artist has transcended the infantile stage in art and has freed him or herself from the constraints of various models, thus tending towards or reaching maturity, and is also the threshold at which a good musician moves from being a virtuoso to a master.

The style of musical performance is therefore the result of hard work and perception, a true connotation of the richness of artistic development. While talent is important in the study of music, it is also essential to work hard. It is only through perseverance and unremitting effort that one can reach the climax of the art of musical performance.

4 Analysis of the characteristics of the performing art of Bel canto

Over the centuries since the European Renaissance, Bel canto has gained widespread recognition and popularity throughout the world. In terms of Bel canto itself, it has been richly nourished and developed, from artistic theory to artistic practice to form a distinct artistic school and a complete artistic system.

4.1. A wide range of smooth and unified

The Bel canto is based on the breath, from the low to middle vocal area and then into the high vocal area that is supported by a strong breath. The breath support point is in the diaphragm, the diaphragm's strong toughness and support make the voice smooth and elastic. The use of the shut-off method after entering the upper register, and the change of voice leave no trace of the change of voice and greatly expands the range of the upper range. Bel canto singers are generally able to sing up to two octaves, flowery sopranos and some singers can sing up to two and a half octaves and even up to three octaves.

4.2. Beautiful tone embodies the metallic penetration

The beautiful and unified tone of Bel canto is pure and flawless, with powerful penetrating power. It is the most representative and beautiful voice in the world of the opera vocal genre, so the title of opera Bel canto is true to its name. We all know that the voice is produced by the vibration of the vocal folds. The vocal folds grow in the laryngeal apparatus, which is made up of three pieces of cartilaginous tissue that singers call the laryngeal chambers. The vocal folds produce the vocal frequencies, and the laryngeal chambers expand the pure sound source. Bel canto requires opening the laryngeal apparatus to bring it into the singing state. The vibratory function of the vocal folds' frequencies is organically controlled with consciousness, giving full play to the beautiful timbre of the vocal folds and the unity of the sound quality. The sound is expanded even more through the resonating organs such as the pharyngeal and oral cavities, which are concentrated in the head cavity to produce this wonderful sound with great penetrating power.

4.3. Stable and smooth breath is a solid foundation

Bel canto has a coherent voice, beautiful tone, wide range, pure tone quality; voice with penetrating power, but these foundations of the more important foundation is the stable and smooth breath.

4.4. The overall resonance makes the sound a very shocking effect

Another major feature of Bel canto is the overall resonance. This is distinctly different from the resonance systems of other genres, where the acoustics produced by partial resonance and overall resonance are completely different. The effect of overall resonance shows the natural charm of the human voice, which is magnetic and extremely powerful.

5 The Application of the Three Realms in the Performance of Bel canto

In terms of singing, breathing is the driving force of the voice and is the foundation of singing. The most important feature that distinguishes Bel canto from other singing styles is, in a word, the fact that Bel canto is a mixed vocal range singing style. Bel canto has influenced not only the art of vocal performance in China but also in other countries.

5.1. Intuitive thinking training in the art of Bel canto performance

The types of thinking used in vocal singing are action thinking, figurative thinking, logical thinking, aesthetic thinking, and creative thinking. Action thinking is a thinking activity in which the brain directs the action and requires the singer to use specific behavioral actions to solve problems. For example, the singer's expressions and stage gestures need to be conceived in advance so that they can be thought of while singing. Figurative thinking requires the singer to have a rich imagination and to be able to express the image vividly, through figurative thinking, the singer shapes the image with which he or she integrates and expresses the understanding of the lyrics and the expression of emotions in a vivid way. Logical thinking is an activity in which the singer makes a rational judgment, recognizes problems, and solves them according to certain logical rules. Through logical thinking, singers can fundamentally realize the problems that arise in singing, master the relevant knowledge, and constantly summarize their singing experience to make their singing more organized. Aesthetic thinking is the singer's ability to express his or her own aesthetic emotions to the audience through singing, intertwining emotions with the song, singing with the heart, moving oneself, and causing resonance. Creative thinking is the singer's constant summing up of singing experience and creating new ways to solve practical problems' thinking activities. Creative thinking is an advanced form of thinking that requires the singer to have some artistic talent and to be able to recreate the art.

Developing learners' mastery of correct thinking methods is a key part of thinking development and is a sign of an individual's maturity. Correct thinking methods include figurative metaphor, analogical association, reverse thinking, and the identification of correctness and error. Figurative metaphor is a more active way of thinking, as it allows the singer to develop their thoughts and understand the principles of things visually. The analogical association method is based on the perception of other learning objects to cause the same type of things, help learners understand new things, conversion of knowledge, and thus learn knowledge more deeply, is the transfer of knowledge thinking method. The reverse thinking method is the use of reverse thinking to help learn and solve problems that arise in the learning process. The identification of the correctness and error method is a logical way of judging whether the actions between objects are correct. In the development process the singers themselves cannot detect the wrong action, using the correct action to demonstrate, so that the singers can compare and identify. The incorrect movements are then corrected.

Cultivate with a solid foundation of knowledge, which is an important part of thinking, and learners with a solid foundation can enhance the ability to think about learning. Educators need to strengthen the singer's professional knowledge and master the basic elements of music in order to facilitate the singer's rationalization and re-working of sung works. As singers continue to accumulate and deepen their knowledge, the singer's knowledge structure will be optimized and adjusted, and the singer's thinking will be more open and detailed.

5.2. Interest development in the performing art of Bel canto

The art of Bel canto and performance is a constant cultivation of the artistic expression of Bel canto so that enthusiasm and interest in music can be constantly stimulated. The prerequisite for the development of interest is therefore the development of artistic expression in Bel canto.

People engaged in the vocal performing arts career should not only listen to concerts, listen to famous singers' singing performances, ponder their singing vocal methods, observe their artistic performance, watch opera performances, and listen to instrumental music, but also often watch films, enjoy paintings, read novels, visit painting exhibitions and various art exhibitions. Because one's direct life is very limited, being good at drawing from other forms of 'sister art' is a good way to enrich one's musical knowledge and strengthen one's musical training. It also requires a lot of practice, from the practice of singing methods and techniques, the practice of singing works, the practice of accumulating repertoire, the practice of singing on stage, especially the process of creating art by communicating directly with the audience, which is an objective stimulus for the actor and can stimulate the actor's potential passion for creating art, so that the singer can make a breakthrough in vocal technique and be sublimated in emotional expression.

It is one of the prerequisites for learning to sing beautifully, and it is as important as mastering the vocal technique of singing. Artistic expression aims to use skilled vocal techniques to vividly express the ideological content of a song and to infect the listener with a beautiful and emotionally appealing singing voice. Therefore, the correct handling of the relationship between voice and emotion in a singing performance is crucial to the development of the singer's artistic performance ability. The voice is the technical means by which the singer brings the emotional content of the work to the listener. The emotion is the subjective emotion of the lyricist, i.e. the direct reflection of the original emotion and meaning given to the work by the lyricist at one point in the composition, and is also the true emotion of the singer.

Only by combining the above factors can the love and interest of the audience be aroused, and only by applying emotion to the voice, strengthening the artistic impact of the voice, and by achieving empathy, can the interest in the art of performance be cultivated. In any case, it is necessary to perform in a prescribed context, based on the personality of the character to carry out the artistic expression, to convey the feelings of the character to the audience, to arouse the audience's empathy, and to achieve the purpose of artistic re-creation is also the basis for interest in the arts.

5.3. Stylistic choices in the performing art of Bel canto

The performing art of Bel canto requires the selection of a suitable style. The difference in styles is the connotation of artistic development and the goal pursued by the performing art. Bel canto requires the singer to understand the lyrics in depth, to strive for natural simplicity, to pursue a high degree of artistic expression, to develop an aesthetic conception of the voice, to improve appreciation, to make singing more in line with the requirements of the Bel canto, to sing with special emphasis on breath control, to emphasize and beautiful tone, to require singing with a varied tone and sincere expression of emotion. The main areas of voice training are correct breathing, accurate voice onset, vocal coherence, vocal flexibility, vocal resonance, diction, and language, etc.

The artistic style of Bel canto is specifically expressed as follows: Bel canto has specific requirements for the singer's singing posture and singing performance, and only when the singer sings Bel canto in a relatively free and natural state can he or she better control the breath and more easily evoke the emotional resonance of the audience. This requires a high level of singing skill and expressive ability. The aesthetic requirements for the singer are also very high, requiring the singer to have rational artistic thinking, to pursue the unity of specifications and concepts, and to attach great importance to the skill and rigor of the vocal style in the singing process, and to pay attention to the application of good resonance cavities. Some Bel canto styles are very closely related to national languages and therefore have different degrees of regional characteristics, which require singers to integrate their singing skills and expressive abilities with the characteristics of different nationalities.

6 Conclusion

In summary, Bel canto is one of the most representative vocal arts in the world of music, and the artistic style and singing techniques of Bel canto have always been an important object of study for other vocal singing styles, and singers must have a fuller knowledge and understanding of the artistic style of Bel canto and master a higher level of singing techniques to be able to perform more beautiful American songs. The development of the performing art of Bel canto is also based on intuition, interest, and the realm of style, allowing the three realms of musical performance art in Bel canto to be better applied in Bel canto performances to enhance one's musical aesthetic realm and literacy.

REFERENCES

1. Xiaojia WU. Popular Literature and Art [J]. Finance and Economics Science, 2011 (04): 107-108.
2. Chunyu MA. How to Shape the Triple Realm of Music Performance Art [J]. 2017(10):83-84
3. Xiaomang DENG. Sanxuan as "After Ethics" [J]. Morality and Civilization, 2021 (03): 5-21
4. Yirong HE. The Origin of Bel canto and its Development in China [J]. Yellow River Sound, 2016 (01): 59
5. Shan JIANG. The Origin and Development of Bel canto [J]. Art Review, 2017(03): 70-72
6. Xiaobo MA. An Analysis of the Origin and Development of Bel canto before the Nineteenth Century and its Dissemination History [J]. The Sound of the Yellow River, 2017(05): 65-66
7. Yan LIU. An Examination and Reflection on the Problem of Communion between Bel canto and Ethnic Singing [D]. Xi'an Conservatory of Music, 2011
8. Jian ZHANG. The Borrowings and Similarities of Chinese Folk Singing to Bel canto [J]. Drama House (First Half of the Month), 2013(11): 133
9. Jingting LI. Analysis of the Characteristics and Vocal Methods of Bel canto [J]. Drama House, 2020(06): 64
10. Tielong HU. On the Three Realms of Musical Performing Arts - Intuition, Interest and Style [J]. Popular Literature and Arts, 2011(02): 100-101

Discussion on the Cultural Differences of Love and Marriage between China and Japan from Their Television Dramas——Taking *the First Half of My Life* and *We Married as a Job* as Examples

Xueying Su [1,*]

[1]Chinese language and literature, Lingnan Normal University, Zhanjiang, Guangdong, China

Abstract. As an important carrier to present the cultural form of a country, teleplay plays a vital role in promoting cultural level and cultural export. To study and analyze the cultural differences between China and Japan is conducive to promoting the understanding of each other's cultures, reducing the misunderstanding and friction caused by the cultural differences, and promoting the exchanges and cooperation in politics, economy and culture.

1 Introduction

The concept of love and marriage is a timeless human topic that often talked about. As one of the embodied aspects of a country's society and culture, it will also be reflected in the language, living habits and other aspects of the country. At the same time, the TV series are a popular form of entertainment in modern society. Therefore, the study and analysis of a specific type or different types of TV dramas is conducive to a better understanding of the different cultural features of Japan. Through the comparison of TV dramas, it is possible to clearly understand the differences between the two countries, such as customs and living habits, the styles of communication, and to contribute to the deepening of communication and understanding between China and Japan, so as to achieve substantial development of bilateral relations.

2 Analysis of the marriage values in *the First Half of My Life* and *We Married as a Job*

2.1 *The First Half of My Life*

The love drama *the First Half of My Life*, which hit the airwaves during the summer vacation of 2017, tells the story of a woman's transformation after a failed marriage. Luo Zijun, in the beginning of the teleplay is a pretentious stay-at-home wife. Since marriage, she has given up her job, the life she lives everyday is privileged. But after she broke up with her husband Chen Junsheng, her life changed a lot. Because of this, she saw through her former life and started a new one. And the concept the drama wants to convey to the audience is, first of all, to be an independent individual. The independence is not only in economy, but more importantly in spirit. Just like Luo Zijun, her husband Chen Junsheng did fulfill his promise of 'I will support you', and it was for this reason that Luo Zijun gave up her economic independence and became a full-time housewife dependent on her husband. Human beings are emotional animals, but any relationship which is not reciprocal, even with great affection, will not last long. Luo's best friend, Tang Jing, is a positive embodiment of female self-reliance, but also promote Luo's awareness of the main body of the important figure. She has never been attached to men, but completely independent, become her own master. So when see finds problems in emotional aspect, she can also keep clam and solve the problem imperturbably. It can be said that her role is a subversion of the identity of contemporary women as 'the other.[1]' The spirit contained in this role is the reflection of thoughts about love and marriage of most women in China in recent years: be independent and be yourself.

2.2 We Married as a Job

The teleplay is an excellent drama about the common existence of unmarried people in Japan and their views on love and marriage. And the main line of it tells such a story: a well-educated woman with a master's degree called Mikuri Moriyama who had a hard time finding a job. Finally, she got a job in a big company but she was fired. Then her father introduced her to do household chores for a man called Hiramasa Tsuzaki to earn money. After

* Corresponding author: 18002788281@163.com

careful consideration, the two young people choose to become a fake couple by signing a contract and Mikuri takes a housewife as her profession. For the plot, the most central element of this teleplay is the discussion about the 'housewife'. This also reflects the traditional Japanese view of love and marriage. As one of the developed countries in Asia, with a higher level of productivity and relatively optimistic employment and survival pressure, most families can afford daily living expenses by being feed by men. As for women, many of them cannot continue to work under the pressure of conventional wisdom and public opinions, can only stay at home to take care of the whole family and run the household chores. These are the norm in eastern patriarchal societies, especially in Japan, where housewives are heavily objectified[2]. But the teleplay clearly argues for breaking gender boundaries. Mikuri gets the corresponding remuneration by doing housework for Hiramasa, that is, equating the marriage relationship with the employment relationship. We can see the idea it wants to convey to the audience: men and women are equal and the housewives are no longer just attached to the male, but also need to be treated with the same understanding and respect. Apart from that, when Japanese people express their feelings, they are extremely subtle and ambiguous. In the emotional aspects, they pay more attention to the inner world and values. From the teleplay we can know that even quarrel is measured, there will not be a fierce fight or argument, they focus more on showing emotional subtle changes. This is also inseparable from its national promotion of tolerance and compromise[3].

3 Comparison of the cultural similarities and differences between Chinese and Japanese views on love and marriage

The views on love and marriage, which means the basic vies, attitudes and opinions of adults on issues related to love and marriage. Choosing the right partner to set up a family, has always been a very important life event for a person, affecting his or her happiness and future development. As for the choice of a partner, due to the geographical location, the same roots of history and culture, the long-term mutual influence and penetration of Chinese and Japanese views on love and marriage have some similarities, but they also have their own criteria.

3.1 The same section

Japanese women's traditional view of marriage and love believe that marriage is a lifetime of happiness for a woman, no matter what they want to get married, so most of them through blind date to meet the marriage partner. After marriage, even if the marital life is not satisfactory: husband cheating, domestic violence, etc., Japanese women will not easily choose divorce, but for the sake of children, for the integrity of the family and insist, and this concept is still recognized by some Japanese women. Such ideas also exist in China: In the feudal society, there was no such word as "love", especially in the marriage, people could not choose the marriage object freely, but usually "arranged by parents and arranged by matchmakers", which was in line with the concept of "matchmaking between families of equal social status and economic ability" at that time. And family atmosphere, lifestyle and other aspects of the close or fit. After all, in that society, marriage was not based on love, but more on the combination of interests of both parties, and catering to the needs of the time to carry on the family line. Although China has already entered a civilized society and people's ideas have changed dramatically compared with the past, there are still some people who hold the traditional view of marriage and love. They believe that once they start a marriage, no matter what happens, they will not easily choose to divorce and insist on maintaining the operation of the marriage, even though the relationship has already deteriorated or caused serious problems.

With the progress and development of society, Chinese and Japanese people have more advanced ideas about whether to keep faithful and monogamous in marriage, which reflects the rational progress in ideology. In ancient times, no matter in China or Japan, polygamy was a very common social phenomenon. For example, the emperor of China could have "three palaces and six houses" and "three thousand concubines in a harem", while the aristocrats or wealthy businessmen could have "three wives and four concubines". Japan has "great Ohashi" -- the Edo era Tokugawa shogun general's birth mother, children, main room, side room and the residence of female officials, is actually the Tokugawa shogunate harem. It is quite normal for a man to have more than one spouse, but in modern society, such a situation is neither legal nor ethical. People insist that the couple need to be sincere to each other, in the love period of both men and women to maintain loyalty and dedication, can not change one's life.

'Marriage is more than being together'(Kefalas et al.,2011). Nowadays, young adults not only marry and have children later than previous generations, but they also take more time to get to know each other before tying the knot[4]. In recent years, whether in China or Japan, late marriage and childbearing seems to have become a trend. There are more and more "non-marriagers", mainly the young generation. Several factors are known to lead to non-marriage or delayed marriage. For example, people may delay marriage to pursue higher education[5]. It's possible that young people are more focused on work and personal life, and don't want to be bound by family; On the other hand, the life pressure is so great that they can not afford to support the family and raise children. However, there is also such a phenomenon in China: when the two sides get married, the man should have a house and a car, which is a "must condition", that is to say, economic ability is highly valued by the women when they get married, often more important than whether the two sides are suitable for living together, have common values and other aspects. But in the final analysis, Chinese and Japanese views on marriage and love will change with the changes of their own economic and social development, which will have an impact on different aspects of society.

In both China and Japan, and in much of Asia, factors such as the low popularity of highly educated women and the harsher employment realities women face than men also play a role in how women view marriage and relationships.

3.2 The differences

3.2.1 China

Under the influence of traditional Confucianism, many Chinese people have formed the concept that 'an individual must be in a marriage in order to become a truly completely person'. 'Take my hand. Take my whole life too' which means a couple grow old together, is the beautiful yearning/aspiration of many people for love and marriage. They hope that sweet love could lead to a happy marriage. As a result, the view 'men and women should get married when they reach a certain age' has also been advocated by most people. On the contrary, nowadays a lot of modern women believe that couples should have common interests, love and tolerate each other , rather than just 'respect each other like guests'; on the other hand, women believe that insisting on a loveless marriage is meaningless and should end it peacefully, instead of making do for the sake of their children or because being influenced by the society[6].

Puppy love has been a common problem in education sector in China, and it is also the germ of many teenagers' views on love and marriage. It is so common for a young person to fall in love with someone at an early age. And lots of Chinese students in junior or senior high schools are burdened with heavy academic pressure and usually choose to relieve the pressure by choosing the way of puppy love, however, this is not accepted by society and most parents and teachers. As far as they concerned, puppy love is a 'great scourges' and all the students should devote themselves to their studies rather than spending precious time to do things which are 'harmful to physical and psychological health'. Such perceptions have an impact on the understanding and views of some adolescent students, some of them still cannot overcome their fear of love after becoming adults.

3.2.2 Japan

Compared with China, Japanese people have not experienced many stages of change in their views on love and marriage. Distinctive features of the Japanese context allow us to shed new light on multiple sources of change in marriage behaviour[7]. For most of them, even if they crush on somebody, maybe they don't show it too much because they are not good at expressing their sentiment adequately, they show their thoughts in a more euphemistic way before being in love with someone and keep observing each other's behavior as a criterion to judge whether he or she could become a life partner.

As for puppy love, Japanese people have a more liberal attitude. In Japan, it is a common phenomenon that students fall into puppy love, many teachers and parents basically do not hold any objection. After all, education about sex in Japan is more open and sound. Parents do not restrict their children's thinking too much, but encourage them to express their emotions, so most students do not shy away from talking about their romance in front of their parents, and parents can better guide their kids by knowing their situations.

Influenced by the idea of gender equality, many Japanese women no longer see marriage as their ultimate goal in life. They increasingly tend to be economically independent women, who believe that when they get married, they will not only do housework, but also have children, take care of them and take on the responsibility of assisting husband and bringing up children. In addition, giving up their current good job and carefree life, which makes them feel that the burden is too big. And when their children is a little older, even if they want to re-employment, it is difficult to find a great job with their energy and state, can only do some short-term or temporary jobs, which cannot reflect their value well[8].

4 Reason for the cultural differences between China and Japan

Although China and Japan have maintained cultural exchanges over the years, the development has led to the formation of Japan's own unique cultural system, which has created differences with Chinese culture.

4.1 The difference in values

China is a socialist society and practices a socialist system. While Japan is a capitalist constitutional monarchy. Therefore, Chinese and Japanese values show obvious differences. After thousands of years of feudal rule and the edification of Confucianism, the concept of family and country has been deeply rooted in the hearts of the Chinese people and they especially pay attention to the consanguineous relationship. It is a symbol of kinship for Chinese people. Besides, it is generally believed that immediate family members and adjacent relatives are the two groups that have blood relation with them.

Japan has fewer nationalities than China. As a single nation, the Japan people have lived on the island for a long time since ancient times and have been greatly influenced by the farming culture. People have developed a strong sense of group in the long-term joint work. Although the hierarchical concept in Japan is very strict, there is a kinship-like force that binds everyone together under the influence of group consciousness[9]. Although Japanese people also have a strong sense of home, the family activities are mainly production-oriented and strictly follow the rules of the family, even those who are not related to the family can also be a part of it. If these unrelated people make a great contribution to the family, their relationship with the family may even exceed that of collateral relatives.

4.2 The difference in cultural development

Chinese culture originated from the Yellow River Basin, and its rise and development mostly centered on there. During the long feudal society, it formed the traditional Chinese feudal culture, which served for the feudal monarch and strengthened the centralization of power, so it had certain class characteristics and limitations.

On the other hand, Japan attaches more importance to foreign cultures and actively and extensively absorbs them. The reason is that since ancient times, Japan has held a strong spirit of learning from countries which are more powerful than itself, just as Japan sent envoys to the Tang Dynasty to learn the prosperous culture of China. After the envoys returned to their country, they would further study and innovate the culture they had learned, so as to keep pace with the times.

4.3 The difference in geographical environment

China's land area and population have reached a certain height respectively. With the use of abundant natural resources, China's agricultural civilization has developed and prospered/flourished. Because of this, the tradition of emphasizing agriculture and suppressing commerce was pursued in the feudal society, and agricultural production was carried out in an orderly way with the state as the leading role, which had an impact on the development and formation of the characteristics of Chinese traditional culture. China's geographical environment was sufficient to provide people with considerable resources for survival, so the nation did not need to choose to plunder in order to survive.

The situation in Japan is the opposite of that in China. Japan is an island nation surrounded by sea on all side, which is more suitable for communication with the outside world through the development of navigation technology. Relying on the convenience of navigation, the Japanese people have developed some excellent qualities such as perseverance, indomitability and tenacious? in the process of fighting against and conquering the sea. In addition, because of Japan's special geographical location, the lack of domestic resources and frequent earthquakes have led to a strong sense of crisis among Japanese people. The feeling of insecurity drove them to seek security by invading foreign countries, and this was the reason why Japan launched aggressive wars against China and other countries.

5 Perspectives on the cultural differences between China and Japan

It is inevitable that there are cultural differences between China and Japan. Only with full cultural exchanges between China and Japan can the world's cultural diversity blossom. Both Chinese culture and Japanese culture have certain redeeming qualities. No matter Chinese or Japanese culture, it has certain merits. After all, China and Japan have different national conditions, different development progress, and different cultural traditions. Therefore, in terms of marriage and love, people of the two countries have different choices.

5.1 Mutual respect and friendly communication

As a close neighbor of China, although Japan is deeply influenced by Chinese culture since ancient times, it is significantly different from China in many aspects, such as living habits and ways of expression. When people from different cultural backgrounds meet, they will usually understand each other's culture from the perspective of their own culture with the values they have formed. One of the important ways to respect other cultures is to break prejudice. Therefore, we cannot unilaterally view the Japanese culture that is different from Chinese culture with inherent concepts. We should understand the different customs of love and marriage between China and Japan, so as to avoid friction and misunderstanding in communication and achieve mutual respect.

5.2 Look at it objectively and take its essence

Any culture has its own uniqueness. We need to look at Japanese culture in a fair and objective manner, take its essence and dross, learn and apply the advantages of its culture, so that it coexist harmoniously with Chinese culture and jointly promote the prosperity and progress of culture.

6 Conclusion

In the process of cross-cultural communication, there will inevitably be the link of Chinese culture transmission and foreign culture introduction. Chinese culture is extensive and profound. Only by exchanging and colliding with each other can different cultures produce gorgeous/brilliant sparks and promote the diversity of world cultures. Whether it is the inheritance of traditional culture, the influence of modern society, or the impact between different cultures, the ideology of people in China and Japan are constantly changing and developing, showing their own national cultural characteristics. The emergence and development of Chinese and Japanese marriage views ultimately take stable and harmonious family relations and civilized and upward marriage and family as the ultimate ideal[10]. Through the two TV dramas of the same type in China and Japan, this paper compares and analyzes the cultural differences in love and marriage between China and Japan to achieve the ultimate goal of respecting the cultural differences, enhancing cultural self-confidence and promoting the friendly cross-cultural interaction and communication between China and Japan.

References

1. Song Xiaohui.An Analysis of female images in urban emotional drama-Taking as an example[J]Media, 2018(05):104-105.DOI:10.19395/j.cnki.1674-246x.2018.05.053.
2. Xia Ying. What's in Japanese TV—Japanese culture and knowledge of the language as seen through [J].China Education Innovation Herald(Mid-ten day issue),2017(23):146-147,DOI:10.16400/
3. Chen Peng. Cultural Differences between China and Japan from Japanese TV Series [J].Shenzhou,2013(21):46.

4. Li,B.,Zhou,C.,Luo,J.(2022) Way of love and after-marriage well-being: Evidence from China. INTERNATIONAL STUDIES OF ECONOMICS,17:394-410
5. Ishiguro,I.(2020)Maximizing Tendency Predicts Delayed Marriage Among Japanese Adults.Japanese Psychological Research,64:53-63
6. Xu Kui, Wang Jin. A Comparative study of Chinese and Japanese women's on Marriage and Love [J].Comparative study on cultural innovation,2019,3(21):62-63.
7. Fukuda,S., Raymo,J., Yoda,S.,(2019) Revisiting the Educational Gradient in Marriage in Japan. Journal of Marriage and Family,1378-1396
8. Liu Xiaodan. A comparison of Chinese and Japanese modern views on marriage and love from the perspective of men. [J]. Journal of Harbin Vocational & Technical College,2013(03):123-124.
9. Han Lei. A brief analysis of the cultural differences between China and Japan and their causes. [J]. YOUTHFUL YEARS,2018(01):230-231.
10. Zhao Hui,2014. A comparative study of Marriage views between China and Japan.https://kns.cnki.net/kcms2/article/abstract?v=3uoqIhG8C475KOm_zrgu4lQARvep2SAk9z9MrcM-rOU4mSkGl_LWfwkZ8XaUAvUnBCm__7OHWwd3lsrBtiJo4DSWbHGAprRM&uniplatform=NZKPT

Overview of Artificial Intelligence Painting Development and Some Related Model Application

Mingyang Liu[1,*]

[1]School of Computing and Data Science, Xiamen University Malaysia, Xiamen, FuJian, 361006, China

ABSTRACT. The application of Artificial Intelligence (AI) in painting has gained significant attention in recent years, promising to revolutionize the digital art world and enhance traditional painting processes. AI painting involves the use of computer algorithms and machine learning techniques to generate and manipulate digital images. This paper provides an overview of current research in AI painting, covering key findings, literature review, and analysis of the field's current status and future direction. Research has shown that AI algorithms can generate unique digital art forms and automate tasks in traditional painting processes, freeing up time and effort for artists. The field of AI painting holds immense potential and further research and development will likely lead to new and innovative applications. In addition, this technology has a long way to go on the road of commercialization.

1. INTRODUCTION

AI painting is a growing field that has the potential to revolutionize the creation and consumption of digital art. This technology is attracting increasing attention due to its potential to expand the creative horizons of artists and designers, and foster the development of innovative applications [1]. With AI painting, even individuals with limited technical skills can produce high-quality works of art in a matter of minutes, greatly lowering the barrier to entry in the field of digital art.

In recent years, research in the field of AI painting has made impressive strides, with a growing number of studies exploring innovative ways to use artificial intelligence to generate and manipulate images. For example, researchers have used advanced techniques like automatic learning and unsupervised learning to segment and analyze image details such as character posture and identity [2]. Additionally, the use of Deep Convolutional Generative Adversarial Networks (DCGAN) [3] has demonstrated the potential for unsupervised learning to outperform supervised learning in certain scenarios. Some advancements are achieved by improving convolution networks with the help of Generative Adversarial Networks [4]. And the development of diffusion model enables AI painting to draw high-resolution and realistic images [5] etc.

Despite these impressive achievements, there are still several challenges that need to be addressed in order to further advance the field of AI painting. Some of the key areas of focus include improving the quality and authenticity of images generated by AI, developing new algorithms that can produce a greater visual appeal and diverse styles, discuss the originality of artworks produced by AI and its ethical issues as a unique personal expression, and exploring the potential of AI to create interactive and immersive digital art experiences.

2. RELEVANT MODELS

2.1. Generative Adversarial Networks

Generative Adversarial Networks (GAN) are a class of deep generative models that have gained significant attention in recent years due to their ability to generate high-quality, diverse and realistic images.

GAN were introduced by Goodfellow et al. (2014) [4], where two deep neural networks, generator and discriminator, are trained in an adversarial manner. The generator tries to generate synthetic samples that are indistinguishable from real ones, and the discriminator tries to correctly identify whether a sample is real or fake. Since then, GAN have been used in various computer vision and machine learning tasks, including image generation, style transfer, and representation learning.

The generator and discriminator are both deep neural networks, and the training process for GAN is accomplished through a minimax game, where the generator tries to minimize the error it makes in fooling the discriminator, and the discriminator tries to maximize its accuracy in distinguishing between real and fake samples. The game continues until a balance is reached, such that the generator is producing high-quality synthetic samples that are difficult to distinguish from real ones, and the discriminator is as accurate as possible in differentiating between the two.

* Corresponding author: a1278336390@gmail.com

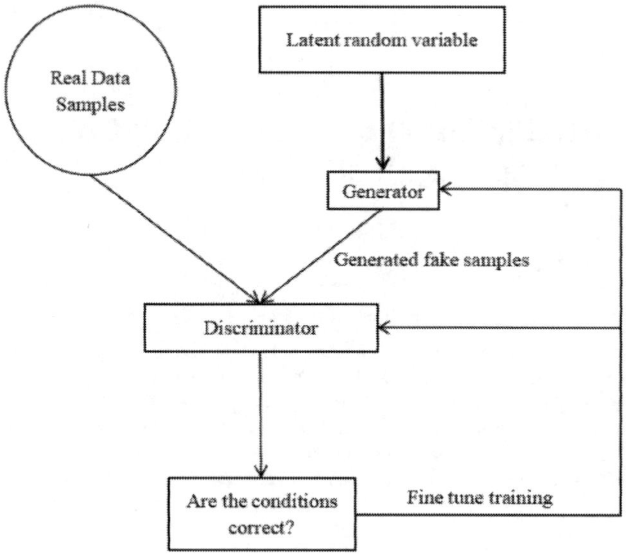

Fig. 1. Generative Adversarial Networks schematic diagram. [6]

$$\min_G \max_D V(D,G) = \mathbb{E}_{x\sim p_{\text{data}}(x)}[\log D(\boldsymbol{x})] + \mathbb{E}_{z\sim p_z(z)}[\log(1 - D(G(\boldsymbol{z})))] \quad (1)$$

The GAN mathematical formula [4] refer to "Equation (1)", where G=generator; D=discriminator; Pdata(x)=distribution of real data; P(z)=distribution of generators; x=sample of Pdata(x); Z=sample of P(z); D(x)=discriminator network; G(z)=generator network [6].

Overall, the principle of GAN is to use the adversarial relationship between the generator and discriminator to learn a generative model that can produce synthetic data samples that are similar to real ones. The process allows for unsupervised representation learning, as the generator and discriminator can learn to extract and use important features from the data to generate high-quality synthetic samples.

However, GAN also have several known disadvantages [7], including:

Instability: GAN can be difficult to train due to their unstable loss landscape and the possibility of mode collapse, where the generator produces limited variations of the training data.

Training data requirements: GAN require a large amount of training data to learn the underlying distribution of the data and generate high-quality samples.

Model collapse: GAN may produce images that lack diversity, with certain modes or features being ignored by the generator.

Evaluation: It can be challenging to evaluate the quality of GAN-generated samples, as there is no objective metric for measuring the similarity between the generated and real data.

Computational resources: GAN can be computationally expensive to train and require significant computing resources, such as Graphics Processing Units or Tensor Processing Units.

GAN is not suitable for processing discrete data, such as text.

Certainly, with these shortcomings, there are also some optimization models. Here, four popular improved GAN models, their advantages, disadvantages and principles are briefly provided.

2.1.1. Four popular improved GAN models

Conditional GAN (CGAN): CGAN was proposed by Mirza and Osindero (2014) [8] to enable the generation of specific types of data, such as images of certain classes or with certain attributes. CGAN extends the original GAN architecture by conditioning both the generator and the discriminator on additional input, such as a class label or an attribute vector and transforms the unsupervised GAN into a semi-supervised or supervised model. This allows for more precise control over the generated samples and enables applications such as image-to-image translation. One disadvantage of CGAN is that it requires the availability of the conditioning information at both training and inference time, which may not always be feasible.

Deep Convolutional Generative Adversarial Networks (DCGAN) are a type of GAN proposed by Radford et al. (2015) [3] that use convolutional neural networks in both the generator and discriminator networks. DCGAN are capable of generating higher quality images than previous GAN models by learning hierarchical representations of the input images. The generator network takes random noise as input and generates images that are progressively refined through deconvolutional layers. The discriminator network, on the other hand, learns to distinguish between real and fake images. DCGAN have been used for a variety of applications, including image synthesis, image editing, and style transfer. However, like Style-based GAN [2], DCGAN also require a significant amount of computing resources and training time. This kind of model may cause the over-generation problem, which is another type of model collapse. So strange pictures may be generated, but the discriminator still treats them as real pictures.

Fig. 2. Strange pictures caused by over-generation. [9]

Wasserstein GAN (WGAN): WGAN was proposed by Arjovsky et al. (2017) [10] to address the instability of the original GAN training process. WGAN replaces the discriminator with a critic that outputs a scalar score, and enforces a Lipschitz constraint on the critic by clipping the gradients of the critic's weights. This leads to a more stable training process and a smoother loss landscape, which allows for better convergence and higher quality generated samples. One of the main disadvantages of the WGAN model is that it can be more difficult to train than traditional GAN. This is because the gradient clipping used to enforce the Lipschitz constraint can cause problems with vanishing or exploding gradients, which can make it harder to converge to a stable solution. Additionally, the WGAN loss function can be less intuitive to interpret than the standard GAN loss, which can make it harder to diagnose problems or make improvements to the model.

Style-based GAN (StyleGAN): StyleGAN was proposed by Karras et al. (2019) [2] to address the limitation of previous GAN models in generating high-resolution and diverse images. StyleGAN introduces a novel generator architecture that separates the generation of high-level attributes from the low-level details of the images, allowing for more control over the generated images. StyleGAN also uses adaptive instance normalization and progressive growing techniques to improve the quality and diversity of the generated images. One disadvantage of StyleGAN is that it requires a large amount of computing resources and training time to generate high-quality images.

2.2. Diffusion Model

Diffusion models are a family of generative models that have been widely used in various applications such as computer vision, sequence modeling, audio, artificial intelligence, etc [11]. The core principle of the diffusion model is to iteratively update a random noise vector until it approximates a target distribution [12]. This technique has proven to be effective in generating high-quality samples that are visually appealing and exhibit realistic patterns [13].

The diffusion model is a type of autoregressive model that generates samples by updating a random noise vector, following a pre-defined set of diffusion steps. These steps are designed to gradually increase the complexity of the noise distribution by adding a small amount of noise to each element of the vector at each step. In this way, the diffusion process iteratively transforms the noise distribution into the target distribution. The key idea of the diffusion model is to sample a set of noise vectors, apply the diffusion steps to each of them, and use the resulting vectors to generate the samples [12]. The algorithm used in the diffusion model is called the Langevin dynamics [14], which is a well-studied algorithm in the field of statistical mechanics. The Langevin dynamics ensures that the updates to the noise vectors are smooth and gradually bring the distribution closer to the target distribution.

2.2.1. Some advantages and disadvantages of diffusion model

The diffusion model has several advantages over other generative models, such as GAN and Variational auto-encoder (VAE). VAE are another popular method for AI painting [15]. Unlike GAN, VAE are generative models that use a continuous latent space to generate new images. The key idea behind VAE is to map an image to a lower dimensional representation, called a latent code, and then use the latent code to generate new images. The generation of new images is achieved by sampling the latent space and mapping the samples back to the original image space. One of the main advantages of the diffusion model is that it does not require any adversarial training or complicated loss functions, which are commonly used in GAN [10]. This makes the training process more stable and less prone to mode collapse, a common issue in GAN. Another advantage of the diffusion model is that it does not require any additional parameters or network structures to generate diverse samples, which is a common issue in VAE [15]. The diffusion model also allows for flexible conditioning, which enables the generation of samples that are conditioned on some input, such as text or images [16]. Finally, the diffusion model produces high-quality samples that exhibit realistic patterns and textures, which makes it suitable for a wide range of applications such as image generation [13].

Despite its advantages, the diffusion model also has some limitations. One of the main challenges of the diffusion model is that it can be computationally expensive, especially when dealing with high-dimensional data such as images or videos. This is because the diffusion model requires a large number of steps to generate high-quality samples, which can increase the computation time significantly [11]. Another limitation of the diffusion model is that it can be difficult to train, especially for large datasets. This is because the Langevin dynamics used in the diffusion model requires careful tuning of the learning rate and other hyperparameters [13]. Finally, the diffusion model may not be suitable for all types of data distributions. One such situation is when the data distribution is highly structured or complex. The diffusion model assumes that the spread of information or particles will be proportional to the density of the medium through which they are traveling. However, in highly structured or complex distributions, this assumption may not hold true.

To solve the problem of slow sampling and excessive calculation consumption, the existing improved algorithms are divided into four categories: Training Schedule, Training-Free Sampling, Mixed-Modeling, and Score-Diffusion Unification. For Data Structure Diversion, there are three types of methods: Continuous Space, Discrete Space, and Constrained Space. There are two methods for Likelihood Optimization, namely, Improved ELBO and Variable Gap Optimization. Finally, Dimension Reduction can be realized through Mixed-Modeling [11].

3. FUTURE PROSPECTS AND PROBLEMS

In the field of art, AI painting provides artists with new technologies and tools for creating art. An important aspect of AI painting is the interaction between the user and the AI system. In some cases, the user may provide high-level constraints or guidance to the AI system, such as a desired color palette or brush style [17]. This technology enables artists to explore new possibilities and improve their creativity. In the future, AI painting can also be used to repair old and damaged artworks.

In the design industry, AI painting can be used to generate graphics and logos. AI systems can learn from previous designs and generate new designs according to user specifications, thus reducing the time and effort required to create designs. AI painting can also be used in the fashion industry to generate new designs and predict future fashion trends. For example, in the future, fashion brand LV may use AI algorithm to design the latest brand trademark, predict fashion trends and optimize the inventory management system.

AI painting can be used to create special effects, backgrounds and characters in movies and video games in the entertainment industry. After several iterations in the future, the AI system may be able to generate realistic 3D models, animations and textures to enhance the visual appeal of these products. In video games, AI painting may be used to create dynamic landscapes and environments to respond to user input.

In the future, the potential application of AI painting is not limited to the art, design and entertainment industries. It can also be used to generate 3D medical images in healthcare, create 3D building models in buildings, and so on.

While the use of AI in creating art may offer exciting new possibilities for creative expression, it also raises significant questions about originality and authenticity in artistic practice, as well as ethical issues related to ownership and authorship.

One of the primary challenges in assessing the originality of AI-generated art is the question of who or what is actually producing the work. While the software and algorithms used to generate the artwork may have been created by human programmers, the specific output generated by the AI may be considered wholly original and not necessarily attributable to any individual creator. This can create a complicated legal and ethical landscape for issues such as copyright and intellectual property, as well as challenges in determining who should receive recognition or compensation for the artwork.

Another aspect to consider is the role of personal expression in artistic creation. Traditionally, art has been considered a unique expression of an individual artist's perspective and creativity, reflecting their experiences, emotions, and values [18]. While AI-generated art can certainly produce striking and visually compelling images, it may be argued that the absence of personal intent or subjective experience in the AI undermines the notion of it as a unique personal expression. This could impact the value and meaning of the artwork for viewers, collectors, and critics, and raise questions about the role of technology in shaping the cultural significance of art. In addition, the development of artificial intelligence may cause large-scale unemployment, technological monopoly and other problems in human society.

As the field of AI-generated art continues to evolve, it is important to consider these issues thoughtfully. Then we can come to engage in critical discussions around the benefits and potential risks of this innovative form of artistic practice.

4. CONCLUSION

In conclusion, AI painting is a rapidly growing field that has the potential to revolutionize the way we create and appreciate art. Through the use of deep learning algorithm and neural network, AI painting has an incomparable material base and efficiency. It can use different color, style, composition and other basic elements to render hundreds of different pictures with the same element style at one time. However, it is also possible to have problems with the authenticity and value of artworks generated by AI, as well as the impact of AI on the role of human artists. In addition, there are also ethical problems in the use of data and algorithms to create and disseminate art. In the follow-up practice, these issues need to be discussed and explored.

In this paper, I mainly summarize the principles, advantages and disadvantages of the relevant development of several model algorithms related to AI painting. Then it looks forward to the future of AI painting and summarizes several possible problems.

REFERENCES

1. Gatys, L. A., Ecker, A. S., & Bethge, M. (2015). A Neural Algorithm of Artistic Style. arXiv preprint arXiv:1508.06576.
2. Karras, T., Laine, S., & Aila, T. (2019). A style-based generator architecture for generative adversarial networks. In Proceedings of the IEEE Conference on Computer Vision and Pattern Recognition (pp. 4401-4410).
3. Radford, A., Metz, L., & Chintala, S. (2015). Unsupervised representation learning with deep convolutional generative adversarial networks. arXiv preprint arXiv:1511.06434.
4. Goodfellow, I., Pouget-Abadie, J., Mirza, M., Xu, B., Warde-Farley, D., Ozair, S., ... & Bengio, Y. (2020). Generative adversarial networks. Communications of the ACM, 63(11), 139-144.
5. Rombach, R., Blattmann, A., Lorenz, D., Esser, P., & Ommer, B. (2022). High-resolution image synthesis with latent diffusion models. In Proceedings of the IEEE/CVF Conference on Computer Vision and Pattern Recognition (pp. 10684-10695).
6. CSDN AIStudio. (2022). Machine Learning Notebook - Generation of an Antagonistic Network (GAN) Overview and Getting Started Example. https://blog.csdn.net/ZwaterZ/article/details/124840602.
7. CSDN TensorFlow. (2019). Principle, advantages and disadvantages of GAN. https://blog.csdn.net/chary8088/article/details/89681869.
8. Mirza, M., & Osindero, S. (2014). Conditional generative adversarial nets. arXiv preprint arXiv:1411.1784.
9. CSDN GAN. (2022). Original GAN (DCGAN) defect. https://blog.csdn.net/weixin_41883450/article/details/127119414.
10. Arjovsky, M., Chintala, S., & Bottou, L. (2017, July). Wasserstein generative adversarial networks. In International conference on machine learning (pp. 214-223). PMLR.
11. Cao, H., Tan, C., Gao, Z., Chen, G., Heng, P. A., & Li, S. Z. (2022). A survey on generative diffusion model. arXiv preprint arXiv:2209.02646.
12. Ho, J., Jain, A., & Abbeel, P. (2020). Denoising diffusion probabilistic models. Advances in Neural Information Processing Systems, 33, 6840-6851.
13. Dhariwal, P., & Nichol, A. (2021). Diffusion models beat gans on image synthesis. Advances in Neural Information Processing Systems, 34, 8780-8794.
14. Roberts, G. O., & Tweedie, R. L. (1996). Exponential convergence of Langevin distributions and their discrete approximations. Bernoulli, 341-363.
15. Kingma, D. P., & Welling, M. (2013). Auto-encoding variational bayes. arXiv preprint arXiv:1312.6114.
16. Kawar, B., Zada, S., Lang, O., Tov, O., Chang, H., Dekel, T., ... & Irani, M. (2022). Imagic: Text-based real image editing with diffusion models. arXiv preprint arXiv:2210.09276.
17. Isola, P., Zhu, J. Y., Zhou, T., & Efros, A. A. (2017). Image-to-image translation with conditional adversarial networks. In Proceedings of the IEEE conference on computer vision and pattern recognition (pp. 1125-1134).
18. The teaching and research team of the art communication direction of the Central Academy of Fine Arts. (2022). Art. https://baike.baidu.com/item/%E8%89%BA%E6%9C%AF/12004323?fr=aladdin.

Content Research Strategy of "Three Agricultural Videos" in the Internet Era – Take the Example of the Three Farmers Food Skills Category

Yimeng Zhang[1,*]

[1]China Coal Technology Engineering Group of Taiyuan, Taiyuan, 030006, China

Abstract. Along with the spiralling development of social media and the rise of short video platforms such as Jitterbug and Racer, there has been a boom of a large number of rural people joining the ranks of short videos, reaping the benefits of media literacy and media opportunities by spreading the daily life of farmers and local culture, while deriving a new form of expression for rural development. Narrative short videos on three rural areas have also become a craze entering audiences and influencing their lives. It is not only the novelty of the video style, but also the incipient categorisation of short videos. As a result, more and more urban groups are learning about idyllic rural life through short video platforms, narrowing the gap between urban and rural areas and improving rural economic revitalisation and development.

1 Introduction

1.1 Simple farmers highlight the local customs, and camera language depicts the field life

The emergence of short videos has made people's lives more fulfilling, and the short video of three farmers is a rising star in the short video category. The advancement of vernacular culture is an important strategy for its dissemination, symbolising [1] a specific cultural system that is an integral part of its cultural heritage, both on a material and spiritual level. Self-media, which has become the main theme of the times, has seen the rapid rise of short videos on short video platforms carrying vernacular culture, gradually turning the past form of communication, which relied on radio, television and paper media, upside down. More ordinary people are using their mobile phones to record their lives, interpreting their ordinary and mundane lives, and then disseminating them to the public through short video platforms, bringing rural life out of the countryside and making it accessible to urban residents[2]. This is the meaning of cultural communication and the reproduction and re-dissemination of rural culture. What we see in the video is not only the warm and quiet life of the countryside, but also the awakening of nostalgia buried in the heart.

The rise of short videos on agriculture has added a colourful touch to the development of rural revitalisation in the new era, leading the development and construction of new rural areas. It has an irreplaceable role in spreading rural customs and folk culture. The content follows the trend of the times, follows the pace of the times, and gives a new meaning to rural culture. It is unique in the way it is disseminated, using technology skilfully to present the whole of the countryside to urban audiences in a new light.

The short videos of agriculture, rural areas and farmers have unique shooting techniques. They use a large number of narrative techniques of panoramic and midrange videos. The three-dimensional space shot from all angles and in multiple directions breaks the inherent template of traditional media videos, and visually expands the spatial dimension. The characters and scenery are integrated, making users feel as if they are in the "scene" when watching videos.So as to get the immersive sensory experience, and put the peach blossom life of the talent into oneself.

When shooting and narrating, the short videos on agriculture, rural areas and farmers are no longer a single closed camera position and shooting frame, but not just a single segment of shot language. In terms of lens manipulation, they describe a variety of scenes of rural life through close ups, close-up shots and panoramic shots, reflecting farmers' hard work in fields, colorful and hot rural life, and warm and harmonious neighborhood exchanges.Let the audience see different rural life in a multi angle and all-round manner.

In the user @ restless Ajun food production short video on agriculture, rural areas and farmers, the short minutes of video usually include several links such as vegetable picking, cleaning process, production process, and tasting food, covering fields, their own yards, rural roads and other scenes. The video shows many characters such as sister, parents, neighbors, and creates emotional expression of family harmony and love of labor,In the process of watching, the audience not only learned how to make delicious food, but also experienced the smart small farmer economy of family

* Corresponding author: 865146063@qq.com

sitting around and hard working in such an atmosphere, and was moved by the simple family feelings in the countryside.This multi angle shooting technique shows the tranquility and happiness of rural life, which makes the audience immerse in the video experience and attract the audience's attention.

1.2 Subjects of the study

In this paper, we are looking at three short videos of the same type by three farmers with more than 20 million followers and more than one million likes in the last 90 days, mainly showing food production and artisan skills. "This is a real achievement.

A user living in rural Fujian province, @A Jun, has been posting short videos on TikTok since July 2021, and as of October 2022, has 23.097 million followers, with the highest number of video views at 1.623 million [3]. His videos mainly focus on food production, where he makes fancy versions of common or uncommon food in rural life, creating colourful and beautiful food videos. User @KangzaiFarmer's videos are mainly about home-cooked food, mostly filmed in his own yard, and often featuring pots and pans and pollution-free vegetables. He gives his audience a sense of originality and simplicity, and his videos are mostly close to life, with the most humane atmosphere[4]. User @PanLouLou is based in rural Anhui province and introduces her rural food life. The videos mostly showcase the food and the making of household utensils, fully demonstrating the confident living conditions of rural people. Through the media platform of Jitterbug, those ordinary farmers with their faces to the soil and backs to the sky are made known to the people, their identities are no longer limited to farmers, they are people who love life, ordinary people who are active, this is the heritage of culture and technology, showing the public the beauty of life and the wisdom of labour, transmitting the power that is lacking in the city.

1.3 Research Methodology

1.3.1 Content analysis method

Through careful observation, research and analysis of the three agricultural short videos on the ShakeYin platform, the most distinctive parts of them were selected for study, and through detailed collation of the content, for example, subdivided according to thematic features, the characteristics of images, language and characterisation in the videos were analysed[5]. By synthesising the analysis and filtering out the eligible video masters of the Sannong genre, it can be concluded that Sannong short videos have rapidly matured and become hotspots, occupying the top of the short video charts.

1.3.2 Case Study Method

Through layers of screening, the three talents who met their requirements were analysed in detail. Each talent's short video style is very different, with unique personal characteristics in terms of screen content and audio-visual language of filming, and different levels of fan groups, seeking common ground while preserving differences in case studies. The style of @IdleArjun's work is mostly personal narration, with a cheerful and light-hearted style that is more likely to be loved by audiences. The style of @KangzaiFarmer's videos is simple, unpretentious and close to life, a reflection of the lives of people at the bottom of the countryside.

2 Case study of content production strategy

2.1 Simple peasants highlighting the human touch in the countryside

The gimmick of the "rural netizen" vividly portrays rural life and evokes the "nostalgic feelings" of society at large, which is why the emotional element in short videos on agriculture is so important. In his book Preface to the Film, Raymond Williams coined the term 'emotional structure' to describe the general feeling of people about real life at a particular time[6]. The love of life, the reverence for nature and the expression of real feelings conveyed in such videos are the focus of the 'new emotional structure' in the context of social transformation.

In @User Idle Arjun's short videos on food-based sanshou farming, the production of each piece usually starts with a story, and goes through the four stages of fetching ingredients from the fields, to cleaning and setting aside, to kitchen preparation, and finally dinner table dining. In a short video of about five minutes, the video brings together several scenes from the vegetable field, kitchen and yard, fully demonstrating the process of making food, so that the audience can have an immersive sensory experience while watching, not only learning to make food, but also being touched by the simple and genuine feelings of the countryside.

2.2 Use of dialect dialogue and integration of local characteristics

In all kinds of film and television productions language is the most important expression of the character's personality, ideas and storyline. In this media platform, language has become an iconic symbol of the Sannong short videos[7]. For example, the short videos of @LiZiQi, @MaiXiaodeng and @Eastern Selection are mainly in Mandarin.

Users such as @KangzaiFarmers and @PanLouLou mainly used dialects for their conversations. Analysing the various short videos of the three farmers in Jitterbug, the majority of ordinary people use dialects for dialogue. Dialects are mainly languages used in fixed areas, which are regional, remote and niche, and this kind of unartistic language can create a strong sense of authenticity among the audience. The regional and differential culture in dialect represents folklore and the lowest level, which makes the characters created in the video more fleshed out and their personalities more prominent. The @kangzai farmer communicates with the old children appearing in the video in what appears to be Mandarin but is in fact a very representative Guangxi dialect. nl pronunciation, qi eggs (eating eggs), etc., give the listener a clear hometown resonance. Such videos make clever use of dialects and continue to spread on the Internet, creating a new trend that triggers imitation and change among netizens and promotes the spread of culture. On the one hand, the use of dialect conversations by groups in the same region can bring them closer to each other, making cultural transmission closer to life; on the other hand, as a communication medium, Jitterbug gives ordinary vegetarian farmers the opportunity to showcase themselves, and users do not have to use their own unfamiliar dialects, but can express their emotions in the dialects of the region, which is not only conducive to breaking geographical restrictions, but also to the exchange of language and culture between regions This not only helps to break down geographical restrictions and promote inter-regional linguistic and cultural exchange, but also helps to increase the self-confidence and pride of the people in the language and culture.

2.3 Showing the "digital" Peach Blossom Garden and building a mimetic environment

With the increasing pressure of urban life, more and more people in the city are yearning for a place like the Peach Blossom Garden where they can slow down the pace of life and enjoy the slow life in the countryside to get their soul ironed out[8]. As a result, the legendary "Peach Blossom Garden" has become a dream destination in people's minds. Many short videos on agriculture use the media platform to showcase the rural landscape and their daily lives, enabling audiences to discover the commonalities between the countryside and the Peach Blossom Garden.

The video by user @KangzaiFarmer has been described by netizens as a "real-life version of Peach Blossom Garden", in which people live in a small courtyard in the countryside, grow pollution-free green vegetables, do their own food, and recreate the life of an ancient small farmer's economy[9]. They make the audience in front of the screen envious and yearn for this kind of carefree life in the countryside. These videos pay great attention to background music, as if the clouds are in a "paradise". The song "Living Well is Living Well" by Zhou Shen is used many times in the video by the expert @Ajun, who can't be bothered, and the song is gentle and quiet, with a slow rhythm, echoing the rural life created by his work, reminding people to love life and conveying It reminds people to love life and transmits positive rural energy. For example, the video released on 5 August uses Mao's "Thinking of You", as well as "An Ordinary Day" and "Murmuring" to create a relaxed and comfortable ordinary life. It builds a picture of a paradise of self-containment and long grass, a desirable mimetic living environment.

2.4 Focusing on body narration, focusing on reflecting the wisdom of working people

As the name suggests, body narrative is to take the body as the narrative subject, form the main narrative process through the dynamic or static form of body language and the ideology of presence or absence, and achieve the purpose of communication and exchange [10].Among all kinds of short videos on agriculture, rural areas and farmers, the most indispensable is physical labor, which is mainly in the form of action, and the charm of characters is also represented by physical performance, becoming a highly personalized visual cultural symbol. Arnheim believes that "sports are the phenomenon most likely to attract strong visual attention" [11].

Whether it's @ Li Ziqi, who focuses on food shooting, or @ Kangzi, a farmer who devotes himself to showing the daily life of farmers. The main body narrative has become the main body in the short videos on agriculture, rural areas and farmers, gradually becoming the representative with specific symbols, giving the local short videos a specific meaning.The user @ Kangzai Nong Ren is an account mainly for recording daily rural food. In a five minute short video, he uses body language to record the cutting, processing, preparation of food materials, and the production of finished products. Every frame of the screen describes the body language vividly.In the video, the old children and Kangzai cut the leaves one by one with scissors, and at the same time, they washed rice and boiled water, making the hand movements seamless.Through such a series of food production processes as cutting, washing, boiling and wrapping glutinous rice, the wisdom and hard work of working people are reflected, and such a desirable real life in the world.

The user @ busy Ah Jun, whose works are smooth and tidy, takes the rural economy and food system as the main body, and the story development content revolves around labor.At the beginning of the video, Ah Jun picked trees and grapes in the field, and introduced his local characteristic fruit grapes through body language. A series of verbs, such as "climbing, picking, washing, squeezing, sawing", and Ah Jun's skillful actions, in this short video, he interpreted the ingenuity of rural people with body language. The short video mainly set off his pastoral style through music.In addition to the colorful behavioral language of the main character, Ah Jun, other characters in the video are also interpreting the freedom and simplicity of rural life with labor.By releasing the potential in the body, the author of the short video on agriculture, rural areas and farmers transmits the free and comfortable state to the audience through the video, and uses the body narrative to show the labor technology. The purpose is to pursue self display and enrich the shaping of self character, and then obtain the sense of social identity, and find the sense of existence that was missing in the past society and was covered by urban culture.

Rural economy is the chain in which the memory of local culture can survive and continue. It not only carries the way of life and production of farmers, but also continues the lifeline of the prosperity of local culture.However, under the influence of social modernization, a large number of farmers went out of the countryside, and the primitive rural civilization was in crisis. The road to continue rural development was bumpy and full of difficulties.Therefore, as a new way of cultural transmission, the short video of agriculture, rural areas and farmers will certainly assume the responsibility of inheriting cultural memory.

3 Summary

To sum up, today's hot short videos of farmers have developed rapidly from a single category, limited channels and low traffic to a place in many platforms, becoming a popular source. Nowadays, short videos on agriculture are not only a form of expression to show the daily life of people, but also a spiritual hub to promote rural revitalisation, economic development and social progress. The high level of user activity on the ShakeYin platform has helped urban and rural residents to progress together, narrowing the cultural differences between them, and providing a channel and platform for the dissemination of short videos by rural farmers [12]. This is an innovative form of local cultural communication, which is based on the farmers' own perspective, making videos with local style.

References

1. Guo Qingguang. Communication Studies [M], Beijing: People's University Press, 2011:29.
2. Jia Shengyun, Dong Xiaoran, Zeng Xiangmin. Research on the Characteristics and Strategies of Full-motion Video Narrative [J]. TV Research, 2017 (06): 10-13.
3. 2022 Tik Tok Data Report https://www.chanmama.com/douyin? BDPC & PP & 3152805483002 2022
4. Flying Melon Data Analysis https://dy.feigua.cn/home/Product/? chl=0111sj-ocpc
5. Ma Le. A study of "three agricultural" short videos in ShakeYin platform, Heilongjiang University, China, 2021.12.4.
6. Gu Lijie, Zhang Qing. The rise of rural online celebrity and the turn of local culture [J]. News Lovers, 2020 (12): 37-40.
7. Xu Xuening. The help of dialect to the bottom narrative of domestic films [J]. Popular Literature and Art, 2013 (21): 195-196.
8. Wang Mingwei. Rural communication of short videos of agriculture, rural areas and farmers from the perspective of mobile modernity [J]. Audiovisual, 2020 (11): 160-161.
9. Zheng Daqun. On Body Narration in Communication Form [J]. Academia. 2005 (05): 183-189.
10. Zheng Daqun. On Body Narration in Communication Form [J]. Academia. 2005 (05): 183-189.
11. Rudolf Arnheim. Art and Visual Perception [M]. Trans. Zhu Jiangyuan, Sichuan: Sichuan Publishing House, 1998: 508.
12. Ma Le. A study of "three agricultural" short videos in ShakeYin platform, Heilongjiang University, China, 2021.12.4.

An Analyze on Intersection of two female figures in the Film In the Mood for Love and an Interpretation of its Metaphor in cultural Identity

Qi LI

Otto-Friedrich University of Bamberg, Bamberg, Bayern, Germany

Abstract: The interpretation of the film In the Mood for Love carried on from the perspective of the female characters: the intersection between Su and Chow's wife, and moreover, the cultural identities, which symbolizes the Hong Kong culture that was bred grew from intersection: different backgrounds assimilated in each other.

1 Introduction

As the inspiration for Hong Kong director Wong Kar-wai's famous film *In the Mood for Love*, the novella *Dui Dao*(《对倒》), written by the Hong Kong writer Liu Yichang, was translated into English as *Intersection*. This is a term of describing two original stamps that look exactly the same but opposite in position to each other. A translation as such reflects accurately the content and plot of novellas, at the mean time brings about its metaphorical meaning to the film as well. As Deppman Hsiu-Chuang described in her book *Adapted for the Screen:* "Stamps on Envelopes Provide a Short-lived but Necessary Connection between Sender and Recipient"[1]. In the novella it defines the meeting an old man and a girl, but when adapted to *In the Mood for Love*, intersection described the secret love between the two protagonists as well as „visually linking them and rhyming their devastating isolation“[2]. Thus we may rise the questions: to what extend does the film covers the novella and how was Wong Karwai inspired from the novella while representing his own memory?

2 Metaphor in the intersected appearance and outfit

As one of the protagonists of the film, Su Lizhen stands in the opposite to Chow Muyun's wife, Ms. Chow. They fell in love with the same men and lived in the same neighborhood. Therefore, the relationship between these two women is intersected to one another. On the other hand, Su Lizhen's background as a migrant from Shanghai can not be ignored, as was emphasized from the beginning of the film, while Ms. Chow speaks Cantonese through the film, which implicates of her being native in Hong Kong. Thus, the contrast of the both female characters can be analyzed as the representations of intersection between the second generation of Shanghai immigrants and the Hong Kong native.

These two female characters appear in opposite ways from figure to identity. Su wears a high-necked cheongsam, a traditional Chinese dress, which is in line with her cautious character, while Ms. Chow prefers western modern dress sleeveless dress with a plunging neckline. Su fulfills "the traditional expectations of being homebound, submissive, and fithful."[3] Already in the first scene, when the neighborhood play Majong together, Su makes room for Ms. Chow (*In the Mood for Love* 00:05:20), which is also a metaphor for the later plot of their positions in relationship.

[1] Deppman, Hsiu-Chuang: Adapted for the screen: the cultural politics of adaptation in modern Chinese fiction and film, Hawaii, 2010, P. 104-105.

[2] Brunette, Peter: Wong Kar-wai, Urbana und Chicago 2005, P.94

[3] Deppman, Adapted for the screen, P. 112

Fig. 1. 00:05:20

The Scene presenting the affair between the intersected two families was nevertheless contrast to the reality, as easily deducted: the absent Ms. Chow and Mr. Chan(Su's husband), with only their voice heard and without their facial expressions shot, is the careless pair, whereas the protagonists, who are observed by the audience, choose to keep their love in secret.[4]

3 Metaphor in social relationships and self identification

The work of Su and Ms. Chow can also be regarded as another metaphorical intersection. Su works in a company that sells boat tickets, while Ms. Chow works in a hotel. Their jobs serve in opposite ways, one is to leave, while the other is to settle, even through briefly, in the city. The same is with their relationship with the neighborhood. Su declines her landlady Ms. Suen's dinner invitation and refused to get close to the neighborhood, who shares the same background with her. In doing so, she escapes form being observed and gossiped by the neighbors - she avoids being judged by the old-fashioned values, which she firmly stick to by concealing her love to Chow. Thus, it can be deducted, that Su's self-identity as the second generation of Hong Kong immigrants is split by the old and modern.

Since the beginning of China's war against Japan in 1937, Shanghainese moved to Hong Kong during the occupation. As the second generation from Shanghai, Su speaks Cantonese and understands Shanghainese, probably from the influence of her parents, just like the director Wong Karwai himself. Su grow up with Hong Kong being influenced by western while being educated by the traditional culture from her family. A cultural background as such results in that, her cultural belonging can be found nowhere, whether in Shanghai or Hong Kong. It can be thus deducted, Su rejected the neighborhood on the one hand to contain her individuality, on the other hand to reject being labeled as a immigrant, or non-local. While rejecting accommodating into the circumstance that she grew up with and familiar to, she can not be counted as a modern soul either. She restrain herself with a traditional moral standard and refuses the liberal modern morality of the 60s, which liberates the individuals from condemned of having affairs or being unfaithful towards marriage. Thus, her isolation in the modern society in the 60s redounds to her being lost in the self identity or belonging to social groups.

This split in identity contradicts to Ms. Chow, a local of Hong Kong, who accepted different culture. She doesn't alienate herself from the traditional Shanghainese or the neighborhood where Shanghai culture prevailed. She doesn't find her affair with Mr. Chan problematic and by no means restricted by the old moral norms. As a native, she finds her cultural identity completely in the modern society, in the openness to all culture, in her presence. She identify herself as a free individual in admitting completely to her desire. Mrs. Chow is more independent and same as Mr. Chan plays the dominating roll in marital relationship: when playing Mahjong with the neighborhood, she replaces her husband and Su companies her husband Chan. doesn't feel shamed for being unfaithful or hurting Su. Nor does she let herself fell into the dilemma of self doubting. In this regard, it can be concluded that the two women are in completely opposite positions when it comes to constructing character or establishing personality. Thus, the identification of the two female figures appears intersected.

[4] Xiao Xia: The complex metaphors: The artistic feature of the film In the Mood for Love , in Movie Review, Beijing 2006(15), P.26-27.

4 Metaphor of cultural backgrounds in individuals and intertextuality

The identities of the two female characters also correspond to the identities of Hong Kong, providing the audience with a glimpse of Hong Kong culture. Since the beginning of the 20th century, the colonial culture grew in China [5] and immigrants with different cultural backgrounds developed Hong Kong's culture in contrast and accommodation. Therefore, the new Hong Kong residents formed a culture in a unique way, which is different from any in its background. Hong Kong itself is not only a geographical concept, but a cultural context in which residents form its won way of thinking.[6] Just like the space in the city, the old and the new blend and converge in the same space, "as ‚old‘ and ‚new‘ are placed together in contiguity and continuity"[7], the unique atmosphere: openness to the to the modern and conservative from the history together constitute the so-called "Hong Kong culture".

During the adaption from the novel *Intersection*, the cultural background in the two cities was hinted in the two contradicted female figures in the movie, while which was represented by the two protagonists in the novella. An old man, Chun Yubai and a young girl, Yaxing walk on the street in opposite directions and their course intersected at an cinema. While the old man with a Shanghai background reminiscent the past in his memory, the young native girl look froward to her future, while feeling anxious about the present. Their meeting represents the meeting of the past and the future, as well as the both cities, Shanghai and Hong Kong. Different from Su, the second generation of immigrant, Chun sees in his past his root and stands firmly in it. However, the culture or the place that he identifies with is actually his own creation, as well as the film In the Mood for Love itself. The whole is constructed by the memories, a heterotopie, as was defined by Foucault, to the real space. This term describes the a space that copies the spaces in reality but doesn't really exist. The space in the mirror for example. In this case, the scenes in the film from the memories can all be seen as heterotopie to reality. Time and space are then melted into one, a combination of time-zone, instead of being split with dichotomy.

Cultural identification located in identifying with the memories of a certain timezone or space and is bound with culture of a certain location. The figures in the novella as well as in the film identify each with intersected time-zones and stagnate in them. However, their memories and expectations don't meet up with the the presence that their life path crossed, which is the reason, why they are isolated. The blossom past of one side and the unknowing future of the other side build the utterly tremendous intersection: when the two protagonists of the novella represents the 30s and the 60s, the film represents the 60s with Wong's memory from the perspective of the 90s. In the constant looking backs, "the nineties to the sixties as the sixties are to an earlier moment, and so on and on." [8] Through out the adaption, the isolation and seeking of the identity of the individuals become the everlasting theme, which is why the film still can be called as the adaption of the novella.

In the adaption from the novella to the film, the reader can observe the identical dilemma of the Hong Konger: the old generation looked back at the past, while reluctantly accept the modern culture; the new generation looked anxiously forward in the future, while facing the current problems of puberty. It is between these two states of living, the unique culture discourse of Hong Kong was formed and became independent from either its origins.

5 The Intersection between identities and the post modern society in the poetical film language

The last intersection in the film and its intertextuality with other pieces lays on the intersected identified time-zones and the era on a bigger plot. By the end of the film, the protagonist Su Lizhen visited the landlady with her son and was told that the neighborhood would be moving to America for the education of their grandchildren. Chow and hs wife are divorced. While Mrs. Chow regained her freedom, Su remains in the most traditional form of family with a child. The Shanghai neighborhood that represents the traditional moral moved into a whole different culture basis, through which the restrict from the past seemed no longer important. In stepping into post modernity the figures returned to their own comfort zones with less struggling. Time-zone as the centre theme of the film became more and more obvious with the ongoing of the plots.

At the end of the film, the time line is drawn to the 70s, with pictures of the protagonist Chow visited a temple of the Kampuchea while working as a journalist and reported the visit of the French president, which signalize the independence of Kampuchea. This event represents the utterly proceed of post modernity. Chow talks then his secrets into an old tree in the temple and hopping that his past would be sealed into the tree. As shown in the screen shot below.

[5] Akbar. p. 68

[6] Zheng Jiawen: The mirror image structure of Hong Kong Culture: From Dui Dao to In the Mood for Love, Guiyang 2007, 216.(12), P. 50-55.

[7] M. Abbas. Akbar, Hong Kong: Missing Culture and Politics, Minneapolis, 2002, P. 67

[8] Abbas, P. 52

Fig. 2. 01:30:42

The loneliness and split identification remain in the 60s and the confusion ended with the end of colonial era. In the scene, the figure of Chow stands not in the centre, but the ruins of the temple. The loneliness of the individuals seems unimportant and insignificant at that point, what truly matters is the time-zone on a bigger scale. The individuals are meant to accept the changing of era.

Even through Su and Mrs. Chow are not present in this scene, but they are brought up to the front. This is also the poetic in the film language of pictures. Their identifications with certain time-zones or the social norms in certain era are deconstructed by the comparison with the ancient temple. The female figures are utterly symbolized as the dilemma of the modern souls between the shackles of the past and the confusion of the future.

6 Conclusion

In the works, I emphasize the intersection of the social identities of the two female characters, Su Lizhen and Ms. Chan, in which the metaphor in the two cities, Shanghai and Hong Kong, was made. Moreover, the intersection is that, what Hong Kong culture is. It is bind together by the difference, accommodating to all while isolating the individual. Cultural identification related to the memories of a combination of time and space, referenced above as "time-zone" that is highly individualized. The intersection of which results to the isolation of the individuals of modern society, and the culture of Hong Kong. In the The process of studying the way the film was adapted itself opens up a artistic window to understanding the Hong Kong culture and the solitary of the individual in it.

References

1. Abbas. Akbar, M. Hong Kong: Disappearing Culture and Politics, Minneapolis, 2002.
2. Brunette, Peter: Wong Kar-wai, Urbana und Chicago 2005, P.94
3. Deppman, Hsiu Zhuang: Screen Adaptation: Cultural Politics of Modern Chinese Novel and Film Adaptation, Hawaii 2010.
4. Xiao Xia: The complex metaphors: The artistic feature of the film In the Mood for Love , in Movie Review, Beijing 2006(15), S.26-27.
5. Zheng Jiawen: The mirror image structure of Hong Kong Culture: From Dui Dao to In the Mood for Love, Guiyang 2007, 216. (12), pp. 50-5

SHS Web of Conferences **167**, 01007 (2023) https://doi.org/10.1051/shsconf/202316701007
CACC 2023

The Value and Character of Music as a Tool for Political Mobilization

Yulin.He*

University College London

Abstract: Music itself has no practical meaning. However, music can be given political meaning by politicians to some extent. This paper will analyze the particularity and value of music in mobilizing political movements from the Iraq War and Vietnam War.

1 Introduction

It is believed that music itself had no practical meaning [2], that is, it is meaningless to do with religion, race and environmental occasions. Besides, Pop music itself has a certain emotional power, can be used as a means to shape the collective atmosphere. Therefore, it is more important to investigate the effect of music on political movements, rather than the specific content and genres of these songs. In the United States, from the point of view of war, music can be a means to increase the cohesion and solidarity of the military, or it can be a channel to get people to resonate and stop the war. From the perspective of political election, music is a means for politicians to convey political strategies, and it can also be a bridge to gain public support through celebrity politics. In this way, the meaning of music in politics is given by politicians. Specifically, music has multiple meanings, both provocative and contradictory, based on multiple themes. Therefore, this article will analyze the specificity and value of music in mobilizing political movements, from analyzing the wars include the Iraq war and Vietnam War and several examples in American presidential election.

2 The value and embodiment of music in American war

Music, as a form of communication, is symbolic, so music, like language symbols, is endowed with multiple meanings. As it was said [10], music composers cannot completely control the meaning of music expression, that is, once music has audience and enters the world of public discourse, the meaning of music will be shaped by the audience. American military history has proved that music has the characteristics of inspiring soldiers to fight. In the Iraq war, music had a close relationship with the U.S. military in Iraq, because the development of new technology and the global popularity of music enabled music to play a positive role in military deployment [11]. Specifically, music could be used to cover up the sound of soldiers' location, such as helicopters and explosive devices, which can protect soldiers from the sight of the enemy to a large extent and has a good combat effect. In addition, the U.S. military has installed audio and communication systems in tanks, allowing soldiers to listen to music while on patrol to relieve the tension and stress caused by the war and become the source of inspiration for soldiers. In contrast, since the 1960s, there has been protest music in the United States, with rock music from the Vietnam War as a typical example. It can be said that the Vietnam War was the first rock 'n' roll war in the history of the United States. Rock 'n' roll was a rebellious force that cured the confused American soldiers and spread anti-war slogans to all parts of the country. In the end, the counterculture, mainly rock music, defeated the political appeal of the American government, influenced the public's perception of the war and social injustice, and to a large extent led to the failure of the United States in the Vietnam War [13].This part will investigate the reflection and value of music as a kind of benign and malignant political mobilization in the American war by specifically analyzing the Iraq war and the Vietnam war.

2.1 Music in the Iraq war: a source of combat inspiration.

In American military history, military radio broadcasts often played songs that were patriotic or inspired to fight, and the U.S. government believed that music could keep soldiers' morale up and ease tensions. During the Iraq war, American soldiers listened to music before leaving their military bases to patrol, and metal and hip-hop were often their preferred musical genres. To some extent, music, as the inspiration for combat, is related to listening to music during patrol, because listening to music during the execution of tasks could make people highly alert and focused. Thus, communication mode between people is more through music, which greatly reduces language communication and enhances attention

*19938278951@163.com

https://doi.org/10.1051/shsconf/202316701007

in combat. Soundtracks In War [14], shows the sense of participation of music among young American soldiers, as well as the positive effect of music on War. At the 0: 40 in this video, one of the soldiers said, 'playing heavy metal music before we go out'. The heavy metal music acts to get rid of the psychological barrier for the soldiers in the combat and stimulates the progress of the battle. Also, two young soldiers in the video were playing folk music with guitars to express their homesickness and arouse their desire to finish the battle quickly and successfully. Some black soldiers used hip-hop lyrics to express their strong desire to win, and the frequent use of bullets, war and enemies in the lyrics was undoubtedly a catalyst for the promotion of combat inspiration. In addition, [11] also describes the positive effect of music on war mobilization through interviews with American soldiers involved in the Iraq war. Among them, soldiers take part in special musical rituals as they inspire the fight. For example, the song I Don't Give a Fuck, written by Lil John, was the popular track when the soldiers left the camp, they also play the song until they lose control and scream. This is reasonable, because war is out of control, and the process of combat requires great vigilance, so soldiers need to be excited. In this way, by playing a song loudly and rudely over and over again, they gain the positive power of the song and are more likely to stay excited and ready for battle. Thus, music could unite a type of people, that is, to increase the sense of team in the army, which is crucial [12]. By performing the particular song together in the same space, soldiers create a common consciousness through their common actions, namely war mobilization.

2.2 Music in the Vietnam war: mobilizing people to protest and stop the war

During the Vietnam War, the radio stations of the armed forces were closely related to the lives of soldiers and the source of their music. During the political and social revolutions of the 1960s, the counterculture in America was revived within the rock and roll genre to represent the political protest of social change. Folk music, with simple materials and straightforward lyrics, tells of real life, difficult political times and public struggles. With the continuing escalation of the Vietnam war, citizens and soldiers in the war of hostility is very strong, rock music as rebel force has become typical of an American counterculture, during the Vietnam war, rock songs reflect the love and sex, to a great extent, restricts the U.S. government launched the war mobilization strength [13]. For example, Pete Seeger, a communist music composer, he created a song called Waist Deep in The Big Muddy during the Vietnam War period which was extremely dark and ironic. The song tells the story of an army crossing a river in 1942. The captain thinks that the river is very shallow, so long as you plunge into it, you can surely wade across. In the lyrics, even if some people doubt, 'with all this equipment/ no man will be able to swim', the captain still opposes, 'all we need is a little determination'. Finally the captain himself was swallowed by deep water, the others were able to turn around in time. This song was already a strong reflection of the Vietnam War, but in 1969, after Pete Seeger sang the song outside the White House, he added, 'every time I read the newspaper/I think of the story', it seems to be a direct expression of my disaffection with the Vietnam War. In addition, as rock music spread the idea of peace to the American masses, millions of young people played the slogan "love and peace" at the Woodstock music festival to express the idea that the younger generation in America hates war. The rock music referred to in [6] as a channel to unite the collective was intended to create a public consciousness during the Vietnam War, as a means of controlling and organizing a series of political mobilization. Finally, in 1973, the American government was forced to sign a series of peace agreements. Rock music, as a channel for the peaceful groups in the United States to spread love and peace, gained the resonance of the general public through music.

3 The value and embodiment of music in the United States presidential campaign

The image of a politician is usually shaped by the voice that is exposed to the public [1], while music is symbolic and therefore given different meanings in different scenes. In addition, the emotional power contained in pop music can bring the audience into a specific scene [4], while compared with other types of music, pop music has a wider audience and familiar songs can build a sense of teamwork in the crowd. Throughout the history of the United States presidential election, President Franklin Roosevelt selected the appropriate music to express his political campaign mode, to attract the public's mood, to improve competitiveness during the campaign, President Barack Obama as an African-American presidential candidate, in the public performances, supporting and representing black rap music, to express his desire to redefine the image of black Americans and identity, attracted a large number of black voters. As a result, the campaign songs became the bridge of the American politicians and voters, expressed the politician's political demands, and also attracted the corresponding to the audience at the same time [13]. Besides, due to the development of modern science and technology and new media, music is not only the approach as a politician express his inner political mode during the election, but also the expression of a celebrity politic, that is, campaign songs contribute to the politicians to build celebrity influence to expand their visibility, to mobilize and guide the choice of voters finally. In other words, American politicians realize the importance of media in self-expression and public opinion formation, and celebrities in music, as objects of media attention and personal pursuit, will become an important part of politicians' election campaign [15]. This part will focus on the background of American presidential election and analyze the value of music, especially pop music in expressing political mode and music as the embodiment of celebrity culture.

3.1 Music in conveying political strategies and mobilizing voters

The music of the American presidential campaign is an important part of a candidate's campaign, containing the political model and history, and helps to present an idealized politician [3]. In the United States, the use of campaign music is also influenced by the leadership style. For example, the campaign songs of the Democratic Party usually reflect the role of the people in national politics, while the campaign songs of the Republican Party emphasize authoritative and capable leaders [8]. Franklin Roosevelt, a democratic candidate, used Milton Ager and Jack Yellen's popular song, Happy Days are Here Again, in his presidential campaign to express his political vision for an era of happiness for the American people after the crisis of the great depression. It is worth mentioning that when the song was released, the country's economy was experiencing unprecedented prosperity. In the lyrics, You Cares and Troubles are Gone, the optimistic mood after encountering setbacks could not make the American voters resonate with it. Therefore, President Roosevelt's election process was not optimistic in the beginning. However, in 1932, when the happiness and joy of the American people were shattered by the economic depression a few years before, the song's lyrics about the troubles of the past and its optimism for the future became emblematic of Roosevelt's political strategy. During his election campaign, the lyrics of Happy Days are Here Again were not altered to highlight the candidate's identity or name, but instead followed him on live shows and radio shows. Starting in 1932, Franklin Roosevelt founded the New Deal coalition, which appealed to urban citizens and intellectual conservatives. In this way, to win over these target groups, Roosevelt's campaign chose to use popular music to convey political messages to members of the New Deal. In other words, the expression of annoyance and the optimistic hope in the lyrics expressed the future of the United States that Roosevelt wanted to portray as the democratic candidate, which to a large extent also resonated with American voters. In addition, regular radio broadcasts of campaign songs enabled Roosevelt's political strategies to be widely disseminated through the media. As a result, when citizens heard the song Happy Days are Here Again, they remembered the democratic presidential candidate. In the end, Roosevelt realized his vision of four consecutive terms as President of the United States, and the campaign song surpassed him and became one of the songs of the entire democratic campaign [13]. After that, influential pop music became an important part of the presidential campaign in order to convey political strategy and mobilize popular support. For example, On the contrary, hip-hop's constant references to violence and male hegemony, and even its defense of the drug trade, have become the point that politicians avoid to some extent. Music, therefore, for the American political mobilization, especially in manufacturing public opinion and attract voters is of promoting significance, however, some content out of the reality in the music will also bring interference for politicians and citizens. Thus, in the use of specific music to convey political ideas and desires of thinking is considered.

3.2 The embodiment of music in American celebrity politics

One aspect of American celebrity politic is that celebrities actively participate in the politics of political parties, which is closely related to the content of campaign songs in the campaign for president. It is mentioned that the importance of building a public image for a successful politician [7], which means that politicians to a large extent need to use the social influence of celebrities and the communication ability of media to enhance their popularity and gain competitive advantage when participating in political activities. According to [5], when John f. Kennedy was President of the United States, he often invited entertainment stars such as Marilyn Monroe to parties at the White House. This is a sign of the beginning of the era of mass media-directed governance. Similarly, popular music plays an important role in campaign music in the United States, where every candidate wants to use the media to raise awareness, and combining the skills of a writer with the activities of a politician is advantageous. Specifically, campaign songs themselves are well-known, and the music establishes a connection with the candidate in a relatively simple mode, while the candidate can use the writer's popularity in the society to indirectly establish a connection with more people. As a result, American politicians' choice of certain types of music will in part target the audience they want to appeal to. For Barack Obama, the first black President in American history, the freedom and struggle themes of his favorite hip-hop music also expressed his political vision for black equality, which also gained the support of some artists. Young Jezzy, a famous black R&B singer, through the lyrics in the song My President, Obama and his preference Lamborghini appeared together, reflecting the significant meaning of Obama's election campaign to the black community, the lyrics repeated we, embodying the respect of rappers for Obama, also trying to create a relationship of equality in voters and candidates, and improving the popularity rating and support rate of Obama voters [9]. The results show that Obama's supporters are mostly young, and given the fact that hip-hop music is particularly popular among young people, it makes sense that the celebrity effect of music has played a vital role in political mobilization. Besides, it is common in the United States for musicians to participate election activities with a presidential candidate or to use their popularity to express their support for that candidate in public performances, from one of the videos about the last day of Obama president campaign in 2012 [14], the famous singer Bruce Springsteen sang to Obama's supporters in Wisconsin. In the video, he mentioned his support for Obama's political strategy of change in 2008 and expressed his support for Obama through a live song. In addition, he flew with the President and held parties together. Therefore, in the process of mobilizing political activities, music can convey the political vision of politicians through specific songs and changing lyrics, and enhance its national influence by using popular music and the popularity and influence of writers. However, according to the reach from [7], celebrity influence seemed to have the greatest effect on people with low levels of political participation, which blurs the direct value of celebrity effect in election campaign.

4 CONCLUSION

This article bases on some academic and media discourses about war and the presidential election in the United States, using the Iraq war and the Vietnam war as an examples to explore the music could act as an inspiration in war and as a tool of prevent war, explore the campaign with the example of that the presidential race music helps America presidential candidates to express political strategy, to attract the support of the target population. In addition, the value of music in mobilizing political activities is that music can attract audiences to a large extent and create common social consciousness and team spirit through specific physical environment. Also, the celebrity politics in the United States conveyed by music also expresses the communication function of music in the special atmosphere of political activities in the United States could contribute to American politicians to obtain public support and mobilize political activities. What is controversial is that the research shows that most of the target groups of celebrity effect are not those who participate in politics, that is, it has little significance to the success of politicians' election. Therefore, in the further research, the specific embodiment and influence of music in American celebrity politics will become the focus.

REFERENCES

1. Brader, T. (2005). Striking a responsive chord: How political ads motivate and persuade voters by appealing to emotions. American Journal of Political Science, 49(2), pp. 388-405.
2. Cloonan, M., & Johnson, B. (2002). Killing me softly with his song: An initial investigation into the use of popular music as a tool of oppression. Popular Music, 21(1), pp. 27-39.
3. Dewberry, D. R., & Millen, J. H. (2014). Music as rhetoric: Popular music in presidential campaigns. Atlantic Journal of Communication, 22(2), pp. 81-92.
4. Frith, S. (1987). Towards an aesthetic of popular music. In R. Leppert & S. McClary (Eds.), Music and society: The politics of composition, performance and reception. pp. 133–150. Cambridge, UK: Cambridge University Press.
5. Gamson, J. (1994). Claims to fame: Celebrity in contemporary America. University of California Press.
6. James, D. (1989). The Vietnam war and American music. Social text, (23), pp. 122-143.
7. Street, J. (2012). Do celebrity politics and celebrity politicians matter?. The British journal of politics and international relations, 14(3), 346-356.
8. Larson, J. (2009). American tune: Postwar campaign songs in a changing nation. The Journal of Popular Culture, 42(1), pp. 3-26.
9. Nielson, E. (2009). "My President is Black, My Lambo's Blue": The Obamafication of Rap? Journal of Popular Music Studies, 21(4), pp. 344-363.
10. Pearce, W. B. (1994). Interpersonal communication: Making social worlds. HarperCollins College.
11. Pieslak, J. R. (2009). Sound targets: American soldiers and music in the Iraq war. Indiana University Press.
12. Pickering, M., & Keightley, E. (2006). The modalities of nostalgia. Current sociology, 54(6), pp. 919-941.
13. Schoening, B. S., & Kasper, E. T. (2011). Don't stop thinking about the music: The politics of songs and musicians in presidential campaigns. Lexington Books.
14. The Telegraph. (2012, November 5). US Election 2012: Bruce Springsteen Barack Obama on final day of campaigning. [YouTube]. Retrieved from: https://www.youtube.com/watch?v=JygWoIeW224
15. Van Elteren, M. (2013). Celebrity Culture, Performative Politics, and the Spectacle of" Democracy" in America. The Journal of American Culture, 36(4), pp. 263-283. Retrieved from: https://search.proquest.com/docview/1852709752?pq-origsite=gscholar

Background of Western Modern Art in the 20th Century

Lei Liao[1*]

[1]University of Toronto, Toronto, Canada

ABSTRACT: In the 20th century, many major events were represented by the two world wars. They not only changed people's living environment, but also affected people's mental state. In the field of arts, modern industrial science and technology have opened up the expressive techniques of arts. War trauma has affected the creative mood of artists, and the inheritance of art value from the previous generation have also profoundly affected the inner spirit of arts. From the perspective of social and cultural contexts, the analysis of the time motivation of the development of western modern art in the 20th century can further understand its influence on art expression techniques, and also assist people better understand the connotation of art works, and better appreciate art works.

1. Introduction

With the development of society, it is not difficult to see a trend which is the change between non-realism and realism (Honour & Fleming, 2005) [1]. Moreover, the field of art has gradually shown many genres characterized by non-likeness. Therefore, many schools of western modern art in the 20th century also reflect this characteristic. For instance, grotesque, abstract, exaggerated, obscure and anti-traditional have become the main impression of most citizens on modern art from different countries in the world. However, artists would pay more attention to personal subjective feelings than the "normal" people when they are working on their artwork, and tend to be more extreme in artistic expression, making their works difficult to understand. I think this is not surprising for me. Because in this period, people have experienced world wars, the new development of industry, science and technology and so on. Also, the communication between different countries had increased to some extent and the artists had started their new adventure of exploring art in different aspects.

Art itself is a metaphor for the world, and its existence means that the world is metaphorized. The recent history and culture of mankind are full of cracks, especially the people at the bottom in the industrial revolution, the two world wars, the Nazi Holocaust, colonialism, racism, terrorism and other events and struggles, and they are in particular in hardship and suffering. Therefore, in art creation, the metaphorical meaning of images is more valued by the art world. When looking at the metaphorical expression behind the paintings, we should not only appreciate the intuitive experience conveyed by the surface, but also explore the deeper, rational and hidden philosophical thoughts through the surface.

2. The development of the techniques of modern industrial science and technology to the artistic expression

The modern industry has affected the techniques of art expression and made it proceed faster than before. The development of science and technology has broadened the artists' sight of understanding the world. Hence, they begun to create a kind of new expression of art. At the same time, the fast pace of life which is brought about by industrialization has increased people's sense of tension and made people no longer satisfied with the traditional forms of artistic expression.

In the 20th century, human beings have created more scientific and technological achievements. The development of numerous materials, the exploration of optics and the development of machinery have enriched the expression techniques of art, and artists have boldly used all kinds of new materials. Take an example in the Art of the Twentieth Century by Harrison (2004) "*The Bride Stripped Bare by Her Bachelors*" which is created in 1915-1923 by Marcel Duchamp [2]. He used to show that he wanted to get away from the physical aspect of painting, so he created this masterpiece. And this must be one of the most important showcases of the creation of industrial infection in the history of art.

In addition, Russian Constructivism is the result of the innovative use of modern materials in the field of fine arts (Wang, 2011) [3]. They used steel wire, glass and other materials to make collages, to show the achievements of modern industrial science and technology which have non-traditional abstract concepts. So, they are also called artistic engineers (Gough, 2005) [4]. In Dadaism, the use of new materials is often reflected. The artwork I have mentioned above also can show this characteristic.

Furthermore, advances in industrial technology and

* Corresponding author: sherryliao48@gmail.com

science have greatly reduced the artists' attention to religious themes. As a result of the abandonment of traditional religion, the individual's search for self-worth became the God of their faith. The inability to locate personal values and the confusion of self-knowledge made art a substitute for faith in its role of spiritual salvation for life. Regarding the spiritual expression of modern art, Motherwell (2007) argues that the social condition of the modern world has given form to every experience, which is a spiritual collapse following technological progress and the collapse of religion [5]. This condition has led to the isolation of the artist from society. The social history of the modern artist is a spiritual existence in a world that loves wealth. He also expresses the true expression of modern art history to modern emotions, each artist needs his or her own religious faith, a faith that cannot be replaced by a view of reality, and the artists themselves are all guardians of the spiritual world. Rothko's connection between abstract forms and spirituality is often strongly mystical. If ancient forms bear a resemblance to our symbols, it is not that we are deliberately trying to imitate them, but that we are trying to express the same state of consciousness and its relationship to the world. If previous abstract art was a scientific and objective expression of the times we live in, then our art is a search for an image that is a metaphor for a more complex, inner self, and a new vision and consciousness. Rothko has always denied that he is an abstract painter; he calls himself a myth-maker, which is inseparable from Rothko's Jewish identity. Disliking concrete imagery and the radical style of Western painting, the Jews' quest for the ultimate and their connection in the field of art has a strange and profound spiritual significance.

Through abstract expressionism, art can only contribute to society if it becomes an alternative to religion. Other forms of communication could not achieve this. In the introduction to the 1947 exhibition Imaginary Word Paintings, Newman's on the discovery that the Kwakiutls artists used abstract shapes to paint on animal skins, and that the shapes depicted were a living thing, a vehicle for abstract thought complexes (Schreyach, 2022)[6]. The purely stylistic language of the Kwakiutls was guided by a ritual will toward a metaphysical understanding. The suggestion of the essence produced by this abstract form is a real, not a pseudo-scientific illusion. Thus, Newman identifies the basis of aesthetic activity as the pure idea, an idea that is complex, linked to the mystery of life, of man, of nature, to the hard, black chaos of death, and to the gray, soft chaos of tragedy. For only the pure idea has meaning; everything else is just everything else. Americans evoke their world of emotion and fantasy through a personal language that has no shape of self-knowledge. It is an act of metaphysics. Led by European abstract painters, we enter their spiritual worlds through self-knowing images, which is an act of transcendence.

The development of society and technology after World War II promoted further economic and cultural renewal, and the development and success of Abstract Expressionism, as a typical representative of American culture, has become an important element for historians to study the economic and cultural development of the United States after World War II. Abstract Expressionist thought had a great influence on the resistance to the bourgeois theory of worship and the idea of commercial interests to obtain individual inner freedom and democracy. As Talebi (2020) states, the rise of abstract expressionism is thought to correspond highly culturally to the overall military, economic, and technological rise of the United States in the post-World War II western world [7]. Abstract Expressionism, with its maturing artistic thought and great achievements in creative practice, won the recognition of mainstream American society, and their works were invited by the Metropolitan Museum of Art, New York's Museum of Modern Art, and the Whitney Museum of Modern Art, the highest cultural and artistic institutions in the United States, to hold exhibitions and collect their works.

Abstract Expressionism has become the most representative American art school. First of all, Abstract Expressionism, representing the American avant-garde, became a classic of the future development of American art, successfully defeating the traditional American realism and the trend of returning to vernacular art. Although located in the elite class, its audience was also expanding, and the idea of finding oneself and discovering one's own self-consciousness of subjectivity also became the belief of self-redemption for the new era American people to overcome the fear psychology brought by the bourgeois technological development. Secondly, the powerful and unique artistic expression of Abstract Expressionism was a great success in the exhibitions in Europe, and the economic and cultural arts in Europe were affected by World War II, which shifted the center of culture and art to the United States. The United States got rid of the long-term domination of European culture over it, won the right to express freely, and became the new hegemon of world culture, opening the outstanding American art era in the 20th century.

While searching for the spiritual world of the inner self, the artist expresses it through the visible form of art. Its expression is a tension between the artist's personal values against the contradictory values of interests in the bourgeois industrial development, a struggle and release of confusion, anxiety, confusion, anger, loss and pain in the heart. Traditional reproduction art can be expressed through the imitation of the real world, where the artist translates the subjectified spirit into the object form of the real world. Reality is visible, and the forms of natural objects are readily available, so art can be used to give reality a storytelling and three-dimensional illusion for empathy purposes. But most works of art express forms that are static and immobile. Due to the advancement of capitalist technology, the invention of the camera replaced the imitation of reality, bringing the realist style to the end of art. But the new problem was that the intuitionist to expressionist portrayal of the emotion of life was to express the vitality and fluidity of life, and therefore the abstract form of modernist painting was required to express the stretches of time through form. How the Expressionists gradually abandoned the depiction of nature to express the temporality of life through the formal elements of color, line and form. Matisse and Kandinsky, for example, including the formal symbols of Surrealism, all expressed the way in which artists unconsciously and freely created form. Kandinsky's production of form paralleled the development of music, which he called the structure of melody.

3. War trauma and the artist's creative mood

The development of trauma theory has expanded and extended from the physical dimension to the psychological dimension and then to the social dimension. In the contemporary academic discourse system, trauma has both mental and physical characteristics. A special variant of trauma is war trauma, and the collective trauma caused by war is another core issue of concern for 20th century artists. The traumatic places they wrote about are extensions of memory, and as the places where war trauma is carried, individual and collective memories overlap and intertwine, telling memories that cannot be forgotten. War trauma, for the first time during the First World War, was medically diagnosed and treated as an epidemic symptom of male hysteria.

The war trauma affected the psychological state of most of the artists who lived in that environment negatively. There is no doubt that both world wars played an important role in this matter. During the Second World War, Europe was extensively damaged. The persecution of some artists by the Nazis and the criticism of modernist art by the Soviet Union led to the concentration of art groups, especially the research groups of art theory in the United States. Abstract expressionism is an art movement that emerged in the United States after World War II (McCloskey, 2005) [8]. Artists who lived in that time experienced a serious crisis of faith after World War II, and then began to be irrational and unconscious. This feature also presents in many other European countries. One of the most famous artists, Pablo Picasso, the representative of Cubism, created the famous "*Guernica*" after being bombed in a Spanish town. This work is mainly in black, white and gray, and uses abstract and exaggerated methods to show the people and animals in the town during the war. The black, white and gray tones are used as the background directly feeling the dark and depressive atmosphere of the war. What is more, the masterpiece "*The Tress Dancers*" which is created in 1925 is also negative either. It reveals that the world war has destroyed people's living environment, and depressed the mood of artists in this environment. Meanwhile, this has become an important source and trigger of artistic creation.

This kind of negative emotions make the expression methods of art works exaggerated, deformed, abstract and absurd, and contain profound connotations that are difficult to possess in the peace time. As a consequence, the state of the characters in the art work is a conveyance of mental damage, revealing the scars left on the psyche by the memory of the war. This trauma is externalized from the psyche into physical manifestations, a direct impact of the brutal battlefield on the human flesh and psyche. Therefore, through the portrayal of the mental state of the combatants, the cruelty of war can be reflected, and the physical and mental trauma brought by war can be metaphorically reflected.

4. The inheritance and influence of the art values of the previous generation

Renaissance painting, Baroque painting and nineteenth-century painting are the three major peaks in the history of world art before the twentieth century (Craven, 1979) [9]. Compared with the first two peaks, nineteenth-century painting embodies the transition from traditional to modern forms, with a variety of genres and styles, experiencing neoclassicism, romanticism, realism, impressionism, symbolism, post-impressionism and other stages, becoming the historical source of modernist painting in the twentieth century. The nineteenth-century French-centered art movement can be called the third climax of European art development.

Regarding inheritance and influence of previous generations of art values. In the 19th century, there was a famous genre called post-impressionism. Artists of this genre had several different styles. The three representatives of the genre respectively derived three new genres with great influence in the 20th century. Cézanne pays attention to the volume, harmony and stability of objects and the performance of the internal structure. He believes that depth is more crucial than the plane, and he has created a lot of still-life works to explore this concept. As suggested by Barr (2019), although he claimed that he was studying the classicism of the past, and his ideas became one of the driving forces for the birth of cubist art in the 20th century [10]. The representative of this must be the famous artist Vincent Van Gogh and Paul Gauguin and they always pay attention to the expression of color and subjective feelings. About the latter one, he used a lot of pure colors in his artwork with strong decorative and philosophical nature, couples of his interest in primitive social themes. Hence, his works are full of primitive wildness.

Van Gogh's vision and expression deeply influenced twentieth-century art, especially Fauvism and Expressionism. According to Birren (1979), as a representative of post-impressionism, Van Gogh's style is similar to Gauguin, but the former's most common use is not a coherent solid color, but a non-coherent brushwork similar to neo-impressionism [11]. In his works, we can often see the use of warm colors, giving people a warm and unrestrained vitality, but also a strong emotional color. This technique continued into expressionism in the 20th century. The representative painter Munch of expressionism presents an absurd and intense sense of picture through exaggerated and distorted colors in the famous abstract style work "*The Scream*", which embodies the human inner spirit and has a strong impact. The "*Self-pictorial with Bandage ear and pipe*" is very similar to Gauguin's style of work.

5. Conclusion

As an important cultural form, the development of art is inseparable from the social context and the environment and complementary to the development of other fields. It can be foreseen that with the development of various fields, the art of expression will become richer and richer.

Based on the examples and analysis above, environment affects art, and art assists people to decorate the environment. It can be seen that the research process of art history cannot be separated from the textual research of the environment and the exploration of the artist's mood. At the same time, this connected vision is of great significance to the creation of artists, and also helps people better understand the connotation of art works, and better appreciate art works.

References

[1] Honour, H., & Fleming, J. (2005). *A world history of art*. Laurence King Publishing.

[2] Harrison, C. (2004) History and histories. In: Jason G. (Eds.), *Frameworks for Modern Art*. Yale University Press. Yale. pp. 10-11.

[3] Wang G. (2011) From the background of the influence of culture on fine arts. *Popular Literature and art, 22*: 147.

[4] Gough, M. (2005) *The artist as producer: Russian constructivism in revolution*. University of California Press.

[5] Motherwell, R. (2007). The Modern Painter's World. In Dore A. (Eds.), *The Writings of Robert Motherwell*. University of California Press. Berkely. pp. 27–35.

[6] Schreyach, M. (2022). Visual Documentation for Barnett Newman's Curatorial Projects, 1944–1946: Part II: Commentary and Assessment. *Source: Notes in the History of Art, 41*(4), 287-298.

[7] Talebi, H. (2020). Universal and Diversified Aspects of Qajar Siyah Mashq and Abstract Expressionism in Painting. *Painting Graphic Research, 2*(3), 151-165.

[8] McCloskey, B. (2005). *Artists of World War II*. Greenwood Publishing Group.

[9] Craven, W. (1979). The Grand Manner in Early Nineteenth-Century American Painting: Borrowings from Antiquity, the Renaissance, and the Baroque. *American Art Journal*, 5-43.

[10] Barr, A. H. (2019). *Cubism and abstract art*. Routledge.

[11] Birren, F. (1979). Neo-impressionism: The most scientific of all schools of color in art. *Color Research & Application, 4*(4), 200-207.

A comparative study of two architectural reconstruction design of Casa Batlló

Yang Lu[1,*]

[1]Faculty of Fine Arts, University of the Basque Country, Basque Country, Spain.

Abstract. Historical buildings are inevitably facing restoration and reconstruction in the context of urban renewal, how to carry out architectural renewal while inheriting the profound heritage of historical buildings? It is a challenge that architects in the new era must face. Based on comparative art, this article compares two architectural reconstruction design for Casa Batlló (remodeled by Antonio Gaudí in 1904-1906 and partially renovated by Kengo Kuma's interior finishes from 2015-2021). The results show that in the two renovation design, the two architects have many similar concepts on the continuation of architectural meaning, the importance of light and shadow design and naturalistic curve shape, but due to the different architectural background, cultural background and design techniques, this famous historical classic building, after the two renovation, is now presented to the world in a more inclusive and diversified form.

1 INTRODUCTION

Casa Batlló is located on Passeig de Gracia, in the downtown area of Barcelona, Spain. In 1904, Antonio Gaudí was invited by the owner of the house, Josep Batlló, a Barcelona textile magnate, to carry out the first "artistic packaging method" of Casa Batlló, and the remodeled building was full of mystery and far exceeded the expectations of the owner. This project won Gaudí the Barcelona Award for Best Architecture in 1907 and was listed as a UNESCO World Heritage Site in 2005.

Historical buildings will inevitably face restoration and transformation under the vicissitudes of time and urban renewal. In 2015, The Bernat family, the current owners of Casa Batlló, invited Japanese architect Kengo Kuma to undertake a second makeover. Historical buildings are endowed with different functions and properties in the context of the new era, and the progress of building technology, scientific and technological development and aesthetic needs, the transformation design will inevitably inject new elements. How to carry out architectural renewal while inheriting the rich heritage of historical buildings? It is a challenge that architects in the new era must face. Taking the two reconstruction design of Casa Batlló as an example, the feasible methods of historical building renovation are studied. It explores how to inject a new vitality into the classic historical and cultural heritage architecture that combines diversification, historical sense and contemporary aesthetics.

There is a lack of documentation on the design of the second reconstruction of Casa Batlló, no books, papers and other literature, only a few online articles for the purpose of promoting tourism. Therefore, the research results of this paper fill the research gap, research and analysis of the two reconstruction of Casa Batlló, and compare them, exploring the deep meaning behind the multiple design of "harmony but difference, different and harmonious", which is very valuable research.

2 METHOD

Taking the comparative research method as the main research structure, the similarities and differences of the two reconstruction design of Casa Batlló are analyzed separately, and the architectural meaning, shape, light and shadow design, architectural background, social and cultural background, and design techniques are analyzed and compared, and the continuation and renewal in the renovation design can be seen more clearly.

Firstly, the first renovation of Casa Batlló is sorted out by means of literature collation and research, and the architectural meaning, shape, light and shadow design, architectural background and other aspects are focused on research.

Secondly, through the web articles and pictures released on Casa Batlló's official website and tourism website, combined with the later field visits, the continuation and innovation of the second renovation design were analyzed.

Thirdly, the designer of the second renovation, Kengo Kuma, is compiled by literature and analyzed the characteristics of his other architectural works, and on this basis, the design concept of the designer's renovation of Casa Batlló is studied and a more reliable comparative study is carried out.

Fourthly, all the sorted data are classified and analyzed to intuitively form the same and different comparisons of the two renovation designs of Casa Batlló.

* Corresponding author: ylu001@ikasle.ehu.eus

3 DIFFERENT BUT HARMONIOUS

Casa Batlló has undergone two architectural renovations, both of which are partial transformations: Gaudí's renovation of the façade and interior parts, Kengo Kuma's renovation of the atrium and the decoration of the new staircase connecting the ground floor, and the creation of additional exhibition space in the former coal bunker in the basement. The two outstanding architects have similar interpretations of architectural meaning, light and shadow design and curvilinear shape.

3.1 Continuation of mysterious architectural allegories and naturalistic curves

Gaudí's transformation of Casa Batlló is full of architectural mystery: the façade is covered with ceramic collage (Figure 1), which is as brilliant as ripples at sea level and like the reflection of clouds in the water at dusk (Thiébaut 2005[6]); Cascading shapes resembling fish scales (Ascensio 2004[1]) and dragon scales (Xue 2013[8]) on the ridge, or the evolution of the characteristics of unicorns in the sea described in the novel 20,000 Leagues Under the Sea (Verne 1961[7]) ; The interior is shaped by plants, animals and waves of the natural world, and the overall design is curved, as if you are in the underwater world (Figures 3, 4).

Similarly, nature and the sea have a profound impact on Japanese society and culture. Kengo Kuma once said, "My architecture is not an epoch-making event, but it is indeed an undeniable fact that the sea has changed my architecture (Kuma 2005[4])." Therefore, when Kengo Kuma carried out the second transformation, he fully respected and continued the architectural meaning of Gaudí's architecture, and the metal mesh chain decoration he designed resembled the beautiful form of the waves fluctuating and undulating, like the scales of creatures wandering in the ocean, and like the dragon scales in mythological stories, shimmering with mysterious and shiny metallic colors. Each unit chain element of the net chain is a delicate curve, and the overall shape connects the visual points to form a curved rhythmic rhythm of the stack, which meets the abstract expression of Gaudí's dreamy architectural allegory and naturalistic curve shape (Figure 2).

3.2. The same homage to light and shadow design

When the first rays of light in the morning are cast on the façade of Casa Batlló, the performance of a rainbow of various lights begins (Tritt, Montes 2003[5]). Gaudí's designs demonstrate his importance and sensitivity to light. The multicolored ceramic collage decoration of the building's façade works with natural light at different times to present different dreamlike colors, and the light of the patio perfectly matches the color of the tiles, from top to bottom blue from dark to light, laying the background for the building's marine theme (Figure 4). In this regard, Kengo Kuma said: "Gaudí's transformation of Casa Batlló is a tribute to the light of the Mediterranean, especially its light, shadow and the colors of the sky and sea. The central courtyard captures the light of the Mediterranean Sea and distributes it vertically to all corners of the house. The color gradient from dark blue to light blue adds light to its vertical march and ensures that it does not lose pigment. In our project, we want to talk about the genius of light. "If Gaudí's project for the renovation of Casa Batlló is a tribute to the light of the Mediterranean, then Kengo Kuma's design is a tribute to the light used in Gaudí's transformation design. The aluminium links of different shades are purposefully organized, starting with brighter tones with more light on the roof and slowly darkening until they reach the depths of the basement and gradually turn black. From light to shadow, the architectural language of everything and the homage to Gaudí and natural light interpret by attaching light to curves (Figure 2).

The two renovations, spanning a hundred years of "dialogue", tacitly preserved the essence of the classics: the mysterious and dreamy architectural meaning, the romantic light and shadow design of the Mediterranean, and the curved shape loved by nature, reflecting the design principle of "different and harmonious" in the transformation of historical buildings. However, we should also see more "harmony and difference", that is, the buildings after the two renovations show richer and more diverse characteristics. Therefore, the following analyzes the differences between the two renovation designs from the architectural background and design approach.

4 HARMONY BUT DIFFERENT

4.1 Comparison of architectural contexts

In the first place, the differences in customer needs. The first renovation of Casa Batlló was due to the fact that the owner, Josep Batlló, believed that the appearance of the building was rigid and wanted his house to be unconventional and superior, so Gaudí carried out the transformation almost without restrictions, and his wild creative ideas and mature naturalistic style were fully satisfied in this transformation. Casa Batll ó , now belongs to the Bernat family (since the 1990s), Commenting on the latest renovation, Gary Gautier said: "Gaudí found a way to combine function, craftsmanship and beauty in an innovative and subversive way... If we don't want to be left behind by this era, we need to adapt to it, and what can we do to make our museum the most interesting place in the city? How can we get people excited--not just some, but everyone (Kai 2021[3])?"In the context of the design that takes into account Gaudí's extraordinary approach and vision, and innovation with cutting-edge technology based on respect and continuity, architects seem to need to be careful when transforming it. As Kengo Kuma says, "It seems like a very difficult challenge (Coulleri 2021[2]). "

Secondly, the differences in the architectural context that Gaudí and Kengo Kuma faced, that is, the difference between ordinary architecture and World Intangible Cultural Heritage architecture. Gaudí is free to transform his talents, and it is difficult for Kengo Kuma to break through and innovate under the pressure and restrictions of the historical buildings of the World Heritage Sites.

4.2 Comparison of design approaches

4.2.1 Materials

Gaudí's whimsy combines a wide variety of materials. The façade of the street-facing building is clad in Mares stone, decorated with colored ceramic shards and glass, wrought iron balustrades, and interior decoration materials such as cement, metal, wood, glass, etc.If the transformation on this basis wants to match each material one by one, it can be said that it is difficult to climb, and it is likely to become an accessory with no new ideas. Kengo Kuma, on the other hand, wisely jumped out of complicated materials and chose only one of them to innovate, which is called "responding to all changes with immutability". In collaboration with Kriskadecor, a custom metal mesh installation was created to abstract the meaning of the building.

As Kengo Kuma says, "In our vision, the space is covered by an aluminum curtain, and the fine materials capture the light like a fishing net and show us the various forms of light: brightness, contours, shadows... This way does not use any other materials... Here, we can talk about light and only about light (Coulleri 2021[2]). "The focus of the renovation design is light, and the materials were chosen to show the effect of light more perfectly.

4.2.2 Form of Expression

Gaudí made extensive use of bionic techniques in the renovation, drawing on the languages of biology, botany, geology and zoology to create architectural decorative elements and design changes. Gaudí's creativity never ends, and in all his works, even the smallest details shine with creativity. In his architecture, you can admire the meticulous work that strives for perfection and subtle details (Ascensio 2004[1]).Faced with Gaudí's difficult to crack "architectural code", during the second transformation, Kengo Kuma chose to intervene in an abstract way by making full use of the inclusive expression of modern design techniques, independent of the material, expression, color and historical dimension of the house. The lighter aluminium created by Krisadecor pours directly from the top, and every link of the chain works like a pixel, with natural and artificial light, light and shadow flow in the unpredictable, immersing people in a fantastical world and adding to the mystery of the building.

4.2.3 Color

Gaudí is very good at mastering color, and the color design of Casa Batlló is more colorful and rich like a "pointillism" painting, and the colorful colors and light work together to create a mysterious classical atmosphere. Kengo Kuma chose to cooperate with Italian lighting designer Mario Nanni to display colors by combining materials with artificial lighting, extracting the main colors in Gaudí's colors to change, and gradient aluminum with different saturation tendencies such as blue, silver, and black were organized, starting from the roof to interpret the boundless color of light and shadow. This gradation of detail continues Gaudí's pursuit of light and shadow, and when you explore it, you can discover the mystery of immersion: the metallic chain changes from top to bottom from shallow to deep, and from bottom to top it is a gradient from deep to shallow. The mystery of light and shadow overlaps with the fantasy of fairy tales, forming a sense of overlap that displaces time and space.

5 CONCLUSION

The same building is transformed by two outstanding architects of different eras, expressing respectful architectural language, identity and inheritance. The first renovation, Gaudí's unique naturalistic style shape, made Casa Batlló a century-old classic World Heritage building; Kengo Kuma's latest renovation design incorporates a design philosophy that pursues harmonious interaction between nature, people, and architecture while minimizing interference with historic buildings.There is a clear difference in the effect of the two renovations, but in the difference can be seen the "dialogue" between two outstanding architects spanning a century. This is not only the limitation of architectural background, the architect's pursuit of their own unique design concepts, the renewal of materials and the development of new technologies and new technological means, but also reflects the cross-era integration between different social and cultural backgrounds, the need for aesthetic evolution and the diversified development trend through the continuous inheritance of generations of architects.

The two architectural renovation designs of Casa Batlló show the architect's tacit protection of architectural principles and the creation and inheritance of architectural meaning, and also show the feasibility of diversified renewal of historical buildings under different social era backgrounds, architectural backgrounds and design techniques. In particular, as a representative classic work of the world cultural heritage, Casa Batlló has an important and far-reaching influence on its renovation design, and the comparative study of the two renovation designs fills the research gap, which is an important piece of the research puzzle of historical building renovation design, which has far-reaching significance, especially for contemporary Chinese architectural designers and the inheritance and

 https://doi.org/10.1051/shsconf/202316701009

transformation design of Chinese historical buildings, and has multi-faceted guiding significance.

Figure 1. Gaudí's renovated building façade, Source: Toutiao; **Figure 2.** Curve shape, light and shadow effect of metal mesh chain, Source: Kriskadecor's official website; **Figure 3-4.** Gaudí's interior effect after renovation, Source: Toutiao.

References

1. Ascensio,P.(2004)Dalí and Gaudí. Shaanxi Normal University Press, Xi'an, China.https://book.kongfz.com/167096/3677324749/
2. Coulleri,A.(2021)New Interior for Casa Batlló Stairs & Atrium.https://www.archdaily.com/967908/new-interior-for-casa-batllo-stairs-and-atrium-kengo-kuma-and-associates.
3. Design a smart Kai.(2021)The new transformation of Casa Batlló.https://baijiahao.baidu.com/s?id=1706939819256426137&wfr=spider&for=pc
4. Kuma,K.(2005)Building construction details. China Architecture and Architecture Press, Beijing.http://www.hopebook.net/tushumulu/44966/
5. Tritt,A.Montes,C.(2003)Gaudí. China Architecture and Architecture Press, Beijing.http://item.kongfz.com/book/25251238.html
6. Thiébaut,P.(2005)Gaudí: The Architect of Fantasy. Shanghai People's Publishing House, Shanghai.https://book.kongfz.com/6515/1811543991/
7. Verne,J.(1961)20,000 leagues under the sea. China Youth Publishing House, Beijing.http://item.kongfz.com/book/54161855.html
8. Xue,E.(2013)Antoni Gaudí,Mies van der Rohe, Ludwig.China Architecture and Architecture Press, Beijing.https://book.kongfz.com/191671/4373070129/

The Inner Dynamics of Kidult Culture: Trending Toys and games, Theme Parks, and Animated Movies

Peiying GU [1,a*], Wenrui LIANG [2,b]

[1]School of Art, The University of Edinburgh, 74 Lauriston Place, Edinburgh, EH3 9DF

[2]Department of Media and Communication, University of Malaya, Kuala Lumpur, Malaysia, 50603

Abstract. Trending toys, Disneyland, animation movies ......, which are traditionally connoted as children's synonymous entertainment, have progressively gathered a growing number of elder children and adults, which is called Kidult. This article analyzes and interprets kidult culture from three types of phenomena in an attempt to understand and explore the inner factors of the contributors. Meanwhile, it also provides aesthetic instructions and practical implications for the kidult-derived cultural and creative industries.

1. Introduction

In the multicultural context of contemporary society, an interesting social phenomenon is rapidly becoming a trend. An increasingly number of adults are keen to purchase toys that suit the psychological profile of children as well as the trending toy shop, blind box vending machine and Game consoles experiencing shop is popping up on the streets and shopping malls, products that focus on toy co-branding culture and concepts are being sought after and sold. Moreover, not only toys for adults, but also the world-famous theme parks, such as Disneyland park and the Universal theme park, which have a predominantly young visitor age distribution rather than the anticipated children.

Further research reveals that the popularity of animated films produced by these DreamWorks studios shows that the stereotype that animation is only for children is rapidly dissipating, with adults becoming one of the primary consumers of animated films[1]. Besides, certain animated films targeted solely at adults have seen much higher attendance rates in theatres than those aimed at children. These contradict common perceptions of the concept of child pronouns.

Kidult, also known as Peter Pan Syndrome, is a term used in the West to describe a group of people who enjoy children's products or are eager to participate in childlike entertainment behaviour. In the review of domestic and international studies about Kidult, most of them based on the analysis of design and aesthetics, personality traits and cultural-historical causes and phenomena. The influence of Kidult's fashion products on adult purchases in society is explored, for example, through the dimensions of Kidult fashion emotion to identify the socio-psychological characteristics[2], as well as a focus on the association of personality traits and decisions with Kidult's consumer behaviour based on the Big Five personality model[3].

Author noticed that scholars have mainly studied the media environment in terms of the external context of media communication, mass consumption: 'find out the dimensions of kidult fashion emotion and identify the influence of socio-psychological characteristics'; and the post-modern context, in order to study the media environment's intervention in Kidult's childhood and the formation of consumer values, the transformation of consumer culture on the subjective demands and concerns of people, and the external factors of the new commercial capital operating in response to the psychology of the audience.

This article makes an attempt to interpret and analyze the concept and content of Kidult's favorite products from the perspectives of media cultural studies and psychoanalysis to better understand the inner dynamics of Kidult culture and the attributions inherent in Kidult's choice of such products and content phenomena. To comprehend the reasons for the emergence of Kidult culture, this article will also discuss three thematic tendencies from trending toys, theme parks, and animated movies.

1.1 What is "Kidult"

'Kidult' is a portmanteau of 'Kid' and 'Adult', which first appeared in the *Times* on 11 August 1985 in the article "Coming Soon: TV's New Boy Network". It has used to describe a type of child, young person and young adult[2]. Besides, Wang and Ding believe that Kidult originally had two meanings: a child pretending to be

* Corresponding author: [a]Arielgpy2022@163.com
[b] 1697458702@qq.com

mature and a childish adult. The meaning used later on favored the latter, generally referring to the phenomenon of childish adults. In other words, adults with the mentality, innocent spirit, personality and fun of a child, or adults with a childish, childlike and naive heart. The term used today to describe the prevalent neo-adult condition can be translated as the Kidult[2]. Although this term is more commonly associated with Europe and the United States, in Hong Kong, it refers to a social phenomenon in which a growing generation of adults (roughly described as those aged 30 to 40) enjoys having fun.

Jacopo Bernardini brought up that 'the growing culture of infantilization that seems to have become one of the most recent hallmarks of postmodernity'[4], and he used proposed 'an infantilist ethos is achieved by correlating the postponing of the canonic indicators of maturity with the global market' which was defined by Benjamin R. Barber[5]. Drawing on scholars, he points out that the Kidult phenomenon is mostly found in the genarations X (born 1965-1980) and Y (born 1985-1995). Considering a contemporary perspective, what are the differences and further implications of the Kidult among the Z generation (born after 1995), which is now active as the mainstream?

Kidult is more than just a style. It is a way of life and a mentality[6]. Kidult culture has emerged due to an increasing number of young people recognizing and adopting this lifestyle attitude. From perspective media phenomena and cultural contexts[7], the understanding of the motivations of the Kidult crowd's behaviour is merely an interpretation and commentary on the phenomena generated in the operation of commercial capital to appeal to the audience, an external attribution from the mass media and postmodern contexts. There is still scope for theoretical research on the internal factors that motivate the Kidult population's behaviour. The behavioural paradigm facilitated by Kidult buying trending toys, visiting theme parks and watching animated films has certain theoretical connotations(as figure 1).

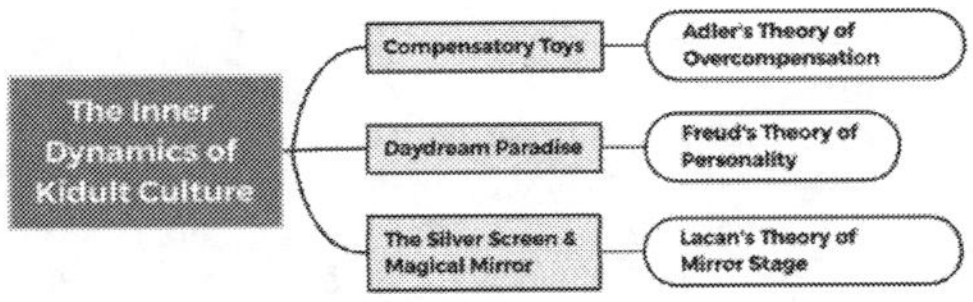

Fig1. The argument structure of Kidult''s Culture and CCI

2. Toys, Theme Parks, and animation movies

2.1 Compensatory Toys

Blind boxes, BJD (Ball-jointed Doll), Switch, PS4s, and other toys that suit children's psychological profile are becoming increasingly popular among adults. Additionally, products that focus on toy co-branding culture and concept are also being pursued and sold, such as a power bank with the figure of Subor Famicom, Lego*IKEA peripherals, and cosmetics with anime co-branding. These childhood memories have been re-engineered and upgraded back into the hands of adults, partly as a factor of fondness, but another important reason may be the mechanism from Alfred Adler's theory of overcompensation. Alfred Adler (1879- 1937) joined the psychoanalytic school that inherited and developed Freud's Personality philosophy. However, Adler's essential vision ran counter to Freud's and he then created individual psychology, introducing the concept of overcompensation. Adler assumed that if an adult's needs were not met as a child, they would continue to seek satisfaction of those needs throughout their lives, to the point of no return. If a child grows up in a poor or exceptionally materially deprived environment, he or she will spend the rest of their lives searching for material fulfilment[8]. The nostalgia for a certain childhood memory of a game even develops into an important deriving option of relaxation for adults. It is a common ground of reasoning that the call from deep in the brain makes people attached to and even dictates their behavior in the long term.

Other psychoanalysts have also made observations. The endowment effect proposed by Richard Thaler[9] argues that if an individual owns an object, they will value it more than other objects of the same kind. They would be more cherished these things and consider them to be a part of themselves. Familiarity is more easily recognized and accepted, and thus judged by subjective intuition as 'good' and 'right'. This is perhaps a result of the fact that those born around the millennium have a richer experience of childhood toys than previous generations. The chance to experience familiar childhood toys again in a contemporary context has the appeal of returning to the beauty of dreams.

Donald Woods Winnicott, a leading British psychoanalyst, also introduced the concept of Transitional Object in 1953, which refers to the child's perception of particular natural material objects as extensions or substitutes for the mother, mediating the child's transition from complete subjectivity to total objectivity[10]. A teddy bear can be used as a new anchor as an emotional projection for the mother throughout childhood, helping to alleviate the anxiety of separation from the mother as one learns to bond with the outside world. Adults who purchase many toys for this purpose may be compensating for unmet childhood needs and gaining similar emotional experiences from consumer behaviour and managed possession as they did as children. Besides, addictive consumption is gradually forming behind the popularity of blind boxes, as a high rate of personal reorder is a significant feature. Many consumers' retaliatory toy consumption behaviour may be a means of recalling childhood, maybe a painful flashback to the subconscious, or a projection of attachment to past experiences to bring the ego-personality full circle.

 https://doi.org/10.1051/shsconf/202316701010

2.2 Daydream Paradise

'Adults are just grown-up children', Walter Disney once said. Based on LBS data sources such as GPS and WIFI probes from Quest Mobile[11], a prominent data service provider, the researcher selected the visitors who visited Beijing Happy Valley, Shanghai Disney park, Guangzhou Chime-Long Paradise and Shenzhen Windows of the World in July 2017(as figure 2), and analyzed the user profiles of the visitors, which were mainly aged 25-35. In particular, Shanghai Disneyland has a high percentage of visitors under the age of 24, indicating Shanghai Disneyland''s popularity among younger visitors.

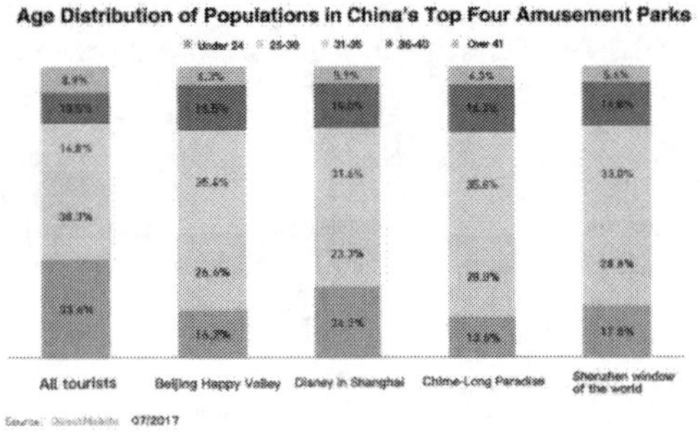

Fig2. Gender distribution of tourists at China's top four amusement parks[11]

The founding figure of psychoanalysis, Freud's Theory of Personality[12] confirms one point that a utopia where people can unleash their subconscious, imagine themselves as a princess or a magical figure, break free from bondage, expose their reality-repressed selves, and pursue the pure and simple pleasures of childhood. In *Creative Writers and Day-dreaming*, Freud describes creative fantasy as replacing for adults the playful activity of childhood, which like daydreaming, fulfils wishes. Therefore, the Kleinian school developed the theory of subconscious fantasy, arguing that creativity is often driven by symbolic and tinkering mechanisms. The subconscious mind is the uncontrolled behaviour and thinking of the human brain, of which the reasons for its formation have much to do with behavioural patterns formed by innate and childhood experiences. If the human psychological structure is compared to a floating iceberg, it is a massive part of it that lurks beneath the surface. The ego is the personality of primal desires represented by the subconscious, which in Freud's opinion(as figure 3), is one of the most primitive and authentic personalities a person can have. Moreover, dreams are a great way to reveal the subconscious in the Self. Kidult's appearance is partly a replay of a latent inner daydream. This will be converted into reality through artistic creation, and an amusement park with an immersive experience is an excellent way to recreate the dream.

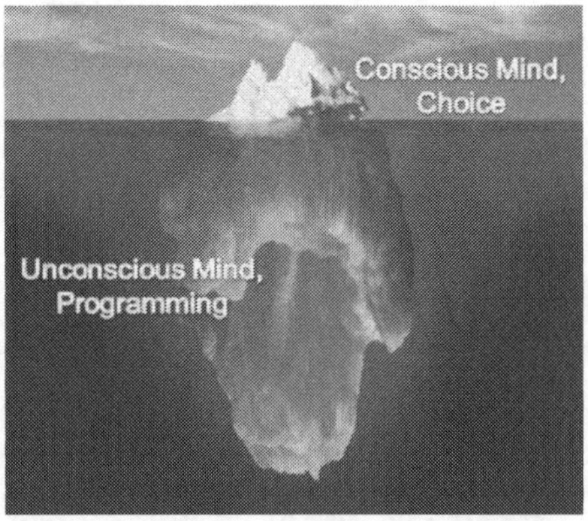

Fig3. Freud's topographical map of the subconscious and consciousness

2.3 The Silver Screen and Magical Mirror

In the book '*The Disappearance of Childhood*', American scholar Neil Bozeman argues that the dividing line between childhood and adulthood is rapidly blurring. Moreover, as television's medium pushes literacy to the margins of culture and replaces it at the centre of culture, a new concept of young adulthood begins to emerge[13]. Disney, the world's leading brand of animated films, released its first animated film in 1937, Snow White and the Seven Dwarfs, which took three years to produce and has become an iconic classic in the history of animated films as well as a memory for a generation. With the subsequent Princess series and the gradual rise of animation studios such as Pixar Animation Studios and Sony Pictures Animation, animated films are no longer the preserve of children, and more and more adults are becoming attracted to them. Besides, personal media tastes are generally difficult to change once formed in childhood and youth. Because cartoons or animated films heavily influenced adults as children, the patterned character images, scene styles, narrative styles, thematic ideas, and spiritual connotations of the films were unconsciously accumulated and preserved in their memories. After sifting years, separation and integration, they formed the perceptual schema and viewing habits of this generation[14]. Animated films made for Kidult became a trend, and the fact that adults became the new audience for animation was gradually accepted, a phenomenon closely related to the mirror stage theory proposed by Jacques Lacan in psychoanalysis.

The film is the art of dreaming. Based on the Jacques Lacan's mirror stage theory, Christian Metz has creatively applied psychoanalysis to film studies. He considers film as a valuable text that serves as a mirror for the full range of human consciousness and unconscious impulses, and a mechanism of identity for human self-knowledge[15]. Three stages of development from infancy to adulthood are explored in Jacques Lacan's mirror theory, and proposed three concepts of the real, imaginary, and symbolic realms. The baby gradually builds up a new perception of self by seeing its reflection in the mirror, up to a complete personality concept. Jacques Lacan uses the real world of the unconscious, the imaginary world of the preconscious and the symbolic world of consciousness to express the psychological structure of the subject, while Christian Metz borrows the distinction between the real, imaginary and symbolic worlds for the study of film[15]. This stage is a fluid transitional process in which people, in their dealings with the outside world, project themselves onto the relative other, which is not necessarily a person but can also be media forms, literature and art, ideology, etc. The process of watching animated films is a process of adult projection of the self, similar to seeking self-identity as a child. The process of watching animated films is also the process of reliving childhood dreams, the form of watching the screen is similar to looking in the mirror, as Snow White stands in front of the magic mirror, so that people walk into the magic mirror in this way with animated films, forgetting the pressure of reality for a short time to be in the mysterious world of fairy tales. It is a mirror image of the human psyche and unconscious instincts[15] and a mirrored expression of human psychological and instinctual desires [15]. Arnheim's visual form dynamics also supports the view that film and television images show co-occurring force patterns and ephemeral force patterns. When adults go to the cinema to watch an animated film that resembles their inherent perceptions, they develop a sense of heterogeneous and homogeneous viewing pleasure[14]. This is the reason why they choose animated films for their strange perceptual experience.

The impact of digitalization, technology and media advancement also grounded thinking about Kidult culture. In the book No Sense of Place: The Impact of Electronic Media on Social Behavior, Merovitz explores that how electronic media blurs the distinctions between adults and children, explaining how the adult state appeals to children's behavior and how adult minds affected by media retain childhood preferences. Thus, the movement of technology and new media has given the fractured memory a more vibrant and vivid character of presence, which is the era and technological trigger that catalyzed Kidult culture.

3. Conclusion

In contemporary art and cultural communication, there may be a possibility that in the current age of artificial intelligence, new forms of Kidult's culture will emerge from new technologies such as virtual reality, NFTs, deep learning and Generative Adversarial Network. The three behavioral modes represented by Kidult mentioned above, in which members buy trending toys, visit Theme parks and watch animated movies, can be interpreted from a psychoanalytic approach to explain the connotations of their formulation also had critical meaning for us.

When dealing with the concerns and consumption habits of the Kidult people, it is not just a matter of focusing on superficial seasonal fashion and replicating what existed in the past, but also on the deeper spiritual needs. It is possible to divide the user group and profile from the missing core of Kidult people's childhood, to develop concern from understanding and meet them in caring.

Turning from the study of the phenomenon to the deep psychological constructs can also provide a new orientation and direction for cultural innovation. Focusing on deconstructing Kidult culture from the audience's perspective, constantly identifying themes, tracing development tracks, exploring and recreating scenarios in will provide creative and inspiring thinking on the product design and development of toys, the concept and conception of parks, and the narratives and filming of animated movies. It also helps to provide aesthetic instructions and practical implications for the emergence and prosperity of the creative industry model based on Kidult extension.

References

1. He, S., Dan X. (2009) Research on the Consumption Value of Animated Movies for Adults. *Movie Literature* 52, no. 9: 26-28.
2. Cha, J. H., & Hong, K. H. (2007). A Study on the Characteristics and the Buying Behaviors of Kidult Fashion Purchasers-Kidult Fashion Emotion and Socio-Psychological Variables. *Journal of the Korean Society of Clothing and Textiles*, 31(9_10), 1373-1383.
3. Vandenhaute, A., & Venken, S. PERSONALITY TRAITS AS DETERMINANTS OF KIDULT MEDIA PURCHASE INTENTIONS AND LIKABILITY.
4. Barber, B. R. (2007). *Con$umed. New York: Norton & Company.*
5. Bernardini, J. (2014). The infantilization of the postmodern adult and the figure of kidult. *Postmodern Openings*, 5(2), 39-55.
6. Wang C., Ding J.(2010) Cultural Interpretation of Kidult Phenomenon.*Youth Studies* 33, no. 3: 37.
7. Wang C., Ding J. (2010) Cultural Interpretation of Kidult Phenomenon, *Youth Studies* 33, no. 3: 41-45.
8. Yang, X. (2006) On the Guiding Significance of Adler ' s Compensation Theory to Individual Adaptation to the Environment. *Journal of Adult Education College of Hubei University* 24, no. 3: 75-77.
9. Liu, T. (2010) The Psychological Mechanism and Influential Factors of Endowment Effect. *Advances in Psychological Science* 18, no. 4 : 646–654.
10. Wang, G. (January 2012) The Progress of Object Relation Theory in Psychoanalysis. *Journal of Nanjing Normal University(Social Science)* 58, no. 1 : 114-119.
11. QuestMobile. (2017). QuestMobile: Amazing! Have you seen the analysis of user portraits in amusement parks? An Insight Report on Tourist Population of China's Four Great amusement parks in 2017. https://www.questmobile.com.cn/research/report-hist

orical/70.

12. Feist, J., Feist G.J.(2005) Freud: Psychoanalysis. In *Personality Theories*, 14-51. Translated by Ru Li. Beijing: People's Medical Publishing House.
13. Postman, N. (1985). The disappearance of childhood. Childhood Education, 61(4): 286-293.
14. Wang, Q. (2017) Study on the Phenomenon of American Animated Film of "Whole Age". Soochow University,.
15. Shu, Y., Li, R. (2015) From Textual Analysis to Mirror Image Analysis——The Analysis Strategies of Psychoanalytical Film Theory.Journal of Renmin University of China 29, no. 2 : 151-156.
16. Meyrowitz, Joshua. (1986) *No Sense of Place:The Impact of Electronic Media on Social Behavior.* Oxford: Oxford University Press.

A preliminary exploration of the cross-cultural production, performance and dissemination models of contemporary Chinese adaptations of Shakespeare's plays (1990-2020)

Huangfu Shanghua*

Sichuan Film and Television University, Chengdu, Sichuan,611311

Abstract: The cross-cultural performance of Chinese adaptations of Shakespeare's plays abroad has gone through two stages: the 'introduction' of Shakespeare's plays and the 'going out' of Chinese adaptations of Shakespeare's plays. The paper aims to analyse the correlation between the two stages. By analysing the process of initiation, dissemination and reception of Chinese adaptations of Shakespeare's plays [see Table 1], which have been invited to perform abroad for many times and have attracted a certain level of response from Western society, the paper attempts to build a network model for the performance and dissemination of Chinese cross-cultural theatre works carrying traditional Chinese culture in the West.

1.Introduction

As cultural exchanges between China and the West intensify in the era of globalisation, the role of intercultural theatre as an important bridge between Chinese and Western cultures is gradually becoming more apparent. Foreign classical theatre works have been localised in China and adapted by Chinese theatre practitioners to the world in the form of cross-cultural performances. Among them, the plays of William Shakespeare[1] (hereinafter referred to as Shakespeare) are among the most representative Western classics and have become an important vehicle for cultural exchange between Chinese and Western theatre. Chinese theatre practitioners have recreated Shakespeare's classics on stage through national and pioneering adaptations and promoted the spread of Chinese adaptations of Shakespeare's plays to the West.

In recent years, there has been no lack of discussion on performance effects and Chinese-style adaptation strategies for foreign performances of Chinese Shakespeare plays. Since the beginning of the 21st century, more and more Chinese Shakespeare plays have been invited to perform abroad, which has attracted the attention of scholars at home and abroad. However, with more and more diversified ways of spreading Chinese culture abroad, the process of the successful performance of Shakespeare's plays in foreign countries and its cultural and social significance have not been paid enough attention. Therefore, on the basis of reviewing the domestic artistic production of Chinese Shakespeare plays since the 1980s, it is of epoch-making significance to investigate the mode and influence of Chinese Shakespeare plays in cross-cultural performances abroad since the beginning of the 21st century.

2.Contemporary Chinese Adaptations of Shakespeare's Cross-cultural Performance Background

The period between the introduction of Shakespeare to China from the West and the arrival of Chinese adaptations of Shakespeare in the West has gone through five stages: the first appearance of performances faithful to the original Shakespeare in the 1930s, the development in the 1950s and 1960s, the silence during the Cultural Revolution, the rise of local adaptations in the 1980s, and the large-scale invitation to perform abroad from the 1990s to the present. In the process of cultural exchange with Western theatre, they began to try to interpret Shakespeare from a Chinese perspective, and from a multicultural perspective, they tried to explore the contemporary value of Shakespeare beyond the classical value of Shakespeare on the basis of active translation and performance, and at the same time participated in the practice of constructing the contemporary Chinese theatre culture, creating a series of Chinese adaptations of Shakespeare with national, contemporary and pioneering characteristics. Since then, Chinese adaptations of Shakespeare have been attracting attention both at home

[1]William Shakespeare (1564-1616) was an English Renaissance dramatist, whose masterpieces include Hamlet, King Lear, Othello and Macbeth.

*huangfu_shanghua@163.com

and abroad. [1]They have gained increasing attention abroad, not only for their Chinese interpretation of Western classics, but also for their cross-cultural presentation of Chinese national cultural history on the international stage. The following performances have been staged.

Until recently, Shakespeare has been one of the main targets of cross-cultural theatre adaptations, and Chinese adaptations of Shakespeare have continued to receive widespread attention from the international community, and the process of performance and dissemination of Chinese adaptations of Shakespeare abroad, as well as their cultural and social significance, are worthy of in-depth study. [2]This study aims to take a holistic view of the entire process of performing the plays from China to the West and the various action points in the artistic field[3] . On the basis of reviewing the artistic production of Chinese adaptations of Shakespeare in China since the 1980s, this study aims to conduct a preliminary study on the cross-cultural production, dissemination and performance patterns of Chinese adaptations of Shakespeare that were successfully staged abroad from 1990 to 2020 and had a certain social influence from a sociological perspective. This is a preliminary exploration of the production, dissemination and performance models of Chinese adaptations of Shakespeare that were successfully staged abroad from 1990 to 2020 and had some social impact.

With the increasing communication between different cultures, since the 1990s, on the stage, Chinese theatre directors, who are at the heart of the production, have gradually made attempts to construct the Western classics in the local Chinese context in the process of the cultural clash between the East and the West, interpreting the contemporary thinking of Shakespeare from a unique local perspective and integrating a large number of elements with Chinese national cultural characteristics into Shakespeare performances, trying to create a new way of interaction between the two distinct cultures. On the other hand, the Chinese adaptation of Shakespeare is a re-creation of Shakespeare in a new cultural context and in a new form, which provides Chinese theatre practitioners with more room for reflection and interpretation in their adaptations of foreign classics. In the process of the dialogue between Chinese culture and world civilization in this way, not only the participation of dramatic texts is crucial, but also many non-textual factors play an indispensable role.

A review of the artistic production and dissemination paths and the image of the works in the different periods of domestic adaptations of Shakespeare is beneficial. At present, the domestic research results of Chinese adapted Shakespeare's plays are still mainly based on the acceptance of Shakespeare's plays in China, and some of them are enlightening, which can provide a basis for the investigation of the artistic production characteristics of Chinese adapted Shakespeare's plays in China. However, under the current background, in the process of cultural exchange, Chinese culture needs to "go out" in addition to "introduce".

3.A Preliminary Study on the Cross-cultural Performance and Communication Process of the Contemporary Chinese Adaptation of Shakespeare's Plays

In this exhibition, we explore the journey of Chinese theatre practitioners in constructing their own cultural identity in the changing context of globalisation.

In terms of the dissemination of Chinese adaptations of Shakespeare's plays performed abroad, the main project initiators are: Chinese producers, Chinese producers, Western producers and Western producers.A sociological analysis of the cross-cultural dissemination and performance patterns of Chinese adaptations of Shakespeare's plays abroad

The cross-cultural performance of Chinese adaptations of Shakespeare's plays abroad has gone through two stages: the 'introduction' of Shakespeare's plays and the 'going out' of Chinese adaptations of Shakespeare's plays. Many non-textual elements also play an essential role. Bourdieu sees 'field' as a 'social space with independent rules of operation'[4] , in which different fields are relatively independent of each other; 'habitus'' refers to the habits of thought formed by individuals during their socialisation, and is a kind of 'habitus'. 'Habitus' refers to the habits of mind that individuals develop during their socialisation, a "socialised subjectivity"[5] in which actors with habitus

[1]In 1986, the first Shakespeare Festival was held in China; in 1990, Lin Zhaohua's Hamlet 1990 was staged at the Beijing People's Art Theatre, generating a warm response in China, and in the same year, Wu Xingguo's Kingdom of Desire was first performed in the UK to great acclaim; in 1994, the Shanghai International Shakespeare Festival was held successfully one after another, bringing together theatre practitioners from all over the world to perform In 1994, the Shanghai International Shakespeare Festival was held successfully, bringing together theatre practitioners from all over the world to perform Shakespeare's classic plays, among which the adaptations of Shakespeare's plays incorporating Chinese elements provoked enthusiastic responses at home and abroad.

[2]From Pierre Bourdieu's (1930-2002) theory of the artistic field. The central concepts are "field", "habitus" and "capital". The existence of a field presupposes the existence of actors who, with their own habits, occupy different positions in the field and form a competition for access to capital, which has an important influence on the production process of artworks.

[3]From the actor network theory of Bruno Latour (Bruno Latour , 1947 - 2022). The core concepts of Latour's actor network theory are 'actor', 'translation' and 'network', in which actors recruit other actors in a translational way, building a dynamic and traceable relationship with different actors through four key stages: problematization, interest, recruitment and mobilization. The four key stages of problematisation, interest, recruitment and mobilisation create a dynamic and traceable network with different actors and facilitate its operation.

[4] Pierre Bourdieu. *The Field of Cultural Production* [M].Cambridge: Polity Press, 1993:162.

[5]Pierre Bourdieu. & Wacquant, L. J. D. *An Invitation of Reflexive Sociology* [M] .Chicago: University of Chicago,

interact and compete within a given field; "capital" forms the basis of social life and is not limited to material and economic forms. Actors with habits and various kinds of capital interact in the artistic field, forming an intercultural theatre practice. The domestic artistic production of Chinese Shakespeare plays is mainly influenced by three factors: the relationship between local and global, the relationship between art and society, and the appeal of the actors of cross-cultural drama in China and the West. The process of initiation, production, dissemination and performance of Chinese adaptations of Shakespeare's plays involves not only the artists, but also the various forces surrounding them, such as producers, critics and book reviewers. Among them, the role of the "cultural intermediary" is crucial. "The concept of the 'cultural intermediary' was introduced by Bourdieu in his commentary on the new petty bourgeoisie and is used to explore 'all kinds of professions concerned with presentation and reproduction'. [1] This concept has inspired researchers to focus more on the actions and meanings of intermediaries in cultural production and consumption when examining different industries and commodity chains. This concept has inspired researchers to focus more on the actions and meanings of intermediaries in cultural production and consumption when examining different industries and commodity chains. Bourdieu sees cultural intermediaries as 'merchants of demand'[2] , which are the Chinese tableaux directors who have to deal with audiences of different cultural backgrounds as well as social needs in the various stages of the creation and performance of Chinese adaptations of Shakespeare's plays.

Latour builds on Bourdieu by including both human and non-human actors in the category of actors, who are fluid and circulating and need to be traced in action in order to understand the connections between actors[3] . The dissemination of Chinese adaptations of Shakespeare's plays at home and abroad involves both human actor nodes such as playwrights, theatre directors, critics, media professionals, producers, producers, audiences, domestic and international scholars, and non-human actor nodes such as the social environment, national policies, communication media, theatres, media, websites, theatre festivals and performance companies. The artistic production of Chinese adaptations of Shakespeare's plays in China and their performance and dissemination abroad cannot be achieved without the combined action of human and non-human actors at home and abroad. This provides a new perspective on the process of cross-cultural performance and dissemination of Chinese adaptations of Shakespeare's plays abroad, and the study of its adaptation and performance has thus moved from the level of theatre proper to that of social culture. Based on Bourdieu's "artistic field theory", "cultural intermediary theory" and Latour's "actor network theory", the cross-cultural performance of Chinese adaptations of Shakespeare's plays abroad can be regarded as a combination of symbolic capital, cultural capital and economic capital. The interaction between the local adaptation and performance process of Chinese Shakespeare and the changing political, economic and artistic cultural environment is the result of the joint action of the artistic, economic and political fields in Bourdieu's "artistic field theory". The process of cross-cultural performance of Chinese adaptations of Shakespeare's plays abroad is in fact a complete network of communication involving many human and non-human nodes of action, including theatrical ontology and social factors.

Construction and operation of domestic distribution networks for Chinese adaptations of Shakespeare's plays. The cross-cultural performance of the Chinese adaptation of Shakespeare's play consists of three stages: domestic production, domestic rehearsal and foreign performance.

According to Bourdieu's 'Artistic Field Theory', 'Cultural Intermediary Theory' and Latour's 'Actor Network Theory', intercultural theatre products are objective expressions of cultural capital. In addition to its own cultural value, it also has the properties of symbolic and economic capital. An intercultural theatre product can be a performance in the traditional sense, or a recorded performance video and a related theatre culture venue. In China, when books, reviews and media reports referring to Chinese adaptations of Shakespeare's plays appear in physical or online bookstores, in paper or online media, readers become consumers who may be interested in purchasing tickets, usually based on two basic buying habits, namely impulse buying or planning to buy tickets. Consumers with impulse buying habits are likely to be attracted by the poster cover, the content of the review and the novelty of the adaptation, and may then choose to buy tickets for the performance, while those who plan to buy tickets make purchases based on their experience of reading the original Shakespeare, their long-standing interest in the content and their trust in the book and play reviewers, and eventually become audience members of the performance. In this way, producers, producers, critics, reviewers and audiences should be included in the scope of the study. These are the human actors who shape and disseminate the Chinese adaptation of Shakespeare as a cross-cultural theatre product through their actions in the artistic field. The process of spreading the Chinese adaptation of Shakespeare abroad involves schools and theatre festivals in addition to the aforementioned sites of action, and the range of actors is accordingly extended to professionals, scholars and students in the theatre field. In addition, the artistic field is also influenced by non-human actors such as the social environment, national policies and communication media. In this process, Chinese table directors are both the liaison between the work and the audience in the artistic field and the bridge between Chinese and Western theatre culture through adaptation activities. Ultimately, with the joint participation of

1992:123-126.

[1] Pierre Bourdieu. *Distinction: A Social Critique of the Judgement of Taste* [M]. Cambridge, MA: Harvard University Press, 1984:359.

[2] Pierre Bourdieu. *Distinction: A Social Critique of the Judgement of Taste* [M]. Cambridge, MA: Harvard University Press, 1984:365.

[3] Bruno Latour. *Reassembling the Social: An Introduction to Actor- network Theory*[M]. Oxford: Oxford University Press, 2005:217.

human and non-human actors at home and abroad, a complete network of actors is formed for Chinese adaptations of Shakespeare's plays to be invited for cross-cultural performances abroad, through the entire process of domestic production, domestic and international dissemination and foreign performances. Based on the investigation of the communication process and effect of the Chinese adaptation of Shakespeare's plays in foreign countries, this paper attempts to explore the mode of cross-cultural performance of foreign classic plays adapted from China, as well as the possibility of cross-cultural drama communication and dialogue in the future under the all-media era.

4.Conclusions

Although Shakespeare has been in the eyes of the Chinese people for nearly 200 years, the artistic life of his plays has been constantly renewed. From the introduction of translations to the staging of Shakespeare's plays, from the staging of civilized plays to the performance of plays loyal to the original works, the vitality of Shakespeare's plays has been continued on the Chinese stage in different periods through different ways of interpretation. However, with the gradual enhancement of communication between different cultures, the attitude of Chinese theater workers towards Western culture is not only limited to simple acceptance, cultural resources should be two-way rather than one-way circulation, therefore, the cross-cultural adaptation, performance and communication of Chinese adaptation of Shakespeare's plays is a systematic project that depends on the network operation.

Since entering the era of globalization, with the continuous deepening of cultural exchanges between China and the West, the Chinese theater circle has been actively adapting and performing Shakespeare's plays to try to deepen the communication between traditional Chinese culture and Western artistic classics. The Chinese adapted version of Shakespeare's plays has attracted the attention of the Western academic circle in this process. The history of Chinese theatre practitioners' adaptation of Shakespeare and the process of performance, dissemination and acceptance of Chinese adaptation of Shakespeare abroad can reflect the development trend of China's traditional culture gradually shifting from passive reception to active export. The local adaptation of Western classics represented by Shakespeare and the cross-cultural performance of theatre to bring Chinese culture to the world not only opens up new paths for Chinese culture to go global, but also is an effective way to improve its international communication capacity.

References

[1]Alexander C.Y. Huang, *Chinese Shakespeares: Two Centuries of Cultural Exchange* [M], New York: Columbia University Press, 2009.

[2]Adele Lee , *The English Renaissance and the Far East: Cross-Cultural Encounters Lau, Miriam* [M], London: Fairleigh Dickinson University Press,2018.

[3]Bruno Latour. *Reassembling the Social: An Introduction to Actor- network Theory* [M]. Oxford: Oxford University Press, 2005.

[4]Dennis Kennedy and Yong Li Lan, *Shakespeare in Asia: Contemporary Performance Bishop, Tom . Shakespeare Quarterly* [M], Cambridge: Cambridge University Press, 2010.

[5]Pierre Bourdieu. *Distinction: A Social Critique of the Judgement of Taste* [M]. Cambridge, MA: Harvard University Press, 1984

[6]Pierre Bourdieu. & Wacquant, L. J. D. *An Invitation of Reflexive Sociology*[M].Chicago: University of Chicago, 1992:123-126.

[7]Pierre Bourdieu. *The Field of Cultural Production*[M].Cambridge: Polity Press, 1993:162.

[8]Bourus Terri , *Songs of Lear presented by Song of the Goat Theatre, and: The Revenge of Prince Zi Dan (Hamlet) presented by Shanghai Jingju Theatre Company, and: Piya Behrupiya (Twelfth Night) presented by Company Theatre Mumbai*[J], Shakespeare Bulletin , Vol. 35, No. 3, (2017), pp. 479-510.

[9]Li Weimin , *Shakespeare on the Peking Opera Stage*[J]. Multicultural Shakespeare , Vol. 10, No. 25, (2013), pp. 29-37.

[10]Renfang Tang, *East Meets West: Identity and Intercultural Discourse in Chinese huaju Shakespeares*[J]. Multicultural Shakespeare , Vol.20, No.1(2019), pp. 61-81.

[11]Rea Kenneth, *The Kingdom of Desire*[N], the Guardian,16 Nov 1990:38.

Leaving and death: the awakening of individual modernity in ethnic minority films——Based on the reading of "A Mongolian Tale" and "Jinpa"

Kaixiang Cui[1,*]

[1]Literature of Theatre Film and Television, Communication University of China, Chaoyang District, Beijing, China

Abstract: The relationship between national tradition and modern ethics has become an important proposition of ethnic minority films in the new century. With the development of society, the minority areas are more closely connected with the outside world, and some individuals of the minorities can be exposed to modern ethics and realize their superiority over the traditional national ethics. Due to the connection with the traditional national society, the awakening individual often bears strong tearing and pain in the conscious transformation from tradition to modern. This irreversible transformation process accompanied by strong pain has been condensed into two symbols of "leaving" and "death" in the film text.

1. Preface

"The whole world may feel such a dilemma, how to adjust the inheritance of a nation in the changing world, as well as how to get along and integrate with the present."[1] In the 1970s, Chinese society stepped into the track of modernization development, and the traditional lifestyle and ethical concepts of citizens changed greatly. However, compared with the central and eastern urban agglomerations, the modernization processes of ethnic minority areas tend to be more slow due to geographical and linguistic isolation, and their ethnic cultural traditions tend to be stronger.

With the enhancement of social mobility, some minority individuals are able to walk out of the strong context of traditional minority culture and receive the education of modern civilization, so they have to face the "two-way choice between traditional culture and modern culture" [2]. Many creators of ethnic minority films paid attention to the unique life experience of the awakening individuals in this period, and extended their more profound thinking on national tradition and modern civilization in the film and television texts.

As excellent works in ethnic minority films, both A Mongolian Tale and Jinpa pay attention to the dual recognition of traditional culture and modern ethics, as well as their simple and painful thoughts on the future destiny of the nation. The result of identification and thinking is reflected in the film and television text: "leaving" and "death". Through these two symbols, the creator conveys his complex feelings of affirmation and reflection on both modern civilization and modern ethics in the film.

2. Leaving: the inevitable choice to integrate into modern civilization

For the protagonists of the two works, Baiyinbaolige and Jinpa, leaving is a necessary way for their growth. Individuals first enter into the discourse system of modern civilization through a certain betrayal of national tradition, accept and identify with the superior part of modern ethics, review the traditional national discourse and national life based on the cultural standpoint of modern cities, and finally complete the awakening at the individual level.

In the film and television text, road and truck are the symbols of departure (Figure 1), the material basis of the minority individual departure in the realistic level, and also the ultimate source of individual rebellion against rationality. In "A Mongolian Tale," Bai's two escapes are made with the help of trucks and roads, while Jinpa in "Jinpa," becomes a link between Tibetan culture and modern civilization with his status as a "truck driver". As creations in the industrial era, trucks and roads, on the one hand, break the geographical isolation and cultural closed state of the minority areas, and on the other hand, bring the minority individuals into the discipline of modern society. In the film, this discipline can be attributed to the care for individual life, the respect for the female body in A Mongolian Tale and the reflection on the tradition of homomorphic revenge in Jinpa.

* Corresponding author: 2950693543@qq.com

Figure 1. The road and truck in "Jinpa"

The protagonist of "Jinpa" is a truck driver, who always traveling between cities and small towns.

Receiving the discipline of modern civilization is not only out of the pressure of administrative orders, but also the result of the active choice by members of ethnic minorities. However, not all acts that actively integrate into modern society can be regarded as a conscious choice of modern culture. For example, in The Silent Holy Stones" directed by Pema Tseden, the son of the stone carving old man goes to Lhasa to become a living Buddha, while the little lama is attracted by "The Journey to the West" on TV (Figure 2). The resources and spirit from the city simultaneously enter the ethnic minority areas through the road, bringing more abundant recreational activities, educational resources and judicial forces. The influence of material resources on social consciousness is often imperceptible. People do not have to study or reflect spontaneously, but can enter a new state of life due to the need for rich entertainment and the passive acceptance of authoritative discipline. In the process of becoming a member of modern society, ethnic minority members often do not realize the cultural separation and ambiguity between national tradition and modern society, and their naturalized behavior is more like a comfortable behavior out of sensory perception and profit-seeking psychology.

Figure 2. The little lama with the optical disc of "The Journey to the West"

In "The Silent Holy Stones", the little lama was deeply attracted by "The Journey to the West" on TV, which was a symbol of modern civilization.

On the cultural level, the individual ethnic minorities who can be regarded as being awakened should meet two preconditions: one is to have a relatively deep understanding of the traditional national culture; the other is to realize the differences between the modern civilization and the national tradition and make a choice. Baiyinbaolige first realized the great dedication of women on the grassland in his grandmother and Somia, and realized the value of "birth" in the harsh environment of the grassland, Only the modern society's respect for the female body and sympathy for the tragic fate of women aroused strong ripples in his heart. It is the strong tear between the attachment to the grassland and the recognition of modern civilization that makes the story of A Mongolian Tale so stirring. Similarly, the driver Jinpa also has a deep attachment to the traditional Tibetan culture, which is aroused by the innocent dead sheep and the Avengers with the same name, so that Jinpa can not continue to live in the secular life constructed by modern ethics.

In film and television texts, tearing often ends in the victory of modern civilization. Baiyinbaolige's study broke the grassland women can not go to school tradition, ending the tragic fate of women in the grassland culture; Jinpa "ended a Kangba tradition in a dream and a cycle of revenge, making the killer Jinpa from the revenge of Mazza's son" [3]. The superiority of modern ethics based on the economic foundation of the city in the face of traditional national ethics makes the departure of awakening individuals inevitable, which is the common judgment of Xie Fei and Pema Tseden as the creators of ethnic minority films, and also the inevitable trend of ethnic minorities in the development of The Times.

3. Death: a painful sacrifice free from national tradition

The national tradition has its lag behind the age, but it can not be based on the dross. Both the dedication and sacrifice of women under the grassland civilization, with the religious belief and homorphic revenge tradition in the Tibetan civilization have played an important role in continuing the national blood and maintaining the social order in the long history. Because of this, the individuals bred by the national tradition are often inextricably linked with the minority land in the spirit and flesh, making their departure often accompanied by strong shock and pain. Therefore, in the film texts of A Mongolian Tale and Jinpa, the ritual scenes designed by the creators to represent their complete departure from the national tradition are often related to death.

In A Mongolian Tale, the ceremony that marks the loss of contact with grassland civilization is the death of Gahala. Both Ganggahala and Baiyinbaolige were raised by their grandmother after losing their families. Their grandmother give them the chance to survive. Therefore, Ganggahara became the embodiment of Baiyinbilige. After Baiyin left, She replaced the company in the traditional social network of the grassland, continuing the connection between Baiyinbilige and the grassland, and fulfilling the latter's responsibility as the son of the grassland. When her grandmother died, Ganggahala completed her final response to the national tradition in the last moments of her life, namely the reward of her birth. The death of Ganggahala became the cult of Baiyinbaolige's escape from the grassland, symbolizing the death of the national character in the life of Baiyinbaolige, which has since become a modern man in the complete sense.

In the creation process, "Jinpa" also set the national incarnation for the driver Jinpa, that is, the sheep killed on the road and the killer encountered by chance. The process of the death is the process of the driver Jinpa separated from the Tibetan ethnic tradition. Under the influence of religious culture, Pema Tseden gives a richer presentation of death, and the text of "Jinpa" therefore presents two meaningful "transcendence".

Firstly,Jinpa expiate the sin of the dead sheep, and the innocent dead sheep is the embodiment of the driver jinba body. The sudden death of a sheep while driving a truck on a modern road, is a symbol of a rediscovery of traditional national life. In the modern way of life, Jinpa betrayed the rules of national survival, which caused his inner guilt. Therefore, Jinpa urgently sought religious rescue. After the transcendence of the dead sheep, Jinpa's body could comfortably go into the secular world, and bought mutton and tryst lovers in the secular world. Secondly,the driver Jinpa release the soul of the killer Jinpa from purgatory.The killer Jinba represents the soul of the driver Jinba which attaches to the tradition.. Similar to Ganggahala, the killer Jinpa replaces the driver Jinpa embedded in the revenge network of the Tibetan society, following the tradition to complete the task of revenge. Therefore, when the killer Jinpa continues its revenge action, the soul of the driver Jinpa is still unable to obtain stability in the secular world. The second ceremony in the text represents the driver Jinpa's complete separation from the national tradition. —— Since religion is also a part of the tradition, this separation can no longer be carried out through the rituals provided by religion. Thus, Jinpa turns to his dream, which "breaks through the despair of the imagination and reproduces the repressed emotions of the day in an incredible form" [3]. Jinpa kills Marza in a dream, "bringing the end to tradition, and making it from reincarnation"[4]. Through the killing in the dream, Jinpa surpasses the sin of the murderer, and also freed the responsibility of the Avengers. So at the end of the dream, Maza stands among the sheep, and the vulture flies by, presenting a celestial burial. But when Jinpa looked up again, a plane flies by in the sky, symbolizing that the traditional national law of survival finally gives way to modern ethics.

Images often become a ritualized display of reality. Through the visual presentation of death and excess, "A Mongolian Tale" and "Jinpa" not only write the inevitability of awakening, but also show the pain brought by awakening to the audience. The affirmation and reflection of modernization are combined in the film, which enriches the artistic connotation of the text and provides the audience with a broader thinking space.

The vulture and plane respectively symbolize tradition and modernity in the film (Figure 3).

Figure 3. The vulture and plane

4. Conclusion

The creators of the literary text and film text of A Mongolian Tale are not grassland people, and the protagonist BaiYinolige also has a natural alien temperament.He was born in the city, raised by the grassland, and educated in the city. Devating from the grassland and returning to the city are the inevitable choice of Baiyinbaige. However, in the process of leaving, the rational confusion and emotional pain is also the inevitable fate.As a result, "A Mongolian Tale" focuses more on "running away", which is a display of the process of individual awakening. Xie Fei's unique life experience between the grassland and the city conveys the understanding and thinking of the grassland tradition and modern ethics.

While the hero of "Killing a Sheep" is a Tibetan, which is more the result of individual awakening, that is, the subject of ethnic minorities seeks the difficult exploration of modernization. Therefore, the text uses a larger part of the scene of "death" and "excess".

"Only after the individual awakening, can the awakening of the nation as a whole be possible and the nation have hope."[5]"A Mongolian Tale" and "Jinpa" with the modern rational spirit, build a bridge of communication and dialogue between national tradition and modern ethics.

However, if we treat the minority with an attitude towards a backward group, ethnic minorities will usually be marginalized.

It is worth noting that the modernization of ethnic minorities has become inevitable because the social and economic foundation has undergone major changes. Due to the progress of productive forces and the development of educational resources, women on the grassland no longer need to sacrifice their youth and freedom to ensure the continuation of settlement; due to the perfection of modern legal system, the Tibetan residents no longer need to rely on religious ethics and homomorphic revenge principles to maintain the social order. As we know, at the end of "A Mongolian Tale", there are many girls in the school, and in "Jinpa", the concept of rule of law has penetrated into the daily life of the Tibetans. The changes in these societies are the basis of the modernization of ethnic minorities. In the face of national tradition, we should hold a dialectical attitude, not simply regard it as a symbol of ignorance and backwardness, nor should we regard the members of the awakening ethnic minorities as uncivilized people.

When Baiyinbaolige return to the grassland, girls can also go to school (Figure 4). The lives of women have been changed as the development of society.

Figure 4. The classroom in "A Mongolian Tale"

Both"A Mongolian Tale" and "Jinpa", not only provide the rational speculation of the national tradition and modern ethics, but also show the humanistic care for the minority —— whether awakened or not yet awakened.It is precisely because of the rational reflection and emotional care that the film and television texts have a lofty value in the dimension of history and art.

References

1. Li Hanfang, Fu Shuangqi. Looking back is for a clearer understanding of hometown -- interview with Pema Tseden, director of Talo [J]. Journal of Cross-Straits Relations,2015:12(45-47).
2. Li Miao, Zhou Zemin. Two-way Writing of "Modernity" -- Cultural Interpenetration in Minority Films in the New Century [J]. New Film,2022:2(66-71).
3. Yang Yuqing. Materialization, Desire and Healing of Dream Narrative Dimension in Roadkill a Sheep [J]. Journal of Beijing Film Academy,2019:12(47-50).
4. Pema Tseden, Ye Hang, Tong Luyao. Mirror Image and Fable: Interview with Pema Tseden, director of JINPA[J]. Journal of Beijing Film Academy,2019:06(56-64).
5. Pema Tseden, HU Puzhong. JINPA: The Obsession and Reflection of Tibetan Films -- An Interview with Pema Tseden [J]. Film Arts,2019:03(51-66).

 https://doi.org/10.1051/shsconf/202316701013

The Return of Actors' Body Essence in Theaters

Shiran Zhou*

Jilin University Of The Arts, China

Abstract. Experimental theaters of dance and drama are gaining international attention for finding unique aesthetics core of Chinese traditional philosophy.Via actors' body to embody mans' inner world without lines artists had been seeking more primitive and sheer forms of artistic expression like the ancients.From the perspective of Somaesthetics,this paper discusses the Chinese traditional somaesthetic consciousness embodied in Dance Theater and Physical Theater.Then we recapitulate viewpoints of body in Chinese traditional philosophy that found expression in Somaesthetics proposed by Richard Schusterman.A type of new relationship of actors and spectators will be the field that is worth to develop further in dramatic aesthetics.

1. INTRODUCTION

Freshly being rewarded the Silver Lion in The 2023 Lions For Dance at February 8 , The company TAO Dance Theater blending the traditional dance of China and modern dance has been dedicated to the propaganda of modern dance for public.Focusing on body more intently can be found in more types of arts of Physical Theater,like Theater SanTuoQi,who has been promoted physical theater,a post-modern style of drama.Aside from their novel forms of performance,their focuses on human body intricate actors' cognition towards bodies connected with minds,emotions and sensations of the world.What being mentioned above are the novel examinations in two separated field of arts,dance and drama,and overturn the tradition in their art forms that include the pursue of skillful body and emphasis on lines.As two independent art forms derived from the same source,dance and drama are inner hearts or minds embodiment of people through human bodies,thus in these innovation of their shapes in theater,artists are detecting ways to re-define actors' bodies for a more pure purpose like the ancients. In this research,we analyse the current novelty of body theaters from a body aesthetic perspective to explore the common awareness of modern and ancient human on their bodies and to contrast gaps of body nature between the primitive art and post-modern art for the goal of using our bodies to better engage and perceive the world,and to effectively express our inner minds and emotions through more natural and sheer ways .

2. THE NOVEL EXAMINATION OF ACTORS' BODY IN THEATERS

2.1 Physical Theater Created By Zhao Miao

One of the most representative explorations of physical theater is founded by Zhao Miao who established Theater SanTuoQi in 1996,and began to perform physical theaters since 2003.At the early stage of the introduction of physical theater into China,artists are still followers of western forms of non-verbal performance.The turnaround from followers to creators has emerged along with the integration of traditional Chinese aesthetic culture.Differing from traditional drama originated from the west,Physical Theater pursue the symbolic expression via actors' bodies rather than realistic techniques,which is highly similar to Chinese Opera in aesthetic pursuit. In the Traditional Chinese Opera,whatever actors' performance,costumes and scenes all present in freehand manners,a kind of virtualized means of artistic expression in almost all kinds of Chinese traditional culture.Actors performing through abstract body actions requires audiences mobilize their abundant imaginations to understand and extend meanings of the whole manifestation.As a non-verbal physical performing form,Physical Theater also unfold people inner worlds in the same abstract and exaggerated body motions.With the thoroughly extensive and profound aesthetic tradition of ancient Chinese,the Physical Theater in China have found its own way to set off an original path.The Chinese original physical theater,Shui Sheng came from the classical novel Strange Tales from Liaozhai in Qing Dynasty,integrates elements of Nuo opera breeding in Guizhou province into its' performance."Nuo dance is a kind of dance to offer sacrifices to the gods, which is

* Corresponding author: 1163837653@qq.com

https://doi.org/10.1051/shsconf/202316701013

effectively integrated into the drama. In spite of exaggerated with it, Nuo dance remains it's sense of beauty . At the beginning of the plot, with the aid of background music, the energy information conveyed on the stage feels like watching a sacrifice ceremony, which it is not actually a sacrifice, but the confrontation between the devil and the water ghost in the play."[1]For the parallel purpose,the ancient Chinese dance in ritual ceremonies for sacrifice and worship resembling as the Greeks did at Bacchanalia. The director uses masks to cover actors' faces ,allows actors' physical movements to communicate directly with spectators via amplified and freehand body languages,which makes actors' bodies representations of souls,characters,even components in the scenery.As to the mysterious meanings conveyed by these abstract somatic actions,they also leave spaces to spectators to envision thoroughly for catching the implications.Realistic performance and true-life stage are these dramatists pursuits not any more instead of stylized theatrics and more primitive physical expression of collective human minds.

2.2 Tao Dance Theater In China

Professional dance has promoted the development of dance theater during the last centuries,while,as to other forms of arts,dance was born out of early human experiences,which embodied intrinsic desires of early humans.Established in 2008,Tao Dance Theater is dedicated to the public promotion of modern dance and researches of the theory of " yuan(circular or circling)movement system", a motion that turn dance back to general public to focus more on their own bodies.The original purpose of Tao Dance Theater to establish the public welfare workshop is to let more people understand the real meanings of the modern dance through teaching engagement rather than rigidifying dance.One of founders of Tao Dance Theater,Tao Ye,said that"Dance is inherently a kind of primitive art and fully interlinked with former poem,music and dance." He continued that" It used to be a whole, which could be without gesture, melody or narrative, so it has never been a boundary. For the creator, it had no limits and the only connection between the three was rhythm, but now it is all separate. Therefore, we try to find a correlation between dance and the public at present, which is of course the dream of all dancers in this era, which is relatively distant." [2] "13""14"are the newly created "Numerical series" dance pieces ,and will soon greet our eyes on stages at home and abroad.

"Yuan movement system" is enlightened by the body charm in Chinese classical dance,which is inseparable with Chinese Opera body movement.Featuring the eastern somaesthetics, " Yuan movement system " assimilating styles of Chinese classical dance and Chinese Opera into modern dance have reconstructed physical languages of modern dance in the Orient."We found that the movements in Chinese Opera make up overall language system.They are not separated into isolated pieces but connected,they align with one circle,one breath which are successive."said by Tao Ye.In the meantime,more participants engaged in the teaching workshop have found their new forms of life,where they cannot unfold in real life,some where individuals are free to move their bodies,and from moving via their bodies to moving from souls.Dancers in the theater were able to find ways to communicate with each other and, more importantly, their own bodies in ways they never had before,just as our ancients attempted to communicate with nature and God through sacrificial rituals at the dawn of dance."There is no exceptions that the primitive motion everyone came this place are pure,just moving their bodies ,but then touching their souls."[3]

3. SOMAESTHETICS EMBODIED IN TRADITIONAL CHINESE AESTHETICS

3.1 Symbiotic Soma

Different from the isolation between sensibility and reason featured traditional aesthetics in the west,the Chinese traditional aesthetics of body put emphasis on the integration of body and mind."Chinese classical aesthetics derives from the body and fully affirms the rational desire of the body. In a sense, Chinese classical aesthetics is not only a kind of aesthetics of the body, but also a kind of aesthetics of the desire" [4]Richard Schusterman have proposed "somaesthetics" blended traditional aesthetics of west and east in his book " Pragmatic Aesthetics:Living Beauty,Rethinking Art",which dissolved the tradition of dividing spirit and corpus.Schusterman's exploration on Chinese traditional philosophy came from the preface writing for the Chinese translation of his book,Pragmatic Aesthetics.After in-depth studies of Confucianism and Taoism,Schusterman put consistency of body and mind in Chinese traditional philosophy in the core of Somaesthetics.Symbiotic soma highlight a interactional , transactional entity contained physical and social environment.Relational and interactional soma should be contextualized,which is defined by our conversations with ourselves,each others,and the world."The self is relational and symbiotic. This idea inspires a broader idea of body reformism -- the idea that we are responsible for caring for and reconciling the environmental characteristics of our physical selves rather than just our own body parts."Said by Schusterman in his book:Body Consciousness And Somaesthetics, " Such a universal model of body self-cultivation expresses the Chinese Confucian ideal of body cultivation: 'being one with all things in heaven and earth'." [5]Schusterman believe that artistic creation has been seen as a tool of self-cultivation and communication,which reveals the ultimate goal of arts——a constant improvement of individuals and his life and arts.

3.2 Plastic Soma

The controversial Somaesthetics Schusterman promoted in the west had received encouragements and supports in China.One of the reasons is that the plasticity of body in Somaethtics identify with Self-cultivation theory in Confucianism and Noble Body of Taoism.In other words,body in both Somaesthetic and traditional Chinese aesthetics highlight a learned body which is called "Born and Made" body that means nurturant body follows naturally born soma."Traditional Chinese philosophy is more than just observation and cognition based on pure theory, but also in the form of a body and mind can be applied practicing the personal experience of spiritual practice, and on the basis of this experience, which obtain the super higher level knowledge of everyday experience."[6]This plasticity embodied in Somaesthetics is body consciousness through numbers of types of somatic education and therapies.These emphases on the soma not only consider body as a way of communication to human themselves and community,but put body on the starting point for shaping and cultivating oneself inner world.The other consistency in Confucianism and Taoism on discussions of body is all shapes of artistic cultivation.Bodily representation in Confucianist thought displaying one's virtue which is directly linked to political reconstruction values Propriety for refined and elegant art,and practical art and techniques are mainly concerned by Taoist thought.Despite the differences,they are the embodiment of Chinese philosophical pragmatism.

4. A NEW SPACE IN THEATERS BETWEEN ACTORS AND SPECTATORS IN DRAMATIC AESTHETICS LENS

As a form of art with the body as the material, drama is the essential embodiment of human life consciousness, and this kind of performance of human body is by no means the body image, it is an unity of physical and mental, which is closely connected with consciousness and emotion. And this uniformity is just ignored by the contemporary drama teaching and drama stage . Fortunately, theater artists have been rediscovering the original essence of the human body in experimental theater. If body movements are objective symbols used to express subjective consciousness in dance,, then in physical drama, the body is a complete carrier that organically integrates the body language of the actor and the inner language of the audience's imagination. Contemporary drama pays more attention to the inner level of human beings. In the school of physical drama, restoring the body's expression of life consciousness and diluting the social attributes of human beings is not so much a pioneering exploration of postmodern drama as a return to the nature of the body in primitive drama. On the basis of showing the essence of human spirit, contemporary drama gradually attaches importance to the narrative expression of the body, intending to convey more philosophical and rational content through the actors' bodies. In all kinds of explorations, the essence of "body" is not limited to the human body itself, but integrates the objective and subjective space, the imagination and philosophy of the audience.

The exploration of somaesthetics in drama field in China is still at the primary stage. Hu Penglin, the earliest scholar who discussed drama from the perspective of body aesthetics, connected the body aesthetics with Antonin Artaud's Theatre of Cruelty, traced the development of the "unity of body and mind" body view reflected in Artaud's drama, and analyzed that the unprecedented attention of cruel drama to the body level was due to its breakthrough in the traditional western philosophy of the body view. Integrate the aesthetics of western drama and Eastern drama and put them into practice, Xiao Ying, professor at Shanghai Theatre Academy, further interprets the body aesthetics in traditional Chinese aesthetics directly from the body aesthetics in opera performances. She believes that Taoist culture directly influences the body aesthetic ideal in Chinese opera, directly appeals to the body aesthetics in traditional Chinese culture and the relationship between body and mind, and builds a different body aesthetics discourse system from the West. In the newly published article from Applied Drama to Dramatic Body Aesthetics, which focuses on Somaesthetics, Hu Xuyue formally discusses applied drama, namely a drama form with more practical significance, and introduces foreign practical cases of applied drama. They present the healing and educational significance of drama through the medium of the body, and applied drama can treat individuals and groups. And even the political implications of social change. It shows us the greater potential of drama development.

Under the influence of defamiliarization effect,Physical Theater had set up a kind of new relationship between actors and spectators reflected on post-modern drama.Expressive body movement had replaced locutionary act,which reserve more expansive margins for reasonable judgment and emotional imagination of audiences.Thus connotations actor's had not literally performed gain more possibility for audiences to interpret it.One of the signs that primitive drama was separated from original dance ,narrativity of human body,has showed a trend of gradual weakening.It is a breakthrough to the stereotype of passive reception for audiences' habits.The thinking process draw people closer to the essence of things.Actors' bodies became semiotic existence leave spectators isolate from dramatic plots.This characteristic of abstraction is also one of Chinese aesthetic pursuits.

5. Conclusion

As the most ancient language of humanity,physical movement has been embodied people pure spirits and minds until the appearance of the characters.Expressed in highly sociable persons,words had already dwarfed body languages.Fortunately,the encouraging signs had showed that experimental theaters of dance and drama in both original forms of arts are re-finding the primitive essence of human body.The presentation of Somaesthetics by Schsterman draws our attractions on soma as the source of conscious activity,which is valuable in whatever philosophy,aesthetics,sociology or art.Attentions on mans' emotional,sensorial and interactional bodies had been witnessing a return in theaters.With more researches about Somaesthetics,the new horizon of Drama Aesthetics of body needs our values to be explored further.

References

1. Luo Ling. The Development Practice and Investigation of Chinese Form Drama [J]. Sichuan Drama,2022,No.262(06):22-25.
2. 3. Tao Ye, CAI Shuo. Tao Body Classroom: "Pay attention to more people's bodies" -- Interview with Tao Ye [J]. Journal of Beijing Dance Academy,2022,No.154(02):25-33.
4. Zhang Zailin, Li Junxue. On Shusterman's Thought of Body Aesthetics -- Also on the Study of Chinese Classical Body Aesthetics [J]. World Philosophy,2011(06):61-71.
5. Quoted from Richard Schusterman, Body Consciousness and Body Aesthetics, translated by Cheng Xiangzhan, The Commercial Press, 2014, pp. 297-299.
6. WEI Shuanxi. Chinese Classical Aesthetics from the Perspective of Shusterman's Body Aesthetics [J]. Sichuan Drama,2015,No.180(08):36-40+47.

Film Literary Criticism and Image Aesthetics：around Film Criticism

Yunning Wang[1,*]

[1]Multimedia Television and Film, Sungkyunkwan University, 00382, Korea

Abstract: Film criticism has been carried out in the form of activities since its inception, in which many outstanding intellectuals and artists have participated.Especially in this respect, European intellectuals can contrast with American intellectuals, who have long held films in contempt."In Europe, intellectuals have been fascinated by filmmaking since the beginning of their media, " writes Tim Bewart and Thomas Szobak.It is not surprising, therefore, that in the 1920s, throughout Europe, Paris, Berlin, Moscow, a group of intellectuals dedicated to all corners of the art and other artists on the same basis as art, came together to discuss and write about film."Unlike the U.S., where movies were excluded from intellectual reflection until the late 1950s, film criticism was a battleground for intellectuals in France.

1 Introduction

Film criticism is the academic practice of applying the principles and methods of humanities and social sciences and film theory to analyze film works and film phenomena.At the beginning of the 20th century, avantgarde film theorists advocated film criticism, promoting various avantgarde films such as impressionist, pure, abstract, Dadaist and surrealist.Musinac's On the Rhythm of Movies (1923) and Leschel's A New Realism (1926) reflect the original concept of film criticism.With the development of film creation, film criticism has experienced the process from the early scattered impression criticism to the establishment of film criticism system.Traditional film criticism incorporates the concepts and categories of aesthetics, psychology and philosophy ontology, reviews film works and guides film creation practice.The Chinese left-wing progressive film criticism in the 1930s and 1940s, Andre Bazan's film criticism articles in the 1940s and 1950s and the film criticism activities in the French Film Manual magazine are examples of film criticism leading to film creation.Film critics include films and film phenomena.Film criticism is fundamentally different from film theory and film history in functionally.Criticism must be combined with the actuality of the film, with specific relevance and practicality.But sometimes the theory must be separated from the concrete work to a certain extent, and with the concrete pertinence and practicality to maintain a certain distance.This limits the direct significance of theory to art practice.Critics are more likely to appreciate the guiding significance of theory than practitioners.To some extent criticism is the medium through which theory directs practice, so the significance of theory to criticism is more direct than that of practice.The emphasis on combining theory with practice often leads to the development of theory towards criticism.The theory itself may not develop at all.Theory and criticism have two important functions: practical function and cognitive function.Although the ultimate goal of both is for the practicality of society, there are still criticisms that attach importance to the function of practice and theory to the difference of cognitive function.

2 Film and Aesthetic Art System

The artistry of film, as a call sign of "art", is derived from the unique visible form of film (i.e., "mobile image") and the particularity of its unique technical equipment.As part of Lonser's critique of modernism, it deduces film as an artistic norm from film's unique material of expression and "production mode", especially the non-historical and utopian ideas of avant-garde aesthetic origin in 1920s and 1930s.According to Lonsill, film is "art" because it belongs to a particular era or system, that is, "aesthetic art system", which refers to unique craft products as art and gives them artistic identity. Lanser analyzes the possibility of the aesthetic art system through the most impure characteristics of film, namely the contradiction story and the contradiction story, and how to restore the classical art system through the discord with the system itself. Through the form of "distorted story", film not only shows the characteristics of the art system to which the film belongs, but also the contradictions and paradoxes of the system are the most distinctive art [1].

*dearmydorothy@gmail.com

3 Discussions on Film——Film Criticism and Film Aesthetics

Film talk came about at the same time as the invention.With the development of film media in various directions, the discussion about film has also developed after countless experiments, differentiation and integration.The development of film discourse can be compared with the development of other forms of art discourse.For example, literary discourse, which originated in the same period as literary works, has been formed by clearly delineating two directions.Annotations, explanations or analyses of individual literary works constitute one direction of discussion of literature, and reflections on literature itself constitute another direction.Philology and literary criticism are extensions of the first discourse, while literary theory or literary aesthetics inherit the second discourse.The two kinds of discourse centered on literature can be distinguished by the difference between the former and the latter, which focuses on abstract and collective literature itself.Jean-Marie Scheffer pointed out that words of these two nature are usually unfriendly to each other, but each has an indispensable nature.This form of discussion of literature can almost be applied to discussion of films.As with literature, discussion of films is distinctly different from one subject to another.One is the whole film, or as a whole film (cinéma) as the object of discourse, can be referred to as "image aesthetics" (or film aesthetics).The other is the discussion of individual films, which range from mere informative writing to serious analysis of works, but can be called film criticism in a comprehensive sense:

First of all, taking image aesthetics as an example, it is almost impossible to comb the trend of image aesthetics in different contexts.So they later applied the results of other disciplines such as psychology, semiotics, philosophy to their reflections on the film itself.In other words, Rudolf Arnheim is an art psychologist interested in the application of Gerstatt's psychological achievements to art or film, Christian Metz is a cinematographer who laid the foundation for the possibility of semiotics.On the other hand, Zigprit Krakkorna Bellabaraz, André Barjón and Jean Mitri have for a long time served as outstanding critics (later or at the same time) of individual films.

Secondly, as far as film criticism is concerned, at the beginning of the invention of film, it has the nature of recording and describing the new technological characteristics of film.After that, it gradually developed into the identification of the aesthetic nature of the works.According to Claude Bailey, the first film criticism beyond simple plot summaries or advertising came after 1908 in France.Film criticism was originally born in the market when it developed into a period of clarification of the aesthetic nature of the works.Later, film criticism was also the work of many film directors, who tried to use film as an art.The new director, who once played the role of a prominent critic, appeared even before his appearance.In fact, good critics play a big role in positioning the film itself as a serious object of thought [2] .

In any case, it is important not to confuse the objects and areas of video aesthetics and film criticism.Just as the discussion of individual literary works is not a substitute for the discussion of literature itself, so is the contrary.On the contrary, the film criticism as a whole can only be applied to individual film criticism in a very comprehensive and indirect way through many detours and media.Of course, this does not mean that an impassable abyss can be created between these two areas and objectives, which can permeate at any time.We will then explore this possibility by taking the French film criticism tradition as an example.

4 The Object and Role of Film Criticism

In fact, not all films need to be reviewed.If film criticism plays a certain role as a medium between the audience and the work, then there are some films that do not need such a medium.In theory, all existing films can be the object of criticism, but in practice, not all films are the object of criticism, nor must they be the object of criticism.Because some films first do not have the need to criticize themselves.Criticism is effective only if it is based on the idea that there is a décalage between the creator (and the person who needs to create it) and the viewer (and the person who needs to see it) and that the deviation is valuable.Furthermore, it makes sense to put a little writing between the movie and the audience whenever they can't face each other properly.If writing is born out of a certain need, criticism is born out of a need to be criticized, or must be criticized.According to Darnay, this need arises first in a certain potential "gap" or "biased" between the audience and the work.The conclusion is as follows: "When a spot immediately touches the crowd it is targeting, it no longer requires additional media.This is the morality of visible sex itself, which is completely worthy of respect [3].On the other hand, films that have room or need to be criticized and films that do not need to be criticized can be distinguished by the quality and manner of time the audience experiences while watching the film.Movies that have a time to treasure, whether through criticism or through physical preservation in museums, become films worth cherishing.This can also be achieved through silence if criticism cannot be excluded from a certain amount of value- évaluation.It is not necessarily the role and task of film critics to add insult to injury.Rather, film critics play a role in reviewing and tracking what happens in movies that achieve great aesthetic achievements.

A critic is a passeur who guides the way between poles.There is something between the creator and the observer.What must be known here is where the priorities are.It seems to me that critics will first write an open letter to the writer, which will then be read by an audience worthy of watching the film.Following the critics represents the "creator" understanding of the non-creator.He is a lawyer in this field.In my view, this is normal and moral.But for others, the opposite is true.Critics target creators and represent the audience's understanding.In this regard, he is a judge.According to the fashion critic's model, he can choose between "popular" and "unpopular" and in some cases, he has worked well enough.Whether it's true or imitative, whether it's true or imitative, these are just consumer guides.

5 Film as a Research Object—— Characteristics of Film Text

Scientific research must first encounter the characteristics of the film text. As an object that exists only in the time of appreciation and passes in front of the eyes, the film text shows a unique aspect compared with other art. Compared with other artistic fields such as literature, drama, art and music, the reason for the relative lag in film research is the characteristics of film texts.

At the human level, it is difficult to quote. First of all, the score shows strong heterogeneity to the means language of the object studied, but it can be quoted at all.The problem is that "the score is not performance, we can't quote sound" and "the score is fixed, but the performance is changed".In some modern music, the differences between score and performance tend to grow, but the problem is much more complex.In this regard, Beirou noted that the textual nature of music is weaker than that of literature and painting.As a performing art, drama—music and drama are largely similar to music—has also appeared as undescriptable and invocable as music.But when it comes to citability, the most serious thing is the film text. According to Beiru, scripts are only avant-texts and should therefore be largely excluded from the issue of references. Because the nature of the script is completely different from the musical score or drama, and it is more like a drawing.He divided the five forms of expression that make up the film according to Christian Metz's division.It is mainly in the form of le sonétique, captions, music, les brutes, and l'image photography mouvante. Phonetics and subtitles do not raise much questions about citation, but in the case of phonetics, "tone, intensity, tone, pitch, pitch" may be raised as in theatrical performances.If noise can be distinguished between motivated noise and arbitrary noise, and the former can be evoked by words anyway, then the latter cannot be put into words at all.In addition to the original questions raised by the music text, the music in the film also presents extremely complex patterns based on the combination of different images.

Of these elements, the essence of the film is the image.The most complicated aspect of film image is that it unfolds in time and space first.The particularity of the image of this movement makes the writing of the film take on the form of content or theme analysis to a large extent, but it is out of the essence of the film, so it cannot be said to be a serious criticism seriously.Whether out of a critical or aesthetic interest, it raises questions that film researchers have to face in advance, reminding them of the problems they should constantly be aware of.

6 Take the French Film Criticism Tradition as an Example

The relationship between film and film criticism began and developed in different ways from other arts.It was at the same time that the modern film, which began to appear in the media, was born that it attracted media attention.On December 28, 1895, two days after the Lumière brothers announced the birth of the film, Le Radical and La Poste published an article focusing on the wonderment of the world as depicted in the Atlas of the Movie and its incorporation into a new dimension of human existence.This can be said to be the starting point for movies to establish a strong relationship with the media and magazines.Shortly after the birth of the film, magazines such as Misofuann Miseaupoint 8 (1897) and Le Fascinateur (1903) marked the beginning of the professional film magazine.In proportion to the increasing influence of film as a mass media, the number of film magazines is increasing, and the influence of film on the public is increasing. Since Ricciotto Canudo, one of the earliest film theorists, named the film "The Seventh Art", formal writing has focused on the promotion of film to art.From then on, it was no longer a literary magazine or a daily film section, but a magazine in the form of a periodical.Then there is Louis Deluc, the birthplace of French film critics and film directors who founded the Ciné-club, which laid an important foundation for film criticism.Since then, countless film magazines have appeared in the French film industry.

From this point of view, it is not difficult to deduce that without the accumulated results of French film criticism, the broad philosophical reflection of the film itself would not have been easy.The role of film criticism in the establishment of video aesthetics can be paid attention to from many aspects.

Above all, film criticism plays a big role in moving movies to serious humanistic objects.This is because, as has been pointed out, film criticism has the nature of "intellectual competition" in Europe, which involves many intellectuals and artists.Film criticism plays an important role in describing the problems raised by film as problems and transferring them to thinking objects.

7 Conclusion

Although film criticism and film aesthetics have different specific concerns and research objects, they play a complementary role to each other.Image aesthetics opens up a whole, systematic and fundamental world of thought, which is rare in the examination of individual films.However, this kind of image aesthetics work is based on the serious contact with individual film works and the numerous film criticisms that follow the happenings of film works keenly.The role of film criticism in the establishment of video aesthetics can be paid attention to from many aspects.Above all, film criticism plays a big role in moving movies to serious humanistic objects.This is because, as has been pointed out, film criticism has the nature of "intellectual competition" in Europe, which involves many intellectuals and artists.Film criticism plays an important role in describing the problems raised by film as problems and transferring them to thinking objects.It can be pointed out that film criticism has a certain media nature.Image aesthetics is related to the outstanding film works that constitute the core of the world film history, although it is only an indirect way.Although the relationship between individual film works and film aesthetics is rare and direct, it is usually indirect and media.Film criticism can be said to play a role in the media of audiences and works on the one hand, but also in the media of individual films and video aesthetics on the other.Image aesthetics and film criticism have a relatively clear distinction between objects and fields, but not insurmountable relationship. Because, as mentioned above, although the two words of image aesthetics and film criticism are not friendly to each other, there is inevitable tension in this respect, but fundamentally, they are indispensable to each other. If film criticism is defined as bottom-up aesthetics and image aesthetics as top-down aesthetics, then at any rate, the two trends meet in one place, and the real concrete reflection and the real abstract reflection can be said to agree. The place where both energetic and creative thinking begins is where the two kinds of words are actively communicated. There is no doubt that both film and television aesthetics and film criticism are based on the intense humanistic spirit. This humanistic spirit includes not only a comprehensive vision of film and art, but also a comprehensive basic knowledge of philosophy, history and social sciences, and above all, the ability to think about the image of film. But in film aesthetics and film criticism, exposure to the humanistic spirit of poverty risk is relatively large film criticism. Here, we decided to replace the conclusion by looking at the inappropriate relationship between film criticism and so-called "grands discours" of which image aesthetics certainly forms a part.Film criticism, which is based on the perspective and reflection of individual films, presupposes a careful face-to-face approach to the works that have become the subject of study.Such a job is not to blindly apply "fashionable" grandiose words to the interpretation of individual film and television works and other methods can replace.As a general discourse of the whole film, image aesthetics is relatively susceptible to the influence of other humanities such as philosophy. In fact, some works are written under this influence. For example, Bellabarraz's film: It is not difficult to read the influence of limpidism from the nature and development of new art (1948).The problem arises when these words are applied directly to the analysis and interpretation of individual film and television works. This attitude of looking not for the foundation of criticism within the work but for externalities and applying them to individual applications is fatal to criticism itself.

References

1. Zhang Zhaocao. Narrative Aesthetics and Cultural Criticism of 'Desktop Cinema'[J]. Film Literature, 2021,000(004):43-47
2. Yang Chijing. Critical reality in the space-time alternation of films[J]. Journal of Sui Institute of Chemistry, 2021,041(008):86-88
3. Cao Jingbo, Dong Yuhong, Liu Wei. Aesthetic Traceability, Automatic Mechanism and Aesthetic Reflection of Screen Films[J]. Film Literature, 2021,000(013):34-39
4. Li Lei. Bella Barraz's Classic Theory of Film Aesthetics and Formalism [J]. Film Review, 2021,000(015):62-65
5. Ma Pengcheng, Wang Chunqiao. French Film Education: Filmmakers for All[J]. Invention and Innovation Educational Informatization, 2018,000(010):17-18

A Narratological Study of Streaming Science Fiction Films: The Case of "Dune"

Wei Zhou*

School of Art, Northeastern University, ShenYang 110819, LiaoNing Province, China

ABSTRACT: As technology continues to advance, science fiction films are adopting new narrative forms. The Hollywood-produced film "Dune" follows the traditional narrative model while also absorbing elements from popular culture, thus becoming a new type of secular mythology. It takes the traditional narrative model one step further and becomes what Deleuze describes as a new paradigm of cinematic narrative, known as "pure optical and sound situations". By studying the narrative logic and style of "Dune" through the theoretical paradigm of narratology, we can not only focus on the changes in the narrative content and form of science fiction films in the era of streaming media but also explore the aesthetic meaning of science fiction embedded within it.

1. INTRODUCTION

"Dune" is a science fiction film directed by Denis Villeneuve. With this work, he was nominated for Best Picture and Best Adapted Screenplay at the 94th Academy Awards, as well as at the 79th Golden Globe Awards, and grossed over $300 million at the box office, receiving both commercial success and critical acclaim. In an interview, Villeneuve stated that his creative techniques and ideas were influenced by directors such as Steven Allan Spielberg, Stanley Kubrick, and François Truffaut. Throughout his career, "Dune" can be considered the culmination of Villeneuve's work in using science fiction to tell stories and analyze social issues. The film touches on topics such as religion, philosophy, and environmentalism. As a science fiction film, it provides a "daydream" that allows people to escape the reality of the rapidly advancing modern world, while also reflecting the conflicts viewers encounter in reality and providing an imaginative solution. This makes science fiction movies possess a secular mythological attribute, attracting us to interpret them using the theoretical paradigm of narratology.

With the rapid pace of technological advancements and the ever-increasing complexity of our world, it is no surprise that more and more people turn to science fiction theme as a means of exploring potential futures and reflecting on the present issues we face. As such, "Dune" stands out as a prime example of a film that offers a unique blend of fantastical storytelling and real-world themes, making it a compelling case study for examining the narrative structures and themes within science fiction films. The purpose of this research project is to delve deeper into the significance of the narrative structure and thematic content in "Dune" and analyze how these elements contribute to its commercial success and critical acclaim. Through a detailed examination of the film's use of narratology, we aim to gain a deeper understanding of how science fiction films tell stories and how they reflect the cultural and social issues of our time. The research seeks to demonstrate the value and relevance of science fiction films as a means of storytelling and social commentary, providing insights into how film directors can use narrative techniques to convey complex ideas and emotions to their audiences. By analyzing "Dune" in this context, we hope to provide new perspectives on the power of science fiction theme as a genre and its ability to captivate and inspire our culture. Ultimately, the study will contribute to a deeper understanding of the role of science fiction films in shaping our cultural and social consciousness, and their potential to offer new and imaginative solutions to the challenges we face in our rapidly changing world.

2. SECULAR MYTH NARRATIVE TYPE

"Dune" is a science fiction story about interstellar space in the future, as well as a secular mythological story of modern society. According to Claude Levi-Strauss, the "father of structuralism", even though the stories we read today are constantly changing, we can still identify a common "core" integrated with Greek mythology – a structural unit that is constantly repeated in various mythical stories with different appearances [1]. The significance of this research is to highlight the common social functions that different narrative behaviors possess:

*Corresponding author: zhouwei0021@163.com

through various and diverse stories, they collectively present certain contradictions that are latent in their social and cultural structures, and attempt to balance or provide imaginative solutions.

Science fiction stories offer viewers a transcendent experience on the surface, but in actuality, they present contradictions based on reality, preserving their fundamental properties through similar plots. In Vladimir Propp's study of Russian folktales, "Morphology of the Folktale", he discovered that although Russian folktales have countless characters and varied plots, there are undoubtedly basic and unchanging elements [2]. Propp proposed seven categories of roles and actions, including the villain, benefactor, helper, princess (or person or object to be found), dispatcher or sender, hero or victim, and false hero. A character in the story can play different roles in different paragraphs and at different times during the story. Additionally, he summarized the numerous folktales into six narrative units and thirty-one narrative functions. Based on these narrative units and functions, Propp conducted a comprehensive and specific narrative study of Russian folktales. Propp's research has also been applied by scholars in the fields of popular culture and film studies. Based on Propp's narrative unit and function sequence theory, the film "Dune" can reveal its secular mythological characteristics.

Preparation Unit. At the beginning of the film, we see the young leader of the House Atreides, Paul, living on a planet ruled by his family. The Atreides family receives orders from the Empire to take over the desert planet of Arrakis, which is currently controlled by the Harkonnen family. Paul frequently dreams of a young girl from the Fremen tribe (who is missing a family member around the campfire - Function 1). The Bene Gesserit sisterhood tests Paul with pain and informs him that he is the chosen one, but Paul refuses to accept this fate (Function 2 and 3). When they arrive on Arrakis, they find that the Harkonnens have already destroyed all the mining equipment before leaving.

Conflict Unit. The Empire secretly sends their Sardaukar troops to attack the Atreides base, coordinated with the Harkonnens, and Duke Leto Atreides is assassinated (a family member is harmed by the villain - Function 8). Paul and his mother are taken captive and the Harkonnens plan to abandon them in the desert to die (the hero is forced to leave home - Function 11). Using "the voice," Paul and his mother escape and are lost in the desert. From a distance, they watch as the base is destroyed, and Paul knows that his father is likely dead.

Struggle Unit. Under the influence of the spice, Paul once again sees visions of the future. He launches a holy war against the Harkonnens in his father's name and defeats them in battle (hero faces villain in a showdown and villain is defeated - Functions 16 and 18). Paul dons the ring of House Atreides and vows to revive his family. He plans to have Dr. Kynes inform the six major houses of the Empire's collusion with the Harkonnens to destroy House Atreides, in order to incite rebellion. Duncan saves Paul in the desert, and they make their way to the ecological station where Paul convinces Dr. Kynes to help him. Pursued by enemies, Duncan sacrifices himself to buy Paul time. Paul and Dr. Kynes go their separate ways, with a plan to meet up in Fremen territory.

Return Unit. Paul escapes in a spaceship and travels through the desert to the Fremen camp, but he is not immediately recognized. The Fremen leader decides to protect him temporarily, realizing Paul's potential as a prophet (hero returns but is not immediately recognized - Function 23). However, some of the warriors challenge Paul's identity and call for a duel using the AMATAL RULE (hero faces a difficult challenge - Function 25). Despite the visions of his own death, Paul bravely accepts the challenge and emerges victorious (hero overcomes challenge - Function 26).

Acceptance Unit. Through his bravery and humility in battle, Paul wins over the Fremen people (hero is given a name) and gains the Desert power (hero ascends to the throne).

Although a more detailed analysis could identify the parts of "Dune" that better align with Propp's narrative sequence, the purpose of using narrative theory is to reveal the underlying meaning of the narrative in the "Dune" text. By analyzing the narrative of the text, it can be discovered that "Dune" has a considerable degree of secular mythological features as a science fiction film. Walter Benjamin declared in "The Work of Art in the Age of Mechanical Reproduction" that "film is the most malleable artwork, which is related to its complete abandonment of eternal value."[3] There is a natural gap between film and folk tales conveyed in ancient mythology or agricultural civilization eras, and they cannot be simply equated. If science fiction films still serve the functions of ancient myths or folk tales in the modern industrial era, it means that their social context has undergone significant changes. Therefore, if "Dune" is to be positioned as a secular mythological sequence produced by modern culture, its position in the film map of the streaming media era must be comprehensively examined.

To classify this film, it is necessary to examine it in the context of the entire filmography [4]. Firstly, based on the film's tone, it is a science fiction film with science fiction as its core, which is its essential attribute. Secondly, based on the elements represented in the film, adventure, action, and disaster are all indispensable parts of the film. Thirdly, based on the meaning conveyed by the film, symbols such as religion, philosophy, and ecology bring complex and ambiguous connotations. Fourthly, based on the film's incompleteness, the film presents a "pure event" and "pure time" through prior setting. Through in-depth analysis, it can be found that "Dune" does not belong to any specific type of previous films, but belongs to a composite sequence type that mixes technology, tone, elements, and meaning together. As a film in this composite sequence, "Dune" is a continuation and development of films such as the "Star Wars" series and "Independence Day" to some extent. However, on the other hand, it has far exceeded the boundaries created by the Hollywood film industry in the past and has different effects from the above-mentioned films. It is a type of unfinished narrative that arises in a new context and new technology. As a new type of secular mythological narrative, the textual structure of such films

https://doi.org/10.1051/shsconf/202316701015

typically consists of the convergence of two functional sequences. This contemporary version of "myth" is not a simple story, but rather a complement and overlap of two stories and two meaning structures.

The first sequence is a story of saving the world, in which a formerly peaceful society is disrupted by the attack of evil forces, and the protagonist experiences a struggle between good and evil, a competition between dark and light kingdoms, and a journey of hardship as a hero to regain peace. This is precisely the functional sequence studied by Propp in traditional folk literature. In early science fiction films such as "A Trip to the Moon"(1902) and "Frankenstein"(1910), this secular mythological narrative pattern is evident, and it remains a major component of the narrative. The second sequence stems more from the "creation" brought about by popular culture after the industrial civilization: the self-salvation of a hero with a mission to save the world. That is, there are two levels of story and meaning in a film. One is the social mythological level, in which the protagonist overcomes an external threat projected from some internal social dilemma and integrates his or her own community. The other is the personal mythological level, which involves the myth of autonomous will or the myth of the subject. In this level, a person who has been frustrated in modern society and is experiencing personal lack encounters truelove or healing trauma while saving society and regains the meaning and perfection of life. Examples of such science fiction films in the post-industrial era that exhibit the dimensions of social and personal myths include the "Star Wars" and "Matrix" series.

3. NEW SECULAR MYTH NARRATIVE TYPE

"Hollywood movies spare no effort in weaving various fascinating hero myths, but their spiritual core still embodies the ultimate expression of American values, encompassing the 'American Dream' and the values of the capitalist Protestant ethic. " [5] After the social crisis and cultural upheaval of the 1960s, the Hollywood film industry underwent significant changes. Some film genres gradually disappeared, such as Westerns and musicals, while in many cases, there was an increasing number of 'grafted' or collage-type genres: forms that mixed two or more types. Examples include the 'Star Wars' and 'Matrix' series. Different genres, including their grafted or collage styles, appeal to their own audiences both explicitly and implicitly. Viewers of a certain film genre derive pleasure from the extremely familiar story, characters, and narrative patterns being renewed. Another type of viewer finds pleasure in discovering new variations on old clichés by referencing similar films. As a product of Hollywood, "Dune" exhibits distinct modern mythological attributes. By combining Propp's functional sequence and Claude Bremond's paradigm model, we can analyze the two functional sequences of this new mythological narrative type.

In the first functional sequence, the social mythological level, the family or kingdom in folk stories is usually replaced by a modern national state. In Hollywood movies, this modern national state that encounters crises and overcomes them naturally refers to the United States. In Hollywood science fiction films, the plot generally involves a major battle between the Earth's forces and alien evil forces, with the US government naturally assuming the role of representing humanity - the side of goodness, love, and justice. In "Dune," the symbol of future human society is House Atreides under imperial rule. As one of the only senders and receivers of motion elements in the film, it determines the mythological tone of the first function of the film. The so-called House Atreides is a melting pot of races, accepting ethnic groups from the sky and the sea. As the prestige of the empire develops, it undoubtedly reminds viewers of American society, which is compatible with multiple ethnic groups and skin colors - a reaffirmation of the American capitalist Protestant ethic. It is worth mentioning that as political correctness has become increasingly important in modern Hollywood movies, Dr. Kynes, the judge in the film, appears just and rational when faced with the task of handing over two families. On the one hand, this is a coping strategy adopted by mainstream culture in response to the turbulent black civil rights movement for half a century. On the other hand, Dr. Kynes, in the face of the destruction of House Atreides, resolutely chooses to help Paul, which can be seen as an intention to preach to black people in the film."。

Claude Bremond argued in his book, "The Logic of Narrative Possibilities," that "any narrative follows a basic sequence with dual possibilities." [6] This highlights the development of the surface logic of narrative, wherein a narrative text is a constantly dividing structure. According to Bremond's theory, the initial situation presented by the narrative is that the protagonist faces a certain dilemma, which opens up two possibilities for the narrative: either breaking the dilemma or being stuck in a deadlock. The narrative logic also implies a possibility that originated from ancient Greek drama - delay - which leads to constant changes in the story as it progresses towards its ending.

In "Dune," the peaks of the narrative are stacked up, and the main events of the narrative are constantly delayed. This not only includes the switch between the two narrative functions but also the continuous transformation of the main functional sequence into a sub-functional sequence. The main sequence in "Dune" is the confrontation between Atreides and the Empire. The solution to the problem is that the Atreides family plans to unite with the desert forces, but they are destroyed. The new solution is that young Paul Atreides tells the judge to inform the six major families of the Empire's mobilization to incite rebellion. However, this plan is constantly delayed due to the pursuit of the Harkonnen family. Thus, the narrative enters a sub-sequence: how to seek revenge? One of the solutions is for Paul to become the leader who helps the Freman take control of the spice in the desert. However, to join forces with the hostile outside world, he can only rely on his prophet identity, which leads to an interlude sequence: becoming the prophesied THE ONE of the Bene Gesserit and the prophet of the Freman people. The intersection of these two identities intertwines the first

https://doi.org/10.1051/shsconf/202316701015

and second functional sequences of the film's narrative. All subsequent plots serve these two identities.

In the myth of modern individualism, the second functional sequence of the film narrative, "the missing family member by the fireside," refers to Paul's future wife, Chani, whom he sees in the prophecy. In the personal life history of modern logic and concepts, this obviously indicates a lack of desire, while in the specific context of the film, it also highlights Paul's hesitation towards the mission he carries. For the audience, "Dune" is a modern "space opera". Paul resists learning the Voice from his mother, resists the Bene Gesserit's imposition of the identity of a prophet on him, resists his father making him the only heir of the Atreides, and repeatedly struggles with the destiny he carries. In ancient Greek tragic myths, tragic fate is often used as a model. Aristotle once said, "The object of pity is the person who suffers from undeserved misfortune, and fear arises because the sufferer is a person like us. Therefore, tragedy should show characters moving from prosperity to adversity, not the other way around, that is, from adversity to prosperity." [7] Tragedy is the art of conveying emotions of pity and fear. The story of Paul in The story of Paul in "Dune" evokes compassion and fear in the audience, although it does not exactly correspond to tragedy in the ancient Greek sense. Paul's family was destroyed, his father was killed, he was pursued by the Harkonnens, and exiled to the desert. As a young prophet, he carried the mission of launching a "holy war". The hero resists fate but falls into despair, and accepting fate only leads to a more difficult future. When Paul wins the fight based on the Amatal rule, and earns the respect of the Freman, he achieves a basic victory, growing from a youthful boy to a capable warrior. This victory also means that he is walking the path predetermined by destiny and marks the continuation of this modern personal myth.

4. THE MYTH OF NARRATIVE TYPE

The development of technology and changes in philosophy have made film art a dynamic, polysemous, and unfinished part. Furthermore, changes in viewing habits mean that the solemn viewing experience in a movie theater is no longer the only option. The emergence of streaming services has made it possible to watch and experience films freely, anytime, anywhere, and in any form. In this system, the dimensions of social and personal myths are, to some extent, dissolved, and the boundary between popular culture and elite culture has become blurred. As a result, from the form of humanism to anti-humanism, utopia, dystopia, politics, society, and even ultimate philosophy, there is no clear location, making it an extremely complex part. Therefore, in "Dune," viewers can find a narrative form that exists outside of the two sequences - a dynamic narrative that is not bound by social or personal mythological levels, a type of unfinished narrative. With the advent of new technological revolutions, particularly the great advances in streaming media technology, the art of film has become increasingly difficult to interpret. The narrative paradigms extracted from typified social and personal myths on which movies were originally based may have ended, leaving viewers feeling drained of energy. In this context, French post-structuralist philosopher Gilles Louis Réné Deleuze proposed a possible film narrative paradigm in 'Cinema 1: The Movement-Image'[8] and 'Cinema 2: The Time-Image'[9] namely, 'pure optical and sound situations,' and conducted profound philosophical reflections on film. Starting from this paradigm, it is possible to surpass the limitations of past narrative studies, which only focused on narrative content and form, and examine films from their aesthetic implications.

Deleuze believed that modern America had produced several new image types, and traditional narrative paradigms were no longer able to provide satisfactory answers to them. For example, Stanley Kubrick's '2001: A Space Odyssey' presents the audience with a seemingly illogical story, beginning with the moment the large black stone is erected before prehistoric humans, who have just started to recognize tools. Later, black stones appear in various locations in the universe, standing on the moon, floating in space, carrying a certain mysterious meaning. Deleuze believes that this rebellion against traditional narrative paradigms can be traced back to Italian neorealism films, where 'realism is defined by its specific level, which does not exclude fiction, even fantasy. It includes phantasmagoria, strangeness, heroes, and legends. It can accept the eccentric and excessive, but only those that are within its own eccentricity and excess.' In other words, realism is primarily concerned with a reality that conforms to everyday logic, regardless of whether it is draped in the guise of magical or science fiction. Italian neorealism films, represented by works such as 'Rome, Open City,' 'Bicycle Thieves,' and 'The Tramp,' depict the true face of reality because they do not 'represent a reality that has already been decoded,' but 'hit a reality that is always ambiguous and should be interpreted.' In other words, true reality is uncertain and cannot be integrated into a fixed structure or type by logic. Such a true reality is the 'pure optical and sound situations.'"

"Pure optical and sound situations" refer to original situations in real life that have not been artificially extracted, possessing ambiguity and generativity. Deleuze believed that great artistic expression involves expressing the unspeakable, and that great films use audio and visual elements to convey the unspeakable, telling a story that cannot be told: a story of "pure optical and sound situations." Firstly, "pure optical and sound situations" no longer depict "logical events" but elucidate "pure events." Deleuze believed that a "pure event" is one that breaks the atomistic structure of the event and returns the event itself to its generative original state. This is closely related to the notion of a "pause," where "pure events" always exist in continuity, and the time of the pause is not only before and after the event but also within the event itself. In "Dune," the focus is on the protagonist's anticipation of this "pause," and the "pause" becomes the protagonist, causing a fracture in the sequence of events where "the generative simultaneity escapes the present." In this sequence of events, causal logic is inevitably disrupted. For example, in "Dune," the assembly of events does not follow any necessary or contingent rules but unfolds in a manner that cannot be defined by the potentiality of "pure events," forming the experience of the heterogeneity, interweaving and fusion of reality and memory. Secondly, "pure optical and sound situations" are no longer in the linear time of reality but in "pure time." Deleuze opposes linear time and proposes a new view of time in film art - "pure time," which emphasizes the multidimensional integration and communication of time. Deleuze believed that the "present" or "moment" is actually a flowing "moment," and similarly, the "future" has already arrived. Any "moment" or "present" includes the "past" and "future." The narrative of "Dune" is deeply connected to this non-linear "pure time," as Paul is the trinity of the flowing moment, the present, and the future, living in the triple self-image of the past narrative of Bene Gessrit, the present, and the future. Therefore, we can see that Paul's situation is also a kind of "pure time."

5.CONCLUSION

Narrative research involves analyzing the ways in which stories are constructed and told in films, as well as the effects that these stories have on audiences [10]. By examining how narratives are created, scholars can gain insight into the cultural and social meanings that films convey. In the case of science fiction films, narrative research is particularly important because these films often deal with complex themes and ideas that require a carefully constructed story to effectively convey [11]. "Dune" is a prime example of how narrative research can reveal the deeper meanings and implications of a film. The film's narrative structure is a key component of its success, as it weaves together multiple threads of storytelling to create a rich, multi-layered experience for viewers. The film's story explores themes such as religion, power, and ecology, using a complex narrative structure that keeps audiences engaged and invested in the story. Moreover, "Dune" goes beyond traditional mythological narratives by presenting a dynamic and unfinished type of myth that evolves and changes as the story progresses. This innovative approach to narrative storytelling is one of the reasons why "Dune" has resonated with audiences both commercially and artistically.

In conclusion, narrative research is a crucial method for analyzing the complex and nuanced narratives found in contemporary films. Science fiction films, in particular, require careful analysis of their narratives to fully understand the social and cultural implications of their stories. "Dune" is a prime example of how narrative research can uncover the deeper meanings and implications of a film, and how a dynamic and unfinished type of myth can contribute to a film's success.

REFERENCES

1. Lévi-Strauss, C.(1955)The Structural Study of Myth[M]. Basic Books press, New York.
2. Propp, V.(1968)The Morphology of Folk Tales[M]. University of Texas Press, Texas.
3. Benjamin, W.(2012)The Work of Art in the Age of Mechanical Reproduction[M]. Prism Key Press, German.
4. Chen Xuguang. (2022)What kind of "imagination" is needed in Chinese science fiction films: "Dune" in the context of "imaginative consumption"[J]. World Cinema. 2022(01).
5. Yuan Zhizhong, Sun Wei. (2016) Rethinking the Cultural Ethics of IP Creation in Chinese Cinema[J]. Hundred Schools in Arts, 2016,32(03).
6. Bremond, C.(1973)Logique du récit [M]. Editions du Seuil, Paris.
7. Aristotle.(2006)Poetics and Rhetoric[M]. Barnes & Noble Classics, New York.
8. Deleuze,G.(1986)Cinema 1: The Movement-Image [M]. University of Minnesota Press, Twin Cities.
9. Deleuze,G.(1989)Cinema 2: The Time-Image [M]. University of Minnesota Press, Twin Cities.
10. Hayward, S.(2013)Cinema Studies: the Key Concepts[M]. translated by Zou Zan, Sun Bo, Li Yaoyang. Beijing University Press, Beijing.
11. Hellmann, C.(1988)The History of World Science Fiction Film [M]. translated by Chen Yupeng. China Film Press, Beijing.

Research on Field Investigation Methods of Folk drama

Huhe Qiao[1,*]

[1]PhD Candidate, Literature Department, No.1 Northwest New Village, Lanzhou, China

Abstract. Field investigation is an important research method for studying folk drama. Taking field survey experience of Errentai as an example, this article summarizes the methods and precautions that need to be used in the field survey of folk drama, hoping to provide a substantive case for the research methods of folk drama.

1 FOREWORD

Folk drama is an important part of folk literature, and it is also a form of folk literature and art with the highest artistic achievements, the largest performance scale, and the most complete performance links. Conducting a field survey of folk drama is conducive to comprehensively grasping the current situation and development rules of folk drama performance, and is also an important supplement and detail improvement to the textual research of traditional folk drama. Through participation in observation, questionnaire interviews, and other methods, it is of great practical significance to further understand the psychological state of the audience of folk drama and the latest performance scripts of folk drama in the contemporary society. This article will take the popular Errentai drama in North China as an example, and explore the field investigation methods of folk drama through the author's experience and field investigation practice.

At present, the theoretical research on folk drama and field research has been relatively rich. For example, Khayambashi Shirin[1] and Vetter Jeremy[3] and other scholars discussed the history of field research and the challenges it poses, and some scholars carried out thinking in the process of field research[4]. Steadman Jones Richard[5] provided the case of field research. Liu Hongyan[6], Zhao Q, Lu X[7], and others have conducted relevant research on Chinese folk drama and Errentai. Alshallaqi Mohammad[2] also carried out a useful discussion on field research and research ethics.

2 FOLK DRAMA AND ERRENTAI DRAMA

Throughout China, there has always been a tradition of performing folk drama, and various forms and styles of folk drama performance have been derived from different local cultures. Folk dramas are based on mass life, with language art as the backbone, and incorporate music and dance elements around language art texts. In modern environments, elements such as lighting are added to enhance the overall performance atmosphere.

Errentai is developed on the basis of Mongols and Han folk songs, and is widely distributed in several provinces in North China. Errentai's performance forms not only include singing and dancing, drama, rhymes, but also include various forms such as proverbs and joke sketches.

The content of the Errentai not only includes Mongolian folk songs sung in Chinese, but also the theme of expressing the yearning for beautiful love among the young people of the Mongolian and Han ethnic groups, as well as the content of depicting the daily life scenes of the local people. Props and clothing are relatively simple, and most of them are production and daily necessities, as well as clothing, shoes, and hats.

The research results of Errentai are relatively few, especially the field research. Research on the Errentai often traces its history of development, as well as its artistic expression, language style, singing, and dance aspects. There are also articles on the current situation and future direction of Errentai through on-site investigation, but the exploration is not very in-depth.

3 SELECT FIELD SURVEY POINTS

Errentai is a form of folk drama that is widely spread in Shanxi, Inner Mongolia, Hebei, Shaanxi, Ningxia, and other provinces. Areas where performance of the Errentai is relatively concentrated include Hequ in Shanxi, Fugu in Shaanxi, Kangbao in Hebei, Hubao'e in Inner Mongolia, and Ulanqab in Inner Mongolia.

Errentai drama is widely spread in five provinces and regions. Therefore, the selection of field survey sites will inevitably be concentrated in the above five provinces and regions. According to the literature review before the survey, the Tumet region in Inner Mongolia is the most suitable field survey point. There are three reasons for this:

1. The Tumet region is densely populated and economically developed, with a relatively large number

* Corresponding author: johuhint@outlook.com

of performance scenes for field investigation and observation.

2. The Tumet region has a large number of performers and rich interview objects. Compared with other regions, the cultural integration of Mongolia and Han is the most prominent feature.

3. The author was born in Hohhot City, and has a good grasp of the local dialect and dialect, making it relatively easy to conduct in-depth interviews and observations in the society.

Based on the above reasons, the author chose Tumet Right Banner as a field investigation site.

4 FIELDWORK METHODS AND PROCESSES

Fieldwork is not accomplished overnight, but rather a manual task with numerous links and cumbersome operations. The specific operation process should follow the following process:

1. Establish trust relationships

In the field survey process, the most important and prerequisite task is to establish trust relationships. It can be said that once a trust relationship is established. Fieldwork is half successful. Facing an acquaintance society like China, how to become one of our acquaintances is the first practical issue we face. There are three actual situations of trusting relationships:

(1) Already integrated

Some scholars choose a field survey location that is in itself their hometown or a place where they work, study, and live. I am already an acquaintance in an acquaintance society, and I am extremely familiar with the interviewees. Such fieldwork sites already have mature trust and social relationships, but it is important to point out whether they are in an information cocoon.

Many times, we don't really understand the environment we are in, and we turn a blind eye to many things, unable to discover the underlying principles and problems. Therefore, if we enter a familiar place to carry out field research, we must avoid this situation. We not only need to observe the content of our investigation with a critical perspective, but also constantly reflect on whether our investigation views are biased.

(2) A familiar stranger

Some scholars or field survey workers have already lived in the society where the field survey site is located before entering the field, but due to factors such as studying, they have failed to establish strong social relationships in society. I call this situation a familiar stranger, and it's often the worst. When faced with certain issues, they are often ignored because they are commonplace, and often fall into a situation where the beholder is confused. The way to solve this problem is to compare, see, think, and reflects more.

If you have known about the relevant situation of the field research site, you should avoid blind arrogance and carefully screen the authenticity of the information you have obtained. A little knowledge is the most troublesome, and it is also the biggest obstacle to our in-depth field research. All successful field research should be based on comprehensive understanding and in-depth insight.

(3) How to integrate strangers

The latest situation is that scholars who need to carry out fieldwork know little or nothing about fieldwork sites, and there are no acquaintances or simply strangers in the fieldwork society. This often happens to young students, as being the case with the author. Under such circumstances, how to establish a trust relationship. First, you are required to bring a letter of introduction issued by your unit to the government for assistance. Secondly, it is necessary in order to find ways to integrate into society, such as finding a part-time job locally and establishing preliminary interpersonal relationships. Finally, slowly develop acquaintances and expand the network.

In the existing research results, many field research disguise themselves as teachers, teach foreign languages or find other strengths to enter the community that needs investigation. Only by truly integrating into the community that needs to be surveyed can we observe the overall picture of the entire community, truly understand the operational laws of this society, and obtain more information. Some young scholars, in order to avoid trouble, often adopt the method of paying certain remuneration to interview specific groups, which can lead to the formation of an information cocoon, often obtaining information that is one-sided.

2. Participate in performance activities

If a relatively unchanging trust relationship has been established, the next most important task is to participate in performance practice. The term "participation" here includes not only participating as an audience, but also performing as an actor, fully understanding the artistic connotation that folk literary and artistic performances need to express.

(1) Appropriate Identity

To participate in performance practice, it is necessary to first have an appropriate identity, and sometimes multiple identities are even required in order to close the relationship with the audience. Taking folk drama as an example, investigators can enter the audience as enthusiasts, but this makes it difficult approach the actors. Therefore, investigators often need to learn some professions to communicate with the cast.

(2) Performance timing

To observe a performance, it is necessary in order to understand the performance occasion, specific time and location, and other information. The timing of the performance determines what type of performance occasions the investigator can observe. For example, weddings, funerals, and weddings, different occasions will perform diverse content. The repertoire performed at the temple fair is bound to be different from the wedding, which requires attention.

3. Summary and data compilation

After participating in performance activities or completing interviews and questionnaires, the most critical task is to sort out and classify the collected text, audio and video materials, paving the way for subsequent thesis writing. When organizing materials, it is often to follow the following principles. The first thing is supposed to be timely. After obtaining interview materials, compare them with your plan as soon as possible to see if you have obtained the content and information you plan to obtain, how many discrepancies exist, and if there are any errors or omissions, find ways to remedy them as soon as possible. The second is to be comprehensive, which refers not only to the comprehensive collection of materials, but also to do a good job. After storing the collected materials by category, you should also write field survey notes to reflect on your shortcomings and omissions in this survey. The third is to examine. During the field survey, it is necessary to seek opinions from key interviewees, which are a discussion link. After the field survey, it is also essential to carry out some communication and discussion with your team members or guidance teachers. Such discussions can often find issues that you have missed. As the saying goes, it is nice to see from the sidelines.

5 FOLK TABOOS AND PRECAUTIONS

Throughout China, due to separate folk cultural traditions, different taboos and rules of social and cultural life have also been derived. In the final analysis, our fieldwork activities still involve dealing with people, so we must pay attention to regional folk customs and taboos. Taking visiting villagers as an example, whether it is necessary in order to bring a greeting gift and what kind of gift is appropriate is a question that needs serious consideration.

(1) Meeting Ceremony

Meeting gifts are an important medium for establishing initial trust relationships when visiting the masses. In contemporary society, the form of saluting ceremony is relatively lively. For instance, during a preliminary visit or stampede, it is often possible to initiate a conversation by dispersing smoke or other means. Owing to special living habits in the Tumet region, when choosing a greeting gift, it is often to choose food that is resistant to preservation as the greeting gift. For example, canned food such as canned fruit, canned Babao Congee, and Baijiu is very popular among local people. With the gradual improvement of economic level, contemporary meeting rituals have gradually become rich and diverse. Currently, liquid packaged milk, fruit. Etc. is also common greeting gift. What should be noted is the etiquette of the meeting ceremony. When visiting an old artist, the author chose Baijiu at the meeting ceremony, but because only one bottle was brought, it caused unnecessary misunderstanding. Such minor omissions can lead to needless trouble and obstacles for field investigations. When meeting interviewees, in addition to the greeting ceremony, it is also important to point out that when entering a home, you should not step on the threshold, use the appropriate address appropriately, and not disrupt the seniority. Even when entering the door, you should consider which foot to take first.

(2) Daily etiquette

China's thousands of years of history have spawned a large number of life etiquette norms, especially in rural areas, where many elderly people still strictly abide by these norms. We must also give attention to them at all times during the field survey process. The first is table manners. Small talk during meals can often quickly pull into interpersonal relationships, so being able to grab meals at the interviewees' homes during field surveys is often an important foundation for effective interview activities. In the Tumet area, it is prohibited to insert chopsticks into a bowl at the dinner table, to speak loudly, to drink in order, and to eat in a hierarchical manner. In addition to table manners, there are also many stressed in daily life. In traditional fieldwork methods, experiential investigation is more emphasized, but when investigating folk drama, it is not only about experiencing, but also about learning and integrating into indigenous society. At this time, daily etiquette is the most direct manifestation of whether to integrate into local society. As the saying goes, only by learning and using local cultural etiquette and customs can we better way the local people and provide convenience for field research.

(3) Folk customs

As a form of folk performance, folk drama is inseparable from native customs. This aspect includes not only the local festivals, weddings and funerals, as well as the time and place the organization of temple fairs, but also deeper folk customs such as kinship.

Understanding folk customs, on the one hand, are to avoid unnecessary misunderstandings caused by touching local folk taboo, and on the other hand. It will also provide many clues for our investigation and research. For example, around the 15th day of the first lunar month in the Tumet region, there are about three days of carnival activities, which are often based on folk beliefs. At this time, there will be a large number of minor performances of the Errentai Drama. If we are able to learn about these customs in a timely manner, simply looking for a performance occasion for the Errentai is tantamount to a blind person feeling like an elephant without the full picture.

In general, paying attention to regional folk culture, customs, taboos, and other aspects, on the one hand, is to avoid misunderstandings and create obstacles for field investigation, and on the other hand, is also to find clues for investigation. However, it should be noted that it is necessary to clarify the main body, not to understand folk customs in order to understand them, but to grasp the principal line of folk drama.

Folk drama not only embodies the aesthetic taste of the local people, but also reflects their dialects, local languages, ideological concepts, and even worldviews. Therefore, in the process of field research, recording the information obtained is the most basic work, and we should further understand the world outlook and metaphysical thinking contained in drama.

6 CONCLUSION

As the most conspicuous form of folk literature and art, folk drama is not only widely spread in rural society, but also among migrant workers and the vast youth group. As one of the foremost research methods for studying folk drama, fieldwork is an indispensable and important research method. From the perspective of current theoretical research, scholars studying field survey methods mainly improve and promote field survey methods based on ethnographic methods of anthropology and ethnology. However, folk drama has a relatively timeliness and requires more flexible research methods. Compared to traditional ethnography, the field investigation of folk dramas should pay greater attention to the efficiency of the investigation. The information collected is often complex and requires further processing. In addition, there is more preparatory work to be done in order to study folk dramas, and adequate plans must be prepared.

When carrying out field research, it is necessary to formulate a survey plan and plan adapted to local conditions according to the local specific conditions, so as to avoid emergencies affecting the progress of the entire survey. In addition, the performance content and process of folk drama often contain a large amount of local knowledge. It may be difficult to conduct corresponding investigations and studies without mastering relevant local knowledge, so this further demonstrates the importance of preliminary preparation.

In general, the field research for folk drama needs to be well prepared. In addition, in response to unexpected situations that may occur at any time, in addition to preparing well, there should also be sufficient psychological preparation.

Reference

1. Khayambashi Shirin. Linguistic Challenges of Fieldwork for First-Generation Ethnic Researchers[J]. International Journal of Qualitative Methods,2022,21.
2. Alshallaqi Mohammad. Cultural practices and organizational ethnography: implications for fieldwork and research ethics[J]. Journal of Organizational Ethnography,2022,11(3).
3. Vetter Jeremy. The History of Fieldwork[J]. Histories,2022,2(4).
4. Thornton Sara A.,Cook Sarah,Cole Lydia,Hapsari K. Anggi,Zawawi Norliyana Zin,Page Susan E.. From 'Muddy glee' to muddy reflections on fieldwork and writing[J]. Area,2022,54(4).
5. Steadman Jones Richard. Linguistic fieldwork at the end of empire: British officials and American structuralists in Anthony Burgess' Malayan trilogy[J]. Language & History,2022,65(3).
6. Liu Hongyan. Research on the Inheritance of Local Opera Errentai in Contemporary Context[J]. Frontiers in Art Research,2020,2(3).
7. Zhao Q,Lu X. Investigation and Study on the Ancient Folk Drama Stages in Lantian County of Xi'an City[C]//Singapore Management and Sports Science Institute, Singapore,Information Technology Application Research Association, Hong Kong.Proceedings of 2017 2nd ACSS International Conference on the Social Sciences and Teaching Research (ACSS-SSTR 2017).Singapore Management and Sports Science Institute,2017:133-137.

A Brief Analysis of the Evolution of Audio Language Communication in the Internet Age

Huiwen Zhang*

Associate Professor, Network and New Media, School of Language and Culture, Dongguan University

Abstract. With the popularization of the internet, audio language has rapidly developed into a powerful force. According to data analysis from iiMedia Research, the active user scale of the Chinese audio industry reached 800 million in 2021. This paper analyzes and explains the evolution of audio language in terms of content, technology, copyright, and other aspects, in order to sort out the development context of audio language and explore its communication laws and trends. In the development process of audio language, emphasis should be placed on intelligently presenting audio information, building an audio ecological closed loop, and collaborating to improve market rules to achieve more comprehensive and efficient progress.

1 INTRODUCTION

Language is a social product that serves as a bridge for interpersonal communication and a tool for establishing common understanding. Spoken language includes various forms, such as radio broadcasts, verbal statements, scientific reports, recitations of excellent literary and artistic works, dramas, movies, cross-talk, storytelling, and other forms, as long as they conform to the norms of literary language.

Traditional spoken language is mainly carried by traditional media such as radio and television, while spoken language on the internet is mainly presented in the form of online audio. In this paper, the spoken language on the internet is referred to as "online audio," which mainly refers to network audio forms related to hosts or broadcasters such as audio programs, live audio broadcasts, web radio, and audio books, and tends to comprehensive audio platforms (such as Lizhi, Ximalaya, Qingting FM, etc.), excluding digital music. These audio contents can be listened to through online transmission or downloaded to personal devices for listening.

2 LITERATURE REVIEW

With the rapid development of Internet technology and the popularization of artificial intelligence technology, audio content has attracted extensive attention in the era of the Internet. A search on CNKI platform for "audio content" resulted in 3,536 articles, 860 articles for "online audio," and 90 articles for "Internet audio content." The dissemination and forms of audio content have undergone significant changes and evolution. This review will analyze the dissemination and evolution of audio content in the Internet era from the following three aspects:

2.1 The Transformation in the Era of Intelligent Communication

In the internet age, the forms of audible language communication are becoming increasingly diverse. In addition to traditional forms such as radio, television, movies, music, and audiobooks, there are now many more online audio formats, such as music, radio stations, podcasts, and audiobooks. Yu Ning in his article "Transformation of Audible Language Communication Roles: Media Analysis Based on 'National Trend Dance Steps"[1] , discusses the new era of traditional media and emerging media integration, with traditional television media competing to leverage new media small-screen live streaming to create more high-quality content for cross-screen communication. Xia Xiaochen, in "On the Arc of Intentions in Audible Language Creation in the Era of Intelligent Media"[2], mentions the great challenge that Dreyfus's non-representational intelligence thought brings to the representationalism of artificial intelligence. Audible language creations have a high-level pursuit, manifested in the dialectical relationship among cognition, perception, and intuition. In the era of intelligent media, there will be breakthroughs in audible language creations.

2.2 The "Trap" of Self-Media Communication Channels

A search in CNKI using keywords "self-media," "communication," and "trap" finds 106 articles. Wang Youfeng, in "Characteristics and Discourse Trap of Self-Media Communication"[3], explains self-media, citing a "We Media" report by Bowman and Willis. "Self-media"

* Corresponding author: Huiwen Zhang Email: zhanghuiwen@dgcu.edu.cn

refers to citizen media, including blogs, microblogs, WeChat, and forums. Tacitus proposed the "Tacitus Trap" theory, where society gives negative evaluations when public power loses credibility. In the Internet + self-media era, audible language communication is no longer limited to traditional channels, making it more convenient and free, but also prone to the "Tacitus Trap." Wen Zujun and Chen Fangxin discuss concerns about self-media's rapid development in "How the Government Can Overcome the 'Tacitus Trap' in the Self-Media Era"[4].

2.3 The Evolutionary Development Direction of Audible Language

Bai Guiren and Qingqing's "Trends and Impacts of Audible Language Communication Innovation in the AI Environment" [5] discuss how AI and new technologies accelerate audible language communication, leading to subject, space, and expression transformation. They analyze the manifestations and impacts of these trends on information dissemination laws, cultural connotations, and aesthetic standards. Huang Lei's "Overcoming the 'Tacitus Trap' - Dilemma and Breakthrough of Mainstream Discourse Power in the Self-Media Era" [6]suggests maintaining stance, enhancing credibility, seizing opportunities, setting discourse themes, and optimizing innovative carriers as feasible paths for mainstream discourse power in the self-media era.

These articles have explored and contemplated the breaking of the monopoly of mainstream discourse power due to the rise of self-media and proposed constructive countermeasures for the direction of self-media development. Based on the literature analysis and summary, this paper will discuss the evolution of internet audible language from various aspects, including content quality, technological development, user needs, and copyright issues. Once technology reaches a certain stage, it will ultimately return to the realm of content and user needs. No matter how far "innovation" goes, it will eventually return to the essence of content demand.

3. THE CURRENT SITUATION OF THE DISSEMINATION OF AUDIO LANGUAGE IN THE INTERNET ERA.

In the past few decades, the dissemination of audio language has undergone many changes and evolutions. Around 2017, traditional broadcasting gradually declined from its "golden age," and audio programs carried by audio language began a rapid development path, benefiting from the application of artificial intelligence technology, bringing about great changes to the dissemination of audio language.

3.1 The "Ear economy" bursts with development vitality

The China Business Industry Research Institute released the "2022 Forecast Analysis of China's Online Audio Industry Market Status". The revenue of China's online audio market increased from 2.5 billion yuan in 2017 to 20.6 billion yuan in 2021, with a compound annual growth rate of 69%.

According to data from iResearch, China's online audio market revenue has increased significantly from RMB1.2 billion in 2015 to RMB27.2 billion in 2020, accounting for 17% of the world's total market scale. It is expected to reach RMB54.3 billion in 2022.

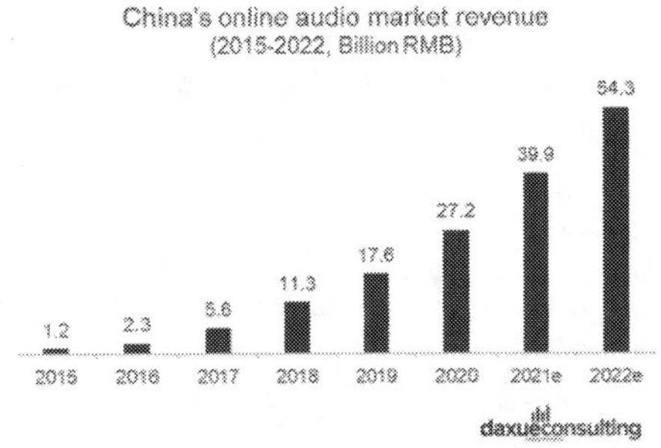

Fig.1 China's online audio market revernue

Data source: iResearch, designed by Daxue Consulting, China's online audio market revenue, 2015-2022

The "China New Media Development Report (2020)" mentioned that the "ear economy" bursts with development vitality, and digital audio has advantages in meeting the needs of specific groups such as children and the elderly due to its accompanying and listening attributes. With the accelerated commercial promotion of 5G technology, all-scenario intelligent audio applications such as in-car audio, smart watches, smart speakers, and smart homes are also growing rapidly, and the audience is developing in a diversified manner.

3.2 The "small universe" of the Z generation explodes

The epidemic deepened the demand of domestic netizens. The user scale of online audio apps continued to expand, and various niche audio apps also experienced certain growth, such as the sleep audio apps Snail Sleep and Xiaoshui Sleep. Taking the podcast platform "Small Universe," which is currently popular among young people, as an example, since its internal testing was opened in March 2020, the cumulative downloads of Small Universe on Android and Apple channels have reached 19.3649 million and 2.3412 million, respectively. According to a survey of Chinese podcast audiences by JustPod in 2022, Small Universe's user usage rate on various podcast platforms or streaming media platforms reached 74.6%, far higher than Apple Podcasts and other veteran podcast platforms as well as large audio media such as Himalaya and NetEase Cloud.

The text is discussing the current state and trends in China's audiobook industry. According to the "2022-2028 China Audiobook Market Survey and Investment Strategy Report" by Co-Watch.com, the industry is predicted to have 420 million users and a market size of 9.37 billion yuan by 2022. The data from iiMedia

Research shows that audiobook users are mainly younger, with a dispersed and spindle-shaped age distribution. The report also found that men make up a larger proportion of audiobook users than women, and that 31-40-year-old users are the main listeners, followed by 26-30-year-old users.

3.3 Mainstream Media Shifts Focus and Makes Efforts

Traditional radio stations have shifted their focus to the internet by settling on platforms such as Himalaya, Qingting FM, and Lizhi. As a result, mainstream media has developed its own mobile internet platforms, such as Jilin People's Broadcasting Station's Mu'er FM and Guangdong Radio and Television Station's Yue Ting client, which offer localized and vertical audio services. In 2020, China Central Television (CCTV) released the "2020 White Paper on 5G Media Applications of China Central Television", and launched the "Yun Ting" app, a sound aggregation platform, to expand its influence in the mobile internet market. Wu Jingyan, deputy director of the Southern Metropolis Daily's integrated media center, stated that audio news is a new path for media integration under the background of comprehensive strategic transformation.

4. EVOLUTION OF THE DISSEMINATION OF INTERNET AUDIO LANGUAGE

The internet era has brought different definitions of how audio language is presented, and with it, changes in user demand. Various audio programs have undergone an evolution from traditional media to digital and intelligent media.

4.1 Diversification of content form and style

With the development of digital technology and changes in user demand, the form and style of audio programs have become more diversified, including news, music, talk shows, education, culture, entertainment, and other types. At the same time, these programs also pay more attention to personalization and customization to meet the needs of different users. With the empowerment of 5G and all-scenario intelligent devices, scene listening devices have also seen a rapid growth trend.

4.2 Cross-platform dissemination

Traditional radio stations could only broadcast audio programs through radio waves, but now audio programs can be disseminated through various channels such as the internet, social media, and smart speakers, achieving cross-platform dissemination.

The car audio market has become the main battleground for audio. According to the Automobile Industry Association, the number of vehicles in China is about 320 million in 2022. With the continuous development of 5G/VR technology, the scale of China's connected car market is expected to increase from 210 billion yuan in 2021 to 800 billion yuan in 2026. Audio language will have great potential in the field of car audio.

4.3 Data analysis and personalized recommendations

Hu Zhengrong, the general editor of China Education Television Station, once pointed out that "voice will be the next huge entrance to the Internet, connecting everything." Audio programs have generated a large amount of user data in the digital age, which can be analyzed and utilized through artificial intelligence technology to help producers understand user needs and preferences and produce programs that better meet user needs. In addition, personalized recommendation algorithms can recommend programs that users like to them, increasing user retention rates and loyalty.

4.4 Real-time interaction and socialization

The development of digital technology has also brought the possibility of real-time interaction and socialization to audio programs. Audio live broadcasts have the same attributes as video live broadcasts, and they do not require revealing one's face, which increases the mystery and concealment and is more popular with listeners. The interactive attributes also increase user participation and interaction. Meanwhile, some applications also provide real-time commenting, liking, and other interactive functions, strengthening the interaction between users and programs.

5. DEVELOPMENT DILEMMAS OF INTERNET AUDIO LANGUAGE

With the advent of the intelligent era, Internet audio language has become an important means for people to obtain information, entertainment, and learning, but it also faces some development dilemmas.

5.1 Uneven Content Quality

In recent years, with the gradual expansion of the online audio market, the problem of uneven content quality has become increasingly prominent. In 2020, Himalaya was interviewed by the Zhejiang Provincial Consumer Rights Protection Committee due to issues such as false advertising for its membership advertisements. One Listen Audio, an online audio platform, has also received complaints and criticisms from users due to issues such as piracy, low quality, and false advertising. Many users feel confused and disappointed when using such platforms. In addition, some users also complain about the recommendation algorithm of some platforms, which recommends content that does not match their interests and needs, further reducing their trust and satisfaction.

5.2 Imperfect Profit Models

In 2020, the global online audio market was approximately 30.61 billion US dollars, and it is expected to reach 59.99 billion US dollars by 2027, with an annual compound growth rate of 10.1%. Online audio mainly includes user subscription services and advertising-supported radio services. In 2019, subscription services accounted for 57.88% of online audio revenue, and subscription fees have become the main source of online audio revenue.

Compared with other content forms such as video and text, the profit model of online audio has not been fully developed. Although some platforms have achieved profitability through subscription services, the profit model is too single. Moreover, for most users, they are more willing to choose free online audio content, which makes the commercialization and profitability of online audio content more difficult.

5.3 Inadequate Copyright Protection and Management

Due to the flexible dissemination methods and channels of online audio, copyright management and protection are relatively difficult. This has also led to a decline in the enthusiasm for creating online audio content, which has had a certain impact on the long-term development of the online audio industry.

In 2019, China's Judicial Document Network released a first-instance judgment document in which Tencent Music sued NetEase Cloud Music for copyright infringement of Jay Chou's songs. The judgment showed that companies such as NetEase Cloud Music had to compensate Tencent Music for economic losses and reasonable expenses for stopping infringement totaling 850,000 yuan. At the same time, NetEase Cloud Music still has problems such as storing pirated songs, lax platform review, and a large number of cover songs. In recent years, this problem has been somewhat improved under the requirement of mandatory real-name registration, but similar events still occur frequently on other platforms, especially some niche platforms, where user flow is frequent and regulation is difficult.

6. EVOLUTION STRATEGIES FOR INTERNET AUDIO LANGUAGE

The evolution of internet audio language is an ever-evolving field that involves multiple aspects such as content production, intelligent audio technology, user service, and copyright management.

6.1 Evolution of Content Quality

In today's digital age, an increasing number of people choose to listen to audio content over the internet, such as paid audio, podcasts, audiobooks, and radio dramas. This trend has become particularly evident in the past few years and is expected to continue to grow. In this fast-developing industry, the quality and innovation of content have become the key factors that determine the success of audio language.

6.1.1 The concept of "content is king" has become increasingly important in the internet audio language industry. Unlike traditional media, internet audio content is not limited by time and space, and people can listen to it anytime and anywhere. Therefore, compared to other media forms, internet audio content requires unique and attractive content to attract listeners. Only by providing high-quality and interesting content can producers win the attention and loyalty of their audience.

6.1.2 The market for audio language content has become more segmented and diversified. Nowadays, different forms of audio content, such as audiobooks, radio dramas, and podcasts, have emerged. This diversification requires producers to pay more attention to content quality and diversity to attract different types of listeners. At the same time, in order to meet the needs of their audience, audio content producers need to continuously improve and innovate based on listener feedback, ensuring that their content remains in touch with their audience.

6.1.3 Innovation and diversity of content are important factors for the success of the audio language industry. Audio content producers need to integrate new technologies and innovative thinking into the production process to break traditional modes and experiences. For example, virtual reality technology can allow listeners to experience the story in an immersive way, while AI voice synthesis technology can improve the production efficiency and quality of audio content. Therefore, with the continuous development and application of new technologies, the future of internet audio language is full of innovation and infinite possibilities.

6.2 Development of Technology

Online audio is a fusion of technology and art. In the traditional broadcasting era, the technology of recording, editing, and playing audio was controlled by a small number of technicians. With the advent of the internet era, this highly difficult technology has become simple and easy to operate, and ordinary audio creators can easily complete the production process.

Currently, speech recognition, speech synthesis, and audio encoding and decoding technologies have gradually matured, enabling internet audio language to achieve more accurate and natural speech input and output. In addition, the continuous development of artificial intelligence technology has also provided a broader space for the application of internet audio language.

6.3 Evolution of User Demands

With the continuous progress and popularization of technology, the user demand for internet audio language is also constantly evolving. Initially, users' demand for audio language content was mainly to be able to listen to their favorite songs or radio programs anytime and anywhere, but today's users not only expect online audio services to provide high-quality sound and smooth playback experience but also hope that these services can recommend audiobooks, knowledge content, and radio dramas that match their own taste. Therefore, some online audio services have started to use artificial intelligence and machine learning algorithms to recommend content that users may be interested in based on their history of listening and rating records.

In addition, users also expect online audio services to provide more social interaction and user participation experiences, such as interactive live broadcasts, online discussions, and user-generated content. Therefore, online audio services need to continuously innovate and improve their user service to meet the evolving demands of their audience.

7. CONCLUSION

With the development of the internet era, audio language communication has become a new and emerging form of media, showing a trend of rapid growth. Audio language communication has highly interactive and participatory characteristics, can create rich and colorful sound content, thus satisfying users' demand for diversified sound resources. At the same time, the convenience, speed, real-time, and global dissemination and reception of audio language communication also make it a popular form of media. The evolution of audio language communication can be traced back to the emergence of network broadcasting, and now has developed into a diversified form of communication including podcasts, web radio, voice social networking, and so on. Driven by voice technology, audio language communication has profoundly affected social life and the media industry, changing the way people obtain information and entertainment, promoting the dissemination of information and the development of innovation.

In the future, audio language communication in the internet era will continue to develop rapidly, presenting trends such as personalized and customized services becoming mainstream, more diverse and refined content trends, and further development of voice technology providing more possibilities for audio language communication. However, audio language communication in the internet era also faces some challenges, such as the balance between the quality and quantity of content creation, copyright protection issues, technical limitations, and data security issues. To better cope with these challenges, the audio language communication industry needs to continuously innovate and explore technology, strengthen copyright protection and information security, and actively explore new business models to better adapt to the changes and development of the internet era.

REFERENCES

1. You Ning. The Transformation of the Main Roles in Audible Language Communication - A Media Analysis Based on "National Trend Dance Steps" [J]. News World, 2023, No. 381(01): 49-52. DOI: 10.19497/j.cnki.1005-5932.2023.01.023.
2. Xia Xiaochen. On the Intention Arc Thought of Audible Language Creation Subjects in the Intelligent Media Era [J]. Journal of Northeast Normal University (Social Science Edition), 2022, 24(02): 145-150. DOI: 10.15936/j.cnki.1008-3758.2022.02.018.
3. Wang Youfeng. The Characteristics and Discourse Trap of Self-Media Communication [J]. Young Journalist, 2014, No. 467(27): 75-76. DOI: 10.15997/j.cnki.qnjz.2014.27.106.
4. Wen Zujun, Chen Fangxin. How Can the Government Overcome the "Tacitus Trap" in the Self-Media Era [J]. Young Journalist, 2020, No. 668(12): 42-43. DOI: 10.15997/j.cnki.qnjz.2020.12.016.
5. Bai Gui, Ren Qingqing. Trends and Impacts of Audible Language Communication Innovation in the Artificial Intelligence Environment [J]. Haihe Media, 2020, No. 003(03): 1-6.
6. Huang Lei. Overcoming the "Tacitus Trap" - The Dilemma and Breakthrough of Mainstream Discourse Power in the Self-Media Era [J]. Journal of North University (Social Science Edition), 2018, 34(01): 56-60.Assistants. Journal of Management Information Systems, 38(3), 736-744.

A Short Film Interpreting the Chinese Characters: 'Trace Back'

Zhe Yu[1*], Szu-Ming Chung[2]

[1]Arts of Digital Media, Quanzhou Vocational and Technical University

[2]Arts of Digital Media, Quanzhou Vocational and Technical University

Abstract. In this study, a group of students worked as a team to create and cocreate a short film in a situated learning environment. The film presents Chinese Han characters in a narrative that employs the poetic audio-visual features of Han characters. We used expert interviews to verify the teaching and learning effects of a short film to interpret the Chinese Characters: 'Trace Back'.

1 Introduction

As ideograms, Chinese characters have natural audio–visual imagery. Therefore, creating a short film with Chinese characters offers a unique opportunity for transforming ideograms, or one of the six forms of Han characters, into images. [1] These characters have cultural elements that can be used as a reference for the style, narrative, and creativity of the image. These elements transform the imaged text into a unique medium for the promotion and dissemination of Chinese character culture. The figurative expression of the audio–visual imagery disrupts the static state of the characters, and presenting this information in a short film adds time elements that transform the characters into moving images. For 'Trace Back', a group of Digital Media Arts students collaborated in a situated learning environment established by the researchers to demonstrate these concepts through creation and cocreation. The following sections present the reviewed literature, research objectives, methodology, production content, and results of expert interviews.

2 Research Objectives

For the production of the animated short film, the following four objectives were established:

(1) The potential and methods for converting Chinese characters into images were explored. The audio–visual connection between Chinese characters and images determines whether Chinese characters are suitable for transformation into images and how they can be transformed into videos.

(2) The ideography and imagery of Han characters can enable creation of film images, and both should be used to determine the style and quality of films and television productions on Chinese characters.

(3) The difficulty of comprehending Chinese is increased by the difficulty of the written language. An animated short film may assist people in comprehending the meaning of Chinese characters at a sensory level and thereby improve the dissemination of Chinese characters.

(4) The visual beauty of Chinese characters is difficult to recognise in daily life or in cultural environments. Images in a short film that are based on the ideography and imagery of Han characters can promote the learning and dissemination of Han characters as well as comprehension of their beauty.

3 Reviewed Literature

Using Chinese characters to tell Chinese stories and transforming Chinese into an international language has become key to promoting and disseminating Chinese character culture. Audio–visual language is the most internationally accessible form of language and can break through regional cultural boundaries.

Montage is the film style that most aptly conveys the feeling of traditional Chinese poetic idioms. The Britannica dictionary definitions of montage [2] are (1) a work of art that is made up of several different kinds of things (such as strips of newspaper, pictures, or pieces of wood, [e.g., a photographic montage])' and (2) 'a mixture of different things'—usually with 'of', as in 'a montage of emotions/sounds/images. The similarities between a montage and traditional Chinese poetic idioms can be observed in the following example from Ma Zhiyuan's Tianjin Sha: Qiu Si.Yuanqu writer Ma Zhiyuan composed the scattered song 'Xiaoling'. 'Qiu Si' ('Autumn Thoughts') is a famous poem that uses a series of concrete concepts to describe natural phenomena and a picturesque scene to elicit a sentimental mood. Ma's poem employs a series of images with consideration for the rhymes and rhythms of Chinese Han characters. The direct translation of the original poem is as follows:

*yuzhe2020@foxmail.com

Dead vines, old trees, dusk crows,
Small bridges, flowing water, people's homes,
Ancient roads, west wind, thin horses,
Sunsetting, broken bowel people at the end of the world

Ma uses the unique, crisp rhythms and ending vowel rhymes of Chinese monosyllabic expressions to compare an old man to dead vines, old trees, and dusk crows and to discuss this person walking through his normal, young life across small bridges, flowing waters, and people's homes and ending his life at the end of the world. This example demonstrates the adaptability of ideograms written in Chinese Han characters to the montage film style.

Xu Sheng's *Shuowen Jiezi* (an ancient Chinese text from the Han Dynasty) employs the principle of sixes (*Liushu*; that is, the six graphs or six methods of forming Chinese characters: pictograms, phonetic loans, ideograms, ideograms plus phonetic aspects, combined ideograms, and transfer characters) to label character entries and provide definitions and examples of deconstructed Chinese characters. We employed ideograms and combined ideograms to elicit student creativity.

Qin Zhanyi's *Ancient Chinese Character Flow and Image Montage*, published in *Film Review* [3], illustrates the relationship between film images and ideograms from the perspective of words, poetry, and the connection between Chinese character flow and montage. In the article, Qin offers many poem examples, including 'riding a horse on a bridge at an angle', in which the word for 'at an angle' is subtle and a visual signifier of a change in perspective. Filming of this line would have irregular composition most likely involving a small, full mirror upward shot. Another example is 'the window contains a thousand autumn snows in the West Ridge, and the door is docked on the East Wu Wanli Ship'. Here, the window, Xiling Snow Mountain, and the snow are figurative instructions; these content words suggest the need for a dynamic push shot that is pushed out of a window. These examples demonstrate the influence of ideograms in Chinese poetry and their connections with imagery.

'Thirty-Six Words' [4] is a colour cartoon released in China in 1984. The cartoon was produced by Shanghai Fine Arts Film Studio, written by Ada, and directed by Xu Jingda. The cartoon tells the story of a father who teaches his son to read and explains the origin of Chinese characters and hieroglyphs. Two other animated films, 'Characters from ['word' with 'since' in Chinese] Meeting You' and 'Five Thousand Years of Chinese Characters', present the Chinese character for 'pen'. The first film discusses the evolution of the form of the 'pen' character and the production of the brush and the history of the brush. The film provides direct visualisation of the Chinese character for 'pen'. The film uses Adobe After Effects (Adobe Systems, USA) software to present an animation of the strokes of the Chinese character for 'pen'. The editing includes light music and commentary. The film explains the historical evolution of the word for 'pen' in a few short minutes. This discussion demonstrates the potential for Chinese characters to be converted into animated images.

4 Production Contents

The participating students were inspired by a traditional text called 'The Book of Songs: Jiangjia Cangcang; Tracing Back: A Road of Hindrance and Longing'. The students created an animated short film called 'Trace Back' and created imagery that told a story related to their study path. The title of the film is presented in calligraphy, which was considered appropriate for the Han characters. The short film tells the story of a girl who practiced calligraphy since childhood because her parents asked her to. She traces her growth through a dream and wishes to break free from what is considered normal and find herself.

The story is loosely divided into three acts. The first comprises the first 58 seconds of the film and presents the backstory. This act is the logical starting point of the story; the protagonist has begun to wither in the shadow of her parents' authority.

The second act continues from 58 seconds to 3 minutes and 3 seconds. The second act presents the protagonist's escape and reasons for escaping her parents. Expectations from parents, relatives, and friends and the pressure of various honors lead the protagonist to issue a cry for help. She starts running, which builds the action to the climax of the story. At 2 minutes and 5 seconds, the pace of the film accelerates.

The third act continues from 3 minutes and 3 seconds to the end of the film and presents the theme. The protagonist retrospect on her self-growth and her parents' influence on the process of her growth. She asks herself, 'What about me? What do I want to be?' The most important thing to her is identifying her interests.

In the protagonist's dream, the depth of field is blurred, the outlines of the characters are feathered, and single-light lighting is used. In addition, a Hitchcock-style zoom is used to visualize the frightened state of the protagonist (Figure 1).

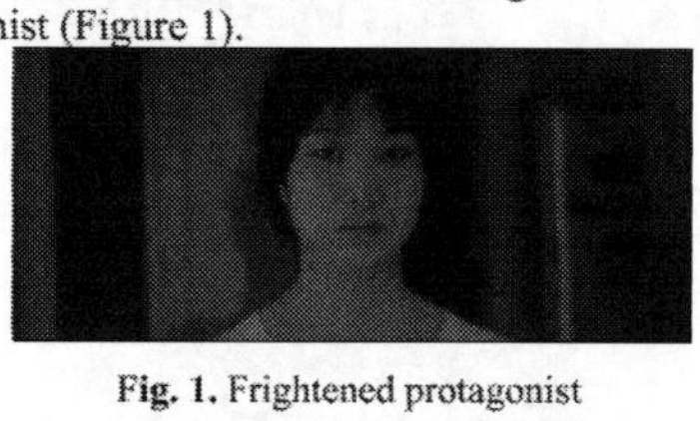

Fig. 1. Frightened protagonist

When relatives and friends discuss the protagonist's lack of clear self-definition, the subjective perspective of the characters is employed (Figure 2).

Fig. 2. The blurry shot

The completed film was uploaded to https://www.bilibili.com/video/BV1hV4y1J7tw/ for viewing.

5 Methodology

In this study, an animated short film was created that conveyed the sentiment of Chinese characters. Analyses and discussions of Chinese characters and images were used to assist students in brainstorming and developing a plot for 'Trace Back'. The students created and cocreated a short film through dynamic actions.

During this collaboration, the teacher played the role of the producer, supervising and assisting in solving technical problems and offering possible solutions.

After the production of the film was completed, interviews were conducted to verify whether the study achieved its research objectives. Table 1 lists the interviewees' background information. The interview questions are listed in Table 2.

Table 1. Coding and background information of the interviewees

Coding	Expertise	Working experience
C1	Documentary film and theory	University teachers, 3 yrs.
C2	Documentary film creation	Director, 5 yrs.
C3	Film production manager	Producer, 3 yrs.
C4	Documentary film creation	Editor, 3 yrs.
C5	Digital Mass media, Documentary film creation	University professor, 23 yrs.

Table 2. Interview questions

Item	Question	Research Objective
1	Do you believe that Han characters can be interpreted through dynamic actions in a short film?	(1)
2	Is this short film closely related to the themes of the poem 'Trace Back'?	(2)
3	Is the imagery of the poem 'Trace Back' expressed appropriately in this film?	(3)
4	What are your suggestions for improving this short film?	(1)(2)(3)(4)

The interview was completed within 45 minutes through telephone, email, and chatrooms. The analyses and discussion of this study are presented in the following sections.

6 Interview Results

After we sorted and summarised the interviews with the five experts, we obtained the following conclusions:

(1) Visualising Han characters in a short film is feasible because of the similar form of creativity of Han characters and images—both are symbolic and can be applied in montage thinking. The five experts agreed that Han characters can be visualised. Two factors reinforced this agreement: the relationship between the characteristics and meaning of Han characters and the symbolic meaning of images. (a) Expert 1 cited an example of a montage of Eisenstein in a movie; he recognised that Chinese characters are theoretically and practically similar to film images. However, the experts suggested that the imagery of Chinese characters should not be limited to surface meaning. (b) The experts reported that the themes and contents of film creation can be diverse and culturally rich. Expert 2 used Chinese character documentaries (e.g., 'Book Briefing China' [5] and 'Words from Meeting You' [6]) to illustrate how Chinese characters can be presented. 'Regardless of whether three-dimensional animation or fonts or paintings are used, film is a vivid means of expressing the appeal of Chinese characters. The aforementioned points illustrate the similarity between Chinese characters and film images.

(2) The structure and style of the short film 'Trace Back' can be considered retrospective, and the film's audio–visual language is mature and effective. The experts understood the film's intended tone from its structure and style and were able to visually identify the meaning and imagery of the poem 'Trace Back'. Expert 1 noted that the short film incorporated dreams and memories into its structure; Expert 2 noted that flashbacks were used to represent the pressure the protagonist felt to practice calligraphy, to identify the cause of the problem, and to investigate the protagonist's past.

(3) Two main points were identified in the experts' answers to question three: The first was that the retrospective imagery was represented in the narrative structure of the film and the audio–visual design contributed to this. The other was that the title and content of 'Trace Back' were somewhat off-target. Expert 5 reported that the short film did not represent specific Chinese characters; it mainly discussed the protagonist's calligraphy and her feelings toward the practice. Nevertheless, the audio–visual design used to express her feelings was effective. Expert 5 believed that 'Trace Back' was not associated with the poem in title or theme; it was associated in style. That is, the imagery of the poem 'Trace Back' was expressed in the audio–visual composition and design. However, the film 'Trace Back' had several problems with its visualisation. The experts presented many suggestions for improving the short film. For example, (a) the film should support the themes of the poem and, therefore, the team should deepen their understandings of the poem 'Trace Back' and focus less on emotions; (b) the design of the ending or coda should be adjusted; and (c) the script and music of the story should be optimised. Expert 5 also suggested that the students adjust the direction of the story and adopt a multilevel and diverse perspective to attract audiences of different ages and classes and thereby improve the spread of Chinese culture through Han characters.

(4) The analyses and suggestions of the experts provided the following potential directions for revising the animated short film: The story should include references to the history of Han characters. This would also reference the protagonist's experience of learning calligraphy, and the theme would become more prominent in the two-line narrative. The number of camera shots of calligraphy should be increased. The experts suggested adding a poem at the end in, for example, song form, as has been done with Charles Bukowski's Bluebird. This would solve the problem of the sluggish end of the story and reference the actions of the protagonist after she retrospects on her experience of learning calligraphy.

7 Conclusion

To motivate a group of students to work as a team and produce a short film, the teacher of this study employed a situated learning environment and introduced the students to a classic text and the meaning of its Han characters through vivid imagery, research, and documentaries. The students successfully produced a film on the subject with a narrative associated with its title. Interviews with experts revealed that (1) the audio–visual connection between Chinese characters and images renders the characters suitable for transformation into imagery and film; (2) the ideography and imagery of Han characters are translatable into video, including film and television; (3) through dynamic actions short films can enable people to understand the meaning of Han characters at a sensory level and can improve the dissemination of Chinese characters; and (4) the images produced in the animated short film of this study were developed with inspiration from the ideography and imagery of Han characters and can play a role in the promotion, dissemination, and comprehension of the beauty of Han characters. Expert 5 suggested that this type of filmmaking can be expanded and enhanced to ensure that all can understand and appreciate such films.

Acknowledgements

This research was sponsored by the Quanzhou Vocational and Technical University.

This film was awarded by Fujian Province Animation and Game Profession Association as an excellent animation short film (2022/06/08).

References

1. Xu, S. and Xu, C.: Shuowen Jiezi (CE 100)
2. Montage: http://www.britannica.com/(20220716)
3. Qin, Zhanyi: Ancient Chinese character flow and image Montage. Film Review, 9, pp. 104-106, (2011)
4. Xu, Jinda (director): Thirty-six words. Shanghai Fine Arts Film Studio (1984).
5. Jin Tiemu and Dai Xiuyang (directors): Book briefing China. Central Radio and Television Station Film and Television Documentary Center (2021).
6. Liu, Huangyan (director): Words from meeting you. Central Radio and Television Station Film and Television Documentary Center. (2022).

Research and analysis of audiovisual art based on the background of new media

Guanwen Diao [1,*], Huiwen Zhao[1]

[1]College of Imaging, Dongguo University

Abstract. In terms of creative concept and audio-visual art, there is still much room for improvement when comparing Chinese TV commercials with American TV commercials, which have a high international standard. How to make up for the shortcomings and form its own characteristics and style has become an urgent problem for Chinese TV commercials to solve. This thesis takes Chinese and American TV commercials as the research object, and adopts a comparative research method to explore the characteristic path of Chinese TV commercials development by sorting out the history, creative concept and audiovisual art of Chinese and American TV commercials, and analyzing the classic cases of Chinese and American TV commercials.

1 Introduction

Since the middle of the 20th century, television advertising has rapidly become an important cultural representation with the unique advantages of both audio and visual, and has begun to influence the public's consumer thinking and purchasing behavior in a subtle way, reconstructing the original system of commercial culture to a considerable extent [1]. TV commercials are about creative ideas and audiovisual art, and in terms of their performance form alone, they should be classified as film and television art. When it comes to television advertising, it is inevitable to start with the television medium. Since 1941, the United States has been embarking on experimental trials of television, which have been a real success, and the commercialization of television has followed [2]. In the 1950s, the world's first color television set was successfully developed in the United States, and viewers received color signals on television in 1954. When it first appeared, television attracted countless viewers with its realistic visual and moving auditory sensations, and at this time, color replaced black and white, and the images became more and more real, vivid and vibrant, together with the original sound, thus greatly enriching the audiovisual experience of the viewer. In addition, color television incorporates a variety of audiovisual presentation elements and is quickly becoming a popular entertainment medium [3].

2 THE CONCEPT AND CHARACTERISTICS OF NEW MEDIA

2.1 The concept of new media

New Media, as a direct translation of the English term "New Media", is a standard foreign cultural language, which was first developed in 1967 when P Goldmark, director of the CBS Institute of Technology, published a project on the product of EVR (Electronic Video Recording). Recording) project in 1967 [4]. Since then, "new media" has become popular in American society and started to spread around the world, and "new media" has gradually become a global hot spot. However, regarding the definition of the term new media, in different periods of the development of new media, various academics at home and abroad have elaborated different descriptions of it, and the following are several more representative views. According to Du Wei, a scholar in China, new media is a new type of media based on digital technology and information technology, which is constantly integrating, evolving and developing [5].

Compared with traditional media, new media is an emerging media based on the upgrading of traditional media such as newspaper, radio and television products, which mainly provides users with various forms of information and entertainment activities through various terminals, using the Internet, satellite, computer and cell phone [6].

Since the term new media has not yet been precisely defined in academic circles, the author believes that a more comprehensive definition of new media should be: new media is a media medium that relies on the Internet, uses computers, cell phones, cable TV, etc. as a

* Corresponding author: 417380386@qq.com

communication medium, and provides resource information and entertainment services to the public with the help of various applications [7].

2.2 Characteristics of New Media

New media, as an information medium with great differences from traditional media, has a variety of characteristics different from traditional media in its information dissemination process, mainly in one of the following aspects:

(1) New media has high interactivity

High interaction is one of the biggest advantages of new media, and plays a pivotal role in the development of new media. Users are able to give feedback to the information based on their own understanding of it, and then export it to the outside world, becoming a disseminator of information rather than just a passive recipient [8].

(2) New media is time-sensitive

With the rapid development of new media, the information dissemination media, mainly cell phones and the Internet, can obtain real-time information at the speed of a thousand miles, and the information obtained is updated quickly and instantly, breaking the time limitations of traditional media [9].

(3) New media is convenient

As indispensable new media electronic products in people's current life, cell phones, tablets and computers provide powerful help for people to obtain information. By using these products, people can break the space and time barriers in accessing information, and they can get answers anytime and anywhere by simply moving their fingers and typing in relevant questions on the search engine, which has improved the efficiency and convenience of accessing information to a certain extent [10].

(4) New media has openness

The globalization of information, as an inevitable trend in the development of new media, has gradually connected the world through the Internet. In the new media, people from different countries can express their opinions and communicate in the same platform, which promotes the collision of ideas and cultures among various countries and nationalities and helps to break the barrier of information exchange between countries and truly realize the globalization and borderless information integration [11].

(5) New media is virtual in nature

As a product of digital information technology, the virtual nature of new media has inevitably become its most essential feature. On the one hand, virtuality allows people to express their opinions, record their feelings and thoughts, and share their lives without worrying about being disturbed by others and avoiding social embarrassment; however, on the other hand, virtuality is often used by people with ulterior motives to spread false information and commit cybercrime on the Internet, damaging the environment of new media usage.

(6) New media has a popular nature

New media has a wide audience. By the end of June 2021, the number of Chinese Internet users had reached 1.011 billion, and the Internet penetration rate reached 70.40%. The lower cost of Internet access and the richness of information media resources have made more and more people start and get used to using new media for socializing, shopping and learning. Anyone can register in the new media platform and create their own account, share and spread their views and feelings, watch new media resources and browse real-time news and information, as long as the network conditions allow.

3 VISUAL ART PERSPECTIVE

Audiovisual art originated earlier as a form of artistic existence in film. After film, in order to enhance the expression and impact of television, the audio-visual art gradually formed theoretical accumulation during its development, and was also continuously explored and summarized until a more systematic methodology was formed to guide the practice of television commercial creation. With the increasing emergence and popularity of new media, audiovisual art has become more widely used in industries other than film and television, such as digital art, interactive networks, and multimedia for film and television. Today, the audio-visual art of TV commercials has become more and more mature, and fundamentally influences and restricts the message and persuasive effect of the advertised product or service.

For TV commercials, in order to pay attention to the application and embodiment of audiovisual art, we must be aware of the most basic components of audiovisual art, that is, visual and auditory elements, specifically: visual elements include scenery and angle, camera and movement, light and color, scene scheduling and temporal relationship conveyed by the picture; auditory elements include voice, music and sound.

Then there are the unique grammatical rules that audiovisual art strictly follows. The grammar of audiovisual art is what makes TV commercials unique and attractive, including the law of shot grouping, the law of sound and picture relationship and the law of audiovisual editing. In other words, the grammar of audiovisual art is the key to combining many audiovisual elements to construct the inner meaning of Chinese and American TV commercials.

Depending on the visual elements chosen and the formal expressions that go with them, television commercials require the creator to coordinate the relationships between the elements and to create a narrative and convey a specific meaning through the final grouping of shots. For consumers, TV commercials rely more on visual art to create a buying atmosphere, to highlight the main artistic image of the product, and to appeal to rational interpretation or emotional resonance. At the level of visual art, the basic elements in TV commercials, such as lines and shapes, light and color, composition and space, and camera and scheduling, all convey information and express meaning in a certain continuous dynamic mode. In other words, the visual art of TV commercials has its own symbolic system or language specification.

3.1 Comparison of spatio-temporal relations

When it comes to the relationship between time and space in Chinese and American TV commercials, it can be said that the two are interrelated and indistinguishable from each other. Specifically, the treatment of spatio-temporal relationships in Chinese and American TV commercials both conform to the grammatical rules of audiovisual art, but there are slight differences in their expression.

3.2 Comparison of lens movement

As a specific way of narrating the content of Chinese and American TV commercials, camera movement has successfully transcended the limitations of simple narrative in terms of thematic ideas and appeal styles, and has become a proven way of expressing emotions and a highly expressive visual language.

In Chinese and American TV commercials, especially those for young consumers and brands or products that explain the true meaning of sports, the use of long shots is also more common, with more changes in angles and scenes to express the richly layered spatial environment, bringing a rich sense of visual form and helping to form a strong lyrical atmosphere. At the same time, the long shot can realize the temporal and spatial continuity and unity of the footage, presenting a complete realistic situation, which can give consumers a sense of rustic reality and natural intimacy when watching the advertising works.

3.3 Contrast of light and color

Light and color are both essential elements for the existence and expression of audiovisual art forms, and both are important components of Chinese and American television advertising graphics.

Overall, the meaning and function of light in Chinese and American TV commercials is mainly reflected in: first, determining the visual style of images, adopting specific light and shadow forms to enhance the visibility and perceptibility of modeling; second, grasping the tone of visual modeling to form the visual style and overall relationship of the picture and other effects; third, depicting and rendering narrative scenes according to the narrative requirements of thematic content, forming a three-dimensional space with great light and shadow effects; and fourth, strengthening and highlighting the characteristics and changes of characters in terms of character and emotion according to the narrative structure of plot development.

Fig. 1. CCTV public service television advertisement "peace and harmony chapter

Take the CCTV public service television commercial "Peace and Harmony" (see Figure 1) as an example. The girl who returns alone at night, riding her bicycle, sings a song to brace herself in the dark hutongs. The good thing is that the old man who sells snacks on the road, a greeting or two, let the girl inner peace a lot. The girl cycled through the intersection, the old man will light up and cast the light in the direction of the girl's travel. It seems to be just a tiny action, but the light that illuminates the girl's way home is deeply touching, and it conveys exactly the care and warmth between people.

Secondly, color is the most external visual expression language of audiovisual art. The color in TV commercials, on the surface, is an objective reproduction and intentional imitation of real life to enhance the sense of reality, but its essence reflects more emphasis on the inner and intention of the characters in the commercials, that is, the color should express the emotion of the characters and the color should portray the spirit of the characters.

Based on the combination of visual and symbolic, emotional and psychological aspects, color has a more unique and deeper meaning of expression. Color in TV commercials can help realize the conversion process from abstract concepts to concrete images and enhance the visual art experience of viewers in terms of picture perception and subjective satisfaction, etc. It can be said that the lasting visual effect of color in TV commercials is crucial because the psychological and spiritual levels of viewers are directly affected by it.

3.4 Comparison of scene scheduling

The scene scheduling mentioned in the audiovisual art of TV commercials, which is used to guide and deal with the actors' performance activities and mutual position relationship, has been borrowed from the early theatrical stage terminology to the creation of TV commercials, and is more reflected in the scheduling of the actors' performance and action lines and the camera positions and changes by the commercial director.

Scene scheduling is a unified global thinking for TV commercials, a conscious pre-arrangement. The reason for this is that:

For one, scene scheduling can elaborate the temporal and spatial relationship of TV commercials, render the narrative theme, and strengthen the artistic infection by controlling the rhythm of TV commercials;

Secondly, scene scheduling can help the advertising director to complete the design and change of the relationship of multiple elements in a single scene, such as the position of the actors and the movement of the camera and their interrelationship;

Thirdly, scene scheduling attracts attention with the help of camera movements to maintain the viewers' continuous interest in the changing images of TV commercials and to enhance their sense of reality as if they were there;

Fourthly, scene scheduling can help the director build his own artistic style to express subjective attitudes related to the theme of the advertisement, thus generating positive narrative effects and audiovisual values.

4 AURAL ART PERSPECTIVE

4.1 Comparison of the ideographic characteristics of sound

The auditory element actively cooperates with the visual element in interpreting the thematic narrative, and exists specifically in the form of sound in TV commercial works. Sound includes three specific forms: speech, music and sound, each with its own attributes and unique functions.

The reason why Chinese and American TV commercials are vivid and concrete and perceptible basically stems from the ideational characteristics of sound, i.e., figuration and imagery, space and reconstruction, deformation and exaggeration, etc. Sound has many different ideational characteristics, which are reflected in Chinese and American TV commercials: firstly, it can create atmosphere and construct reality; secondly, it can link the picture and subtly change the scene; thirdly, it can participate in the narrative and explain the theme; fourthly, it can influence the mood and enhance the experience.

4.2 Comparison of sound and picture relationship

Sound and picture are important components of TV commercials, and the coordination and cooperation of the relationship between the two will directly affect the viewers' reception and understanding of the advertising content. This is because viewers need to process the information of a specific product or service acquired in an integrated manner under the joint action of visual and auditory senses, and produce an all-round perceptual experience. Therefore, the handling of the sound-painting relationship is particularly crucial. For the analysis of Chinese and American TV commercials, we can start from the contrast of two basic forms of paying attention to and dealing with the sound-painting relationship, namely:

4.3 Comparison of the audio-visual art of Chinese and American TV commercials

(1) About the handling of time

The duration of Chinese and American TV commercials is more strictly defined, basically in units of 5 seconds, increasing upward in multiples of 5, as in the case of Chinese TV commercials, which tend to be mostly 5-second, 15-second and 30-second versions.

(2) About the content and screen

Chinese TV commercials are mostly based on life, but they are different from real daily life. In the specific advertisements of different brands or products, there are basically well-trained and beautiful models, props with exquisite texture and craftsmanship, and surreal scenes, which give consumers only a sense of sameness and tedium. In addition, more problems focus on the poor combination of actors' voices and performances with background music, and the incompatibility between celebrities' images and product characteristics.

(3) About Montage

In most cases, Chinese TV commercials design multiple scenes, multiple clues or multiple characters in a single work, and try to show the perfect image of the product or multiple selling points, which causes the whole TV commercial to be straightforward and bland in the articulation of the footage due to the improper handling of montage.

5 CONCLUSION

This dissertation compares the audiovisual art of Chinese and American TV commercials by sorting out the different historical lines of the development of Chinese and American commercials, looking into the changes and evolution of the creative concepts of both commercials in different development periods, and then cutting into the audiovisual art of Chinese and American TV commercials, combining the representative cases of both TV commercials, and closely analyzing the main components of audiovisual art, including the comparison of temporal and spatial relations, camera movement, light and It also includes the contrast of the ideological characteristics of sound and the contrast of sound-painting relationship from the perspective of auditory art, as well as the contrast of the application of montage and the contrast of editing rules and techniques from the perspective of audiovisual editing art. After that, the discussion starts with the influence of creative concept on the content of Chinese and American TV commercials and the influence of audiovisual art on the form of Chinese and American TV commercials respectively, and draws a short conclusion on the dialectical relationship between creative concept and audiovisual art in Chinese and American TV commercials. Finally, the conclusion is drawn in accordance with the starting point of the discussion, and the affirmation of the development path of Chinese TV commercials' characteristics and related suggestions are drawn.

References

1. He Jiajun. The audio-visual art of The Children of Jianghu [J]. Delta, 2023(05):145-147.
2. Lu Di, Sun Yanfeng. Ten years of development and peripheral communication of Chinese audiovisual art[J]. Media,2023(05):89-93.
3. The audiovisual art presentation of Chinese modernization[J]. China Television,2023(02):4.
4. Leng Fei, Chen Hanying. From "silent to silent" to "silent to silent"--The empowerment of Chinese audiovisual art aesthetics for foreign communication[J]. Media, 2023(02):62-64.
5. Wu, Chia-Yi. Radio and television studies: players in the audiovisual arts [J]. Seeking Learning,2022(23):30-31.
6. Leng Wenqi. A study of audiovisual art and design in the movie "Flowery Years"[J]. Art and Design

(Theory),2022,2(07):117-119.DOI:10.16824/j.cnki.issn10082832.2022.07.007.

7. Yang Wen. Reviewing the path taken in the past ten years, leading the peak era to the future [N]. China Press, Publication, Radio and Television Journal, 2022-07-13(007).DOI:10.28907/n.cnki.nxwcb.2022.002354.
8. Gu, Yachi, Yuan, Wenjie. Self-pleasing consumption of "Generation Z" and the practice of "breaking the circle" of audiovisual art[J]. China TV, 2022(07):20-25.
9. Liu Yi, Xia Liang. A comparative study of Chinese and American animation films in the field of audiovisual art[J]. Media Forum,2022,5(12):63-66.
10. Liang Li. Research on the audiovisual art of H5 news [D]. Nanchang University, 2021. DOI:10.27232/d.cnki.gnchu.2021.002775.
11. Jia Yaorong. Research on the communication effect of audiovisual art in grassroots cultural communication[J]. TV Guide,2018(03):185+187.

Research and analysis of interaction in virtual reality films in the context of new media

Long Zhang [1], Zecheng Li [1], and Jahng Sumg Gahb [1,*]

[1] Chung-Ang University

Abstract. With the continuous progress of computer network technology and the burst of new digital media art, virtual reality technology, a revolution in transmitting information, has come into being. As a brand new combination of computer technology and film art, virtual reality films have an interactive experience that no other art genre can achieve. During the movie watching process, the audience can walk into the movie scene, panoramic arbitrary observation of any angle around, and active interaction with people or objects in the scene, creating a personal memory palace, creating a unique storyline, creating a virtual dream space, and re-creating personal life and personal experience in the process of virtual experience. This paper firstly, through understanding the origin and background of virtual reality, analyzing and comparing the development status of virtual reality movies at home and abroad and looking forward to its prospect, so as to arrive at the possible value of virtual reality movies realized under the future digital technology media; secondly, combining the characteristics of virtual reality interactive technology, through analyzing the development process of traditional movie experience, comparing and arriving at the difference between traditional movies and virtual reality movies in Secondly, by analyzing the development process of traditional movie experience with the characteristics of virtual reality interactive technology, we compare and contrast the experience difference between traditional movies and virtual reality movies in four aspects: time and space construction, screen boundary, subjective experience, and interaction mode; secondly, we analyze the experience process from shallow to deep in virtual reality movies, and explore the psychological demand value rooted in the spirit after the intermingling of technology and movie content; finally, we elaborate on the experience value feedback brought to the audience after the practical application of virtual reality technology, and at the same time, we think deeply about this new digital technology medium. Lastly, it is the feedback of the experience value brought by the practical application of virtual reality technology to the audience and the reflection and critique of this new digital technology medium.

1 Introduction

Virtual Reality is "Virtual Reality", VR for short, the Chinese translation of the spirit of the realm, phantom real, is a high-tech emerging in recent years, also known as the spirit of the technology [1]. The "virtual" in virtual reality is a fantasy space generated by computer data, while "reality" refers to any object or environment that exists in the world in a physical or functional sense, which is theoretically realizable, but may also be difficult or impossible to achieve [2]. Therefore, virtual reality is the use of computer simulation to generate a three-dimensional space of the virtual world, the user can use virtual reality equipment to project themselves into this virtual environment, and at the same time provide such as the simulation of visual, auditory, tactile and other senses, and then real-time, unrestricted observation of people, objects, things in the environment and operate, control and other intuitive and natural means of perceptual interaction, to maximize the convenience of the user's operation [3].

2 THE SENSORY EXPERIENCE OF VIRTUAL REALITY MOVIES

Thanks to the creation of interactive systems for virtual reality technology, viewers can have an immersive experience in virtual reality movies [4]. As an individual, the viewer's experience in a natural state rises to a deeper experience in a state of immersion, reaching an immersive state of mind where he or she is convinced of what is in front of them and the environment they are in, and naturally makes interaction and behavioral feedback with the images. The virtual reality space makes the viewer's visual horizon disappear, so that the viewer overlaps with the transformation of the environment they are in and the point of view of the movie character, thus becoming one with the virtual image and being in the image 么, causing the boundary between the virtual world and the real world to become blurred or even disappear. The main significance of the interactive experience given to the audience by virtual reality films

* Corresponding author: makefilm@cau.ac.kr

is that it changes the identity characteristics of the audience in watching the film. In addition to experiencing a real and unreal world in the images, it also establishes a personal view of life and the world, which is the main reason why every director stays true to the aesthetics and artistry of the film. Interactive experiences change the relationship between people and images, images and media, and media and technology [5]. In today's digital network media era, the emergence of virtual reality films is bound to change the public's attitude toward traditional films, but the artistry of film always revolves around the technicality, and what the images ultimately convey is the spiritual dimension that overrides the surface.

People often talk about going to the movies, but actually entering a movie is the highest level of movie experience, and it is also the goal that movie creators have been pursuing, and virtual reality movies achieve this goal, allowing the audience to truly enter a movie for interactive experience [6]. The famous French filmmaker André Bazin's ontology of the image elaborates that all art is based on human participation, and the audience can only realize the value brought by the film experience in the process of participation. The psychological basis for the invention of film is the reproduction of a complete world, a fantasy world of both sound, light and shadow, color, character and environment. Since the birth of cinematic art, every technical innovation has revolved around how to capture the sensory experience of the audience. It has always been the goal of the creators to improve the audiovisual effect and stimulate the sensory stimulation of the audience. In the period of silent movies, directors used montage shots to mobilize the visual experience; in the period of sound movies, sound added aural stimulation to the images and facilitated the expression of actors' emotions: in the period of color movies, the images highly restored the real world and presented a colorful movie world to the audience; in the period of 3D movies, the overlapping and crossed images restored a three-dimensional movie world to the audience, and the surround type The stereo sound effect adds more audio-visual elements to the movie. The sensory experience of virtual reality movies is to use multiple senses to interact with the images themselves without the influence of real time and space, in order to enhance the realism and immersion of the sensory experience [7]. Usually, the human eyes and ears are the most direct way to discern and locate the real or fake space. Virtual reality technology is to take advantage of this, to create an illusion to the audience, the use of interactivity this characteristic, so that the five senses in a state of interaction, when reached a certain time, the audience's senses will be due to the reality of the immediate and the formation of sensory feedback in the interactive process. This interactive process brings the sense of experience, and the feedback process enhances the sense of immersion, and the two are alternately combined, and the multi-sensory experience of the audience is formed.

2.1 Listen to the sound and identify the location

With the emergence of sound films, people realized that sound is what is necessary to constitute the realism of the picture and to convey the atmosphere of the image and the mood of the characters. With the continuous development of audio technology, mono, dual-channel, multi-channel, surround sound, human head music or 3D audio technology have all played an important role in the use of sound effects for films [8]. The human ear is very sensitive to the source of sound signals and can analyze and determine the direction of the source of the sound it hears. When the visual gives an orientation cue, the auditory goes to confirm whether the cue is real or not; it allows the audience to determine the space they are in and make perceptual judgments about the surrounding objects and environment. In normal life, we hear different sounds in both ears, so stereo sound is divided into left and right channels [9]. Now the stereo surround sound in the cinema, is to retain the original signal source direction, the use of cinema surround space to expand the original sound source from the line to the entire three-dimensional plane, producing spatial diffusion and sea sound, giving the audience a realistic sense of presence. The three-dimensional stereo sound effects in virtual reality movies, lock the direction of the source of sound, restore the spatial orientation of the high school and low three audio, that is, the head as the center to include all the angles of the sound, and then based on the special acoustic effect of the human ear, the formation of sound effects law, so as to simulate the sound similar to the real environment.

2.2 Visual Guidance

Virtual reality movies are about immersing the audience in a 360-degree visual experience to see this unreal image space. In virtual spaces it is often easy to get lost, the viewer may not know what point to look at next, and there is a key to influence the visual focus in the experience [10]. At this point, a point of interest is needed to guide the audience and to control the visual direction of the audience. According to Gutenberg's physiological law, people's visual habits tend to move from top to bottom and from left to right, and the first point of view is the upper left corner of the picture, which is the location of the visual center. The upper right corner and the lower left corner are the visual falling blind points that people can easily ignore, but eventually the final line of sight of eye movement will fall on the lower right corner of the picture. According to this law, a perfect visual guidance path can be designed, according to which viewers can effectively choose their personal visual landing point, and can better access image information in order to interact with objects or characters in the environment, helping viewers to effectively choose the main line of the story to interact, preventing viewers from blindly choosing useless scene paths. In today's virtual reality films, the audience is often guided in several ways [11].

3 VIRTUAL REALITY MOVIES CREATE A DIFFERENT STATE OF EXPERIENCE

3.1 Explanation of the state of nature

In virtual reality movie viewing, the natural state is divided into two kinds: one is the natural state between human and senses, which refers to a function of mutual coordination between human body and senses; the other is the natural state between human and image, which means that human is in the image and becomes one with it. To achieve the natural state of the viewing experience, two conditions must be met: first, the audience itself is in the same spatial environment as the image; second, the combination of the body's instinctive response and psychological response to the story plot during the experience. At this time, the psychological response is based on the change of physical behavior response and stimulates deep experience cognition, which not only exists in the brain, but also is closely related to physical behavior, facial expressions and mental activities.

3.2 Enter a state of relaxation in virtual reality movies

To "relax" means to change the attention or control of something from tight to loose, lax. The relaxation state in the virtual reality movie actually means that the viewer's mind is completely detached from the body itself during the movie watching process, and has delivered his or her body to the brain's imagination, the mind is outside the body, allowing himself or herself to make an unconscious virtual trip from the real environment to the virtual movie.

3.3 The state of immersion in virtual reality movies

When the viewer watches a virtual reality movie, the body can act as an interface, forming an inherent pattern through virtual reality devices, virtual reality images and the body. The virtual reality device, as a concretized technical extension, connects to the story content of the virtual image, transports the image information into the device, then connects to the body through the device, shares information with it and receives behavioral feedback made by the body, and finally transmits the feedback information to the image, forming a communication interaction between the body and the image in order to enhance the body's degree of awareness in the virtual environment. Such an immersion mode leads the viewer to discover his or her presence in the virtual image, and this presence acts on the viewer's perceptual-behavioral system. When the viewer tries to break through the illusion, the intense realism of the experience draws you back into the image again and allows the viewer to actively participate in it all. For the real world, this kind of virtual environment with super simulation and similarity will make the audience believe that they are in such a world, a familiar and comfortable environment, a virtual space with personal identity and subjective consciousness as the precursor. What is the real world, it is possible that it is just an illusionary space of personal consciousness after a series of changes over millions of years of generation, evolution, recovery, reproduction and survival, or it is just a shared communication body of all information and emotions. When the audience's pre-consciousness is thinking about everything, it means that it is already in a state of complete immersion in the virtual reality movie.

3.4 The state of lost ego in the real world

Ego is also called self-awareness, which mainly refers to the individual's perception of his or her own existence. The loss of self means that after experiencing virtual reality movies, the audience will not be able to distinguish the "I" in reality from the "I" in virtual reality movies for a period of time, resulting in a loss of self-cognition for a period of time, but it will be restored soon. The phenomenon of sending is very similar to dreaming. When in a dream state, those images will be like those flashing shots in a movie, forming a story in the brain, and the moment you wake up from the dream, the human brain will go back to the scene in the dream, when the memory is still in the dream, but after a period of time, it will be clear to distinguish reality from the dream. For example, when the audience meets a kitten in the virtual reality movie and also becomes friends with it, when they return to real life and see an identical kitten on the roadside, they will unconsciously treat it as a good friend of the virtual reality movie and get close to it. This phenomenon is called the conversion phenomenon, converting and confusing the real world and the virtual world, a short circuit of self-awareness, and a loss of self-cognitive state.

4 THE MULTIPLICITY OF REPRESENTATIONAL INFORMATION AND THE SELECTIVE COINCIDENCE OF IMAGINATIVE THINKING UNDER INTERACTIVE EXPERIENCE

4.1 Factors Affecting the Multiplicity of Representational Information in False-Hot Reality Films

In virtual reality movies, representations are generated, and the influence of different factors will make the representations different, among which the most important factors are the following three aspects: first is the object itself, different virtual objects will hold different characteristics, and the information signal emitted by the object will be different during the interaction of the audience, and the representations received by the audience will be different. The virtual space created in the virtual reality movie makes the audience experience the almost real sensory world again. In this objective space, the difference of each person receiving the apparent information will lead to a different also-ran experience and feedback in the subsequent deep experience, and the value of each person's virtual reality experience will be different at the end of the movie watching.

4.2 The Practicability of Imaginative Thinking in Virtual Reality Films

Imagery" refers to the subconscious thought activity of human, and "image" is the specific performance of the subconscious thought activity, so "imagery" is the active or inadvertent emergence of the picture and its specific content in the human mind. From the level of also science, "imagery" refers to an "image" formed on the basis of the appearance of the inner experience and camp structure. The imagery in virtual reality films is a structure carried by countless images, a synthesis of a collage of representational information combined with the inertia of individual thought. It is linked to a specific perceptual experience and takes over the mode of connection between the information obtained from the sensory experience and the virtual image interaction, which is an advanced information carrier. Thinking is the brain's association of things or environments, a way of thinking. In virtual reality films, imagery stimulates the active consciousness of the mind to interact with imagery, storing the information of sound, light and shadow, things, characters and scenes in the images in the viewer's brain and forming a mental body with the external body, making it an important tool and necessary condition for conducting virtual practice. Viewers tend to interact with virtual reality movies differently according to their own personal experiences and thinking inertia, and the imaginative thinking formed is also a combination of images presented from the viewers' subconscious or reawakened deep memories.

The selective seduction of imaginative thinking under

4.3 The multiple messages of virtual reality films

The so-called "presence" of virtual reality is the wrapping of the sensory system, through the brain's imaginative thinking to confuse the body perception, so that the audience believes that they are in the geographical environment they see in front of them. There are differences in the strength of each person's vision, hearing, taste, and touch. Similarly, there are differences in the human brain's ability to receive and organize information, the power of thinking and the clarity of projecting objective real-world images, and this series of factors leads to different strengths of imagery seduction for each person. Virtual reality movies themselves break the boundary between the virtual and the real, and people, as a bridge between the two, exchange information between the two worlds, an act that itself prompts confusion in thinking consciousness and enhances the possibility of being seduced. On the one hand, in the real space, the body is a symbolic carrier of the existence of life, which is controlled by natural physical laws. And in the virtual space, the body is just an incarnation of behavioral activities controlled by the thinking consciousness, when the brain and sensory system is deceived, the body will also be deceived, it is not controlled by the outside world.

5 CONCLUSION

Film has experienced more than a hundred years of development, with the continuous emergence of silent film, sound film, color film, 3D film, and IAMX film, slowly moving from a primary stage of experimentation to a stage of progress and development to maturity. Nowadays, with the continuous development and progress of science and technology, virtual reality technology has made new breakthroughs in different fields. As an art genre combining technology and art, the marketability and commerciality behind film has become the fundamental reason for the marriage between virtual reality technology and film. The development of virtual reality film needs to go through a long exploration process before it can truly mature, not only to change the traditional viewing mode on the technical level, so that the audience can have a comfortable and immersive interactive experience, but also to break through the limitations of video art on the artistic level, to meet the aesthetic needs of the audience in terms of story content, plot narrative and character setting, and moreover to stimulate the audience to think about their own existence, social phenomena and the real world on the psychological and spiritual level. And today, virtual reality movies create a dream world for the audience. This dream world, not just an illusion in the consciousness, but a near-real image space, where the audience is an independent individual in the space, and any behavioral actions will get feedback, just like living in the movie, the emergence of virtual reality movies finally turns the statement of movies as dream mirrors into reality.

References

1. Xiong Yue. The phenomenal body: perceptual transmission of virtual reality films[J]. Film Literature,2023(04):28-33.
2. Wang Xicheng. The reconstruction of virtual reality narrative on classic film narratology[J]. Journal of North China University of Water Conservancy and Hydropower (Social Science Edition), 2022,38(05):109-114.DOI:10.13790/j.ncwu.sk.2022.090.
3. Li Yingli. Research on spatial narrative and scheduling methods of virtual reality films[J]. Western Radio and Television,2022,43(18):109-112.
4. Song Kepeng. Exploration of spatial interaction in virtual reality films[J]. Beauty and Times(Up),2022(08):116-118.DOI:10.16129/j.cnki.mysds.2022.08.026.
5. Tang Beibei. Research on the impact of VR technology on film in the context of new media [J]. Contemporary Film,2020(10):165-168.

6. Song Jia. Research on virtual reality film [D]. Shandong Arts Institute,2019. doi:10.27281/d.cnki.gsdyc.2019.000048.
7. Che Lin. The 70th anniversary of Cannes International Film Festival:The direction of the festival in the new media context[J]. Contemporary Cinema,2017(07):50-54.
8. Li Fang,Jiang Tao. Application of new media in the context of virtual reality technology[J]. Electronic Technology and Software Engineering,2017(08):74.
9. Hong Shaopeng. A study of virtual reality filmmaking and its narrative language [D]. Harbin Institute of Technology,2016.
10. Liu Qingke. The development of virtual reality technology inspires new ideas of new media animation creation methods[J]. Creativity and Design,2014(05):70-76.
11. Huang Lin,Zeng Zhigang. The aesthetic qualities of new media art[J]. Ethnic Art Research,2014,27(04):105-111.DOI:10.14003/j.cnki.mzysyj.2014.04.038.

https://doi.org/10.1051/shsconf/202316701021

Research and dynamic analysis of interactive methods of film in the context of new media

Jinhao Li [1], Jie Liu[2], and Jahng Surng Gahb[1,*]

[1] Chung-Ang University
[2] The graduate school of Advanced Imaging Science, Chung-Ang University

Abstract. Since the 20th century, art has undergone a radical change in both concept and form under the influence of technology. With the mutual integration of modern technological means and artistic thinking, new media art came into being. Its birth and development is a remarkable novelty, and its rapid development in turn pushes the innovation and transformation of traditional media art, opening a new window for the development of film art. Interactive cinema has grown rapidly based on new media technology, and its all-round interactivity has had an almost revolutionary impact on traditional cinema, thus attracting widespread attention from many researchers in the field of film art. As a new form of image, interactive film has an irresistible charm for audiences. Interactivity, as one of the essential features of interactive film, is also increasingly prominent in terms of expressiveness, and interactivity has become a higher level of interactive film development. The interactive intervention shifts the image narrative from linear to non-linear, and the creative approach from closed to open and diversified, in addition to which the audience deeply experiences the drenching sound of viewing. The interactivity of interactive movies transforms a person sitting passively in a theater quietly enjoying a movie into a person or a group of people actively participating in the creation of the movie, making the movie a real entertainment object for people to manipulate and play with.

1 INTRODUCTION

The 21st century is the era of rapid development of information technology, and in this context, the form of art has changed dramatically accordingly [1]. Interactive film is such a new media art form with interactivity as its basic feature. After the emergence of interactive film, its all-round interactivity has grabbed people's attention tightly. The earliest new media interactive art can be traced back to the 1920s, during which Duchamp's work "Spinning Glass Plate" is the most typical. In the 1960s the concept of new media interactive art became more active, in the 1970s new media artists began to experiment with new media technologies such as video for live performance interaction, and in the 1990s computer interaction technology emerged, allowing humans to enter a new era of art experience. The "human" and "machine" create a unique work for the audience through a dialogue [2]. After entering the 21st century, this interactive new media art form has injected a vibrant element into the development of traditional film art, turning movie watching into movie playing and audiences into real participants or players. The idea that human-computer interaction is an inevitable trend first appeared in the book Digital Survival by Negroponte, who unveiled the prologue of the vast show of digitalization with transcendent courage and insight. It is based on a human-centered approach, and the digital future has another important dimension beyond the technology-only theory that is obsessed with computing speed and display resolution, which is the harmony between humans and computers in interaction [3]. The pioneer of new media art, Roy Ascott, said that the most distinctive qualities of new media art are connectivity and interactivity. There are five stages to understanding the creation of new media art: connection, integration, interaction, transformation, and emergence. This leads to the transformation of the work and consciousness, and finally to the emergence of new images, relationships, thoughts and experiences. When we generally speak of new media art, we are mainly referring to circuit transmission and creations that incorporate computers. However, this media, based on silicon crystals and electronics, is merging with biological systems, and concepts derived from molecular science and genetics. The most innovative new media art will be the combination of "dry" silicon crystal computer science and "wet" biology. This newly emerging new media art has been called "MOIST MEDIA" by Roy Ascott.

There are many forms of new media art, but they all have one thing in common: the user is involved in changing the image, the shape, and even the meaning of the work through direct interaction with it. They trigger the transformation of the work in different ways - touch, spatial movement, vocalization, etc. Whether the interface with the work is a keyboard, a mouse, a light or

* Corresponding author: makefilm@cau.ac.kr

sound sensor, or some other more complex and sophisticated, even invisible "board", the relationship between the viewer and the work is primarily one of interaction. Connectivity is about transcending the barriers of time and space, connecting people around the world. In these online spaces, users can assume various identities at any time, searching distant databases, archiving information, learning about foreign cultures, and generating new communities.

2 The way of interaction

Among these characteristics of interactive films, interactivity is undoubtedly the most attractive, and interactivity is also the basis for the development of other characteristics, which is the characteristic that truly distinguishes interactive films from general art films, so that interactivity is a kind of yearning, an interaction between audience and artist, audience and artwork [4].

In the era of new media art, interactive films, with the help of interactivity, allow audiences to reprocess and recreate their works.

This view of Abrams still holds if the four elements here are changed to the film work, the creator, the visitor, and other visitors in interactive cinema. In interactive movies, interaction can take place between people, between people and devices, and through speech, text, pictures, and video [5]. In general, visitors interact with interactive movies in three main ways. The relationship of these four elements we can represent by a diagram:

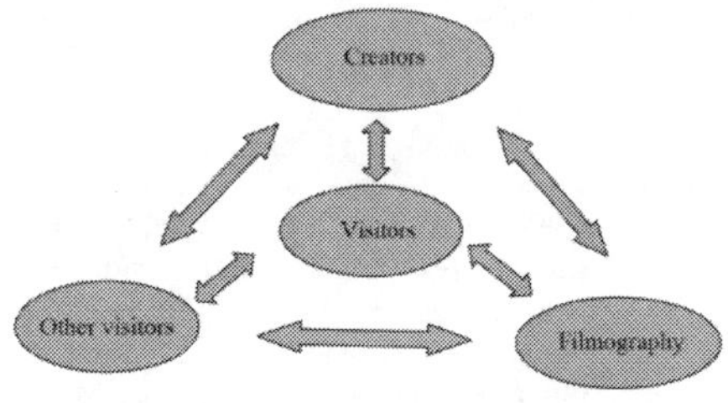

Fig. 1. Interactive movie four elements interaction diagram

From this diagram, we can see that these four elements of interactive film interaction work with each other and their interactions can be complex and varied, but no matter how they change, the visitor is always the core element [6].

2.1 Interaction between visitors and creators

In interactive cinema, we call the viewers who watch the film "visitors," i.e., people who come to visit the film. The people who create the film, such as the writer, director, actors, camera crew, etc., are called "creators," i.e., the people who create the film [7].

2.2 Interaction between visitors and works

One of the greatest features of cinema, and arguably one of the most detrimental to communication with the viewer, is that the playing process and the ending are determined and the viewer can only passively receive information. In interactive cinema, two different ways of interaction exist in the process of interaction between the visitor and the work [8].

(1) Indirect interaction with the work

The participants interact with the work indirectly, mainly through the medium of the film maker during the creation and filming stages of the work [9]. In our country, interactive movies are still in the primary stage. In the creation stage of works, the so-called interaction is mainly for the audience to process and process the relevant information of the movie, and then the demand for the movie is acted on the movie creator through the form of network interaction or live interaction, and then the movie producer acts on the movie afterwards, and this interaction is simple and delayed. China's first interactive movie, PK.COM.CN, was met with the audience in 2008, and the movie was made from this interactive concept. Before the movie was finally produced and presented on the screen, the audience could choose to participate in the movie's subject matter, script selection, director, cinematographer, and actors through an interactive online platform by voting, and the director then finally decided the specific The director then makes the final decision on the shooting plan based on the audience's opinion [10].

(2) Direct interaction with the work

The interaction generated by the audience indirectly with the work is when the audience does not come into direct contact with the work, but uses the creators such as the director as an indirect medium. In contrast, the direct interaction between the audience and the work is when the audience comes into contact with the film work and processes it directly [11].

At the present stage, especially the interactive film works seen in some art exhibition halls, most of them belong to the capture of participants' position, movements, expressions, brain waves and other information through various information collection devices, thus changing the state of the film works. The creator envisions several possible storylines and endings for the development of the film's plot, and shoots and produces film clips with different story directions and endings. When the story is at a crossroads of choices, the viewer only needs to use the remote control or selection button to choose the desired storyline or ending of the movie according to his or her wishes. Alternatively, a higher level of direct interaction involves viewers uploading images in real time as they enjoy the film, so that the uploaded material is integrated with the movie being played. While this approach can produce freer interactions, it is more difficult to achieve under current conditions.

(3) Group interaction between visitors and other visitors

The collective interaction among visitors is mainly manifested in online real-time interaction. When several visitors enjoy a movie together, they are unconsciously divided into subgroups according to their preferences, and the number of subgroups is dynamically changing.

In an interactive movie, several visitors are gathered in one place, and each person can work on the movie according to his or her own preferences, but because there are several people participating at the same time, chaos can occur if there is no effective rule to control it. In order to avoid creative conflicts, we can imagine that there is an auxiliary screen or a small window on the screen or next to it that displays the parameters of the participants' order, number, and deconstruction of the film story. This parameter can be divided into several levels, and each level will have several options, much like a tree menu option, with the first level option being the overall structural part, the second level option being the scene, the third level option being the characters, and the fourth level option being the change of action behavior parameters. Designing in this way allows for greater freedom of choice for the audience on the one hand, and for the visitor to become a collaborative and orderly whole on the other. The similarities and differences in choices between the visitor and other visitors work together to develop the plot and ending of the movie story, which provides entertainment for the visitor while also increasing responsibility.

3 Performance of Interactivity

Interactivity is the most basic characteristic of interactive film in the era of new media art. Interactivity is derived from other characteristics. New media interactive device technology has revolutionized the narrative, creation and viewing experience of interactive films, mainly involving three aspects: the diffusion and interaction of the narrative, the openness of the creation and the innovation of the viewing experience of the participants. In terms of narrative, the traditional passive linear narrative has changed to an interactive non-linear narrative in which the participants are actively involved; in terms of creation, the traditional one-way creation by the author has changed to an interactive creation by the author and the audience; in terms of viewing experience, the audience has changed from individual appreciation to active interaction with other participants, and through collective participation and interaction, they can jointly control the development of the film story.

3.1 Interactive narrative representation

Narrative plays a pivotal role among the constituent elements of cinema. In the glorious history of cinema in the past hundred years, we can find that cinema relies on narrative to present a splendid scene, and at the same time makes cinema transformed and increasingly prosperous. Although interactive films formally rely on new media devices to make the experience immersive, no matter what stage of film development, the basic narrative is inseparable. Interactive film has also become a highly influential and fashionable art and cultural form due to the intervention of interactive narrative.

(1) Choice based narrative

Interactive films appear as different narrative texts depending on the subjective choice of the audience. Traditional narrative events occur in chronological order and focus on continuity. The film that the audience eventually sees on the screen is shot and produced according to the director's intention, and even if the audience has a million different opinions about the storyline and the ending, they can only keep it in their minds. With the support of new media interactive technology, this passive situation can be completely overturned. In the world of interactive movies, different audiences have different interpretations of different movie texts, and viewers make full use of interactive means to choose different characters, different ways of participation, different plot developments and different endings. For example, when the audience chooses the characters for the interactive movie, they can name the characters they choose, set their physical characteristics, and give them their own personal style of speech and behavior. At this point, the "audience" is transformed into a "participant" and "creator", creating their own personal narrative style in the process of selection and creation, thus constructing their own small narrative The film.

In traditional films, the narrative subject of the medium is the director, while in interactive films, the narrative subject of the medium is constructed by the audience, who makes interactive choices in the process of watching the film and reprocesses the film in the choices to construct a new narrative system, which will lead to the instability of the film's narrative structure and the fragmentation and diversification of the narrative meaning.

(2) Diversity of narrative styles

The narrative of traditional films generally develops in a linear mode according to the director's leading ideas, and after the film is produced, it cannot be modified or added to, so the audience can only passively accept the director's arrangement of the storyline and follow the director's thoughts to understand the director's one-way creation. Audiences agree with some of the director's ideas, but also sometimes not very satisfied with the development of the story, but no matter how opposed, the audience must still be good to enjoy the director's planning of the film's content. This kind of narrative is a linear fixed pattern, for the audience, sometimes you can easily guess the end of the story, making people feel too flat. The application of interactive technology in interactive movies can easily subvert this traditional linear narrative, allowing the audience to rely on interactive technology to control the story development and ending of the movie, deconstructing the movie in a non-linear way, making the narrative more open and the ending more unpredictable, with free combinations of narrative styles and presenting diversified possibilities.

3.2 Interactive creative expression

The biggest innovation of interactive film is that it breaks through the traditional fixed one-way creation space of the director and turns into an open space where the experiencer actively participates in two-way creation or even multi-way creation, which liberates the experiencer's mind to a certain extent and opens up his thoughts and vision to feel the dream scenery in the free creation.

The creative expression of interactive film interactivity is mainly the openness of the creative process and the diversity of creative subjects and creative carriers. From the perspective of creative subjects, traditional movies generally rely on professional production teams with professional writers, directors, actors, lighting technicians, cinematographers, choreographers and professional shooting equipment, however, in interactive movies, the creative subjects are diversified, and the creative subjects in interactive movies are professionals and film experiencers, especially with the rapid development of new media device technology, which makes it possible for ordinary movie watchers to become directors.

3.3 Interactive viewing performance

In the era of new media art, many of our audience behaviors have the characteristics described in the "immersion theory". The viewing experience in interactive movies is mainly an immersive viewing experience, which is an uninterrupted behavior caused by human-computer interaction based on the support of interactive technology, in which users truly forget to immerse themselves, enjoy the rich experience brought by it, and achieve a "self-motivating" effect. Csikszentmihalyi believes that immersion has the following eight elements:

(1) Challenging and manageable tasks;

(2) The subject is fully immersed in the task and is distracted;

(3) A sense of subjectivity that can be fully controlled;

(4) Complete freedom of the subject to focus on the task;

(5) The task has a clear objective;

(6) Subjects can get immediate feedback from their actions;

(7) The subject does not feel the slightest change in time;

(8) The sense of identity diminishes and then increases.

Almost all interactive movies have content corresponding to the eight elements of immersion. It is the setup of these movie systems and the arrangement of content that makes it easier for users to generate immersion experiences based on interactive technology, which proves that interactive movies are one of the activities that are particularly likely to cause immersion experiences through human-computer interaction.

4 Conclusion

New media is a form of communication that uses digital technology to provide information and services to users through computer networks, wireless communication networks, satellites and other channels, as well as terminals such as computers, cell phones and digital TV sets. From a spatial perspective, "new media" refers to the media that can be globalized across geographic boundaries, supported by digital compression and wireless network technology, with its large capacity, real-time and interactivity, as opposed to "traditional media". In fact, new media can be regarded as a product of new technology, and the latest technologies such as digitalization, multimedia and network are all necessary conditions for the emergence of new media. After the birth of new media, the form of media communication has undergone a radical change, such as subway reading, office building big screen, etc., are the traditional media communication content transplanted to a new communication space.

As an important part of the new media art form, in order to create a visual breakthrough and better satisfy the participants' all-round three-dimensional physical pleasure, interactive film is always drawing on and absorbing new technologies and new creation methods, and its interactive features are becoming more and more prominent. This paper takes the interactivity of interactive film in the era of new media art as the main object of study. Through the analysis of the connotation and relevant characteristics of interactive film, it is concluded that interactivity is the essential characteristic of interactive film, and the depth and extensiveness of interaction is also the inevitable trend of the future development of interactive film. Then we analyze the specific manifestations of interactivity to gain a deeper understanding of film interaction and lay the theoretical foundation for script creation. The creation of the script is mainly based on the non-linear narrative of the interactive film, the script is discrete, and then the montage principle is applied to these seemingly unrelated fragments, and various arrangements and combinations are made to reflect the interactive features within the interactive film. We hope that the general public can gain an in-depth understanding of interactive film interactivity and take an active part in the process of making their own personal style of film.

References

1. Liu Chen. When film encounters the Internet--film communication in the new media field[J]. Beauty and Times(Up),2022(12):64-67.DOI:10.16129/j.cnki.mysds.2022.12.033.
2. Qi Xin,Wan Shu-Liang. The cultural reflection of new media film[J]. Film Literature,2022(23):82-85.
3. Wang Yan. On the role of new media films in promoting the originality of the content of theatrical films[J]. Tomorrow's Style,2022(22):37-40.
4. Xiao Nuan. Exploring the development path of new media film communication[J]. Tomorrow's Style,2022(06):25-28.
5. Wang Xiaoshang. The construction of mainstream values of new media films under the Internet[J]. Da Guan (Forum),2021(01):62-63
6. Lu Jingren. Analysis of the aesthetic characteristics of new media films[J]. Art Tasting,2019(36):290-291.

7. Zhang Guoyan. Exploring the combination of new media film art and painting art[J]. Global Premiere,2019(08):25-26.
8. Huang Cangsu. Exploring the combination of new media film art and painting art[J]. The House of Drama,2019(20):121.
9. Wei Jun. Analysis of the impact of new media film industry on the development of Chinese film industry[J]. China Media Science and Technology,2019(05):85-87.DOI:10.19483/j.cnki.11-4653/n.2019.05.026.
10. Chen Jie. Comparative analysis of traditional film footage and new media film footage[J]. Media Forum,2019,2(08):37+40.
11. He Xinguang. The construction of mainstream values of new media films in the context of the Internet [J]. Youth Journalist,2018(35):87-88.DOI:10.15997/j.cnki.qnjz.2018.35.056.

Current Situation of TPACK for Art Vocational Teachers: Problem Analysis and Countermeasures

Tao Zhao [1,*], Shijian Li [1], Xiujuan Zhang [1]

[1] Chongqing Vocational college of Culture and Arts, Banan, Chongqing, China

Abstract. The knowledge of Technological Pedagogical Content Knowledge (TPACK) is a new definition of the knowledge structure of higher vocational college teachers in the era of information society. This study investigated the prominent problems in TPACK training of teachers in higher vocational college of arts in our city by sampling. Firstly, in the professional training of provincial or municipal teachers, emphasis was placed on "theory or method propaganda" rather than "guidance of technology integration training"; Secondly, focus on "discipline knowledge", "teaching knowledge" and "technical knowledge" of a single module of training, but lack of "integration and deep integration of the three"; In view of the above problems, it is suggested that educational administrative departments or universities should pay attention to the training of "subject teaching with technology integration" in the professional training of teachers, and pay attention to the deep integration of "Content knowledge", " Pedagogical knowledge" and " Technological knowledge" in the training process. In order to effectively improve the TPACK knowledge structure of the teachers in higher vocational college of arts, promote the high-quality development of vocational education in our city.

1 INTRODUCTION

In the current era of rapid development of information and intelligent technology, The traditional structure of teachers' knowledge and skills has been unable to meet the requirements of today's information-based teaching. Under the premise that teachers have various basic knowledge, such as subject content knowledge (i.e., ontology knowledge), Pedagogical Knowledge (i.e., conditional knowledge), situational knowledge (i.e., practical knowledge) and cultural knowledge, they also need to add teaching technology knowledge to their knowledge structure system. Punya Mishra and Matthew Koehler advocate that teachers be able to use technology to teach effectively. In 2006, they formally proposed the classic theoretical model of TPACK. the ability of teachers in higher vocational colleges of arts to integrate the knowledge of Technological Pedagogical Content Knowledge (TPACK) is directly related to the high-quality development of higher vocational education, the construction of vocational education power, more related to the cultivation of "great craftsmen, skilled craftsmen" and the realization of the goal of manufacturing power. As a new definition of the knowledge structure of college teachers in the information age, subject teaching method knowledge with integrated technology (TPACK) refers to the knowledge complex that integrates information technology into classroom teaching, which is not only the basis of effective teaching of teachers in the information age, but also the embodiment of teachers' information-based teaching ability. It is rooted in three core ingredients: Technological Knowledge (TK), Pedagogical Knowledge (PK), and Content Knowledge (CK),and integrated into other four derived components, that is, PCK, TPK, TCK and TPACK (See Figure 1).TPACK is the most advanced knowledge complex in line with the development needs of The Times[1].

Based on the "14th Five-Year Plan" project of Education science in Chongqing, this study investigated and studied the knowledge of the Technological Pedagogical Content Knowledge (TPACK) of integrated technology among teachers in higher vocational colleges of arts in Chongqing through sampling method, so as to grasp the development situation of TPACK of teachers in higher vocational colleges of arts in Chongqing, and investigate the prominent problems in the improvement process of teachers' TPACK. In order to improve the subject teaching knowledge of integrated technology (TPACK) of vocational teachers, to promote the high-quality development of vocational education to provide reference. The questionnaire adopted by the research takes the questionnaire designed by Schmidt et al for pre-service teachers as reference, and revises and redevelops the questionnaire based on the characteristics of the knowledge structure of teachers in China's higher vocational colleges [2] .The reliability and validity of the questionnaire are tested, and the test results are good. Under the condition of ensuring the good reliability and validity of the questionnaire, 268 teachers were sampled in the city's higher vocational colleges, including 37

* Corresponding author: 517805746@qq.com

professors, 60 associate professors, 125 lecturers and 46 teaching assistants.

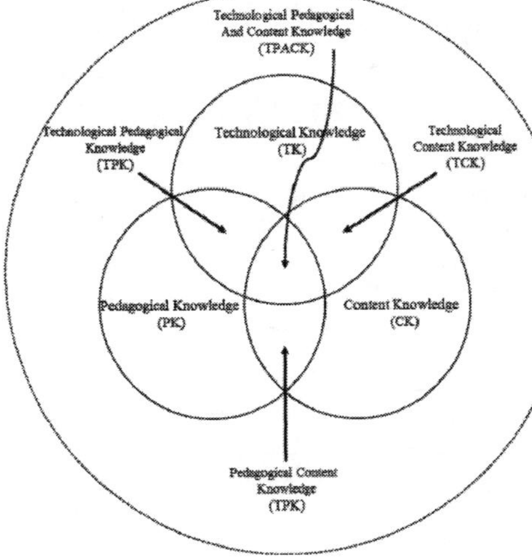

Figure 1: TPACK structure diagram

2 The present situation of TPACK knowledge structure of teachers in higher vocational colleges of arts

2.1 The overall level of technology-related knowledge in the TPACK knowledge structure of teachers in higher vocational colleges of arts is low, and the knowledge of subject teaching method integrating technology has serious shortcomings

The basic information of the sample is that the ratio of male and female teachers is 37.2% and 62.8% respectively; The proportion of students with undergraduate education background of normal university and non-normal university was 32.6% and 67.4%, respectively. 13.80% associate professor and 22.38% professor, 46.64% lecturer and 17.18% teaching assistant respectively. The average values of TK, PK, CK, PCK, TPK, TCK and TPACK of teachers' knowledge dimensions are 3.42, 3.67, 3.92, 3.90, 3.23, 3.36 and 3.16, respectively, in order from height to bottom: CK > PCK > PK > TK > TPK > TPACK.

From the average values, it can be found that the average values of CK, PCK and PK related to subject knowledge and teaching method are relatively high, indicating that teachers in higher vocational colleges of arts have a relatively high level in subject knowledge and teaching method knowledge. However, the TK, TPK and TCK of technology-related knowledge structures are significantly lower, and the average value of PK is the highest and that of TK is the lowest in single dimension. The average value of PCK is the highest and the average value of TPACK is the lowest in multidimensional aspects, which indicates that there are serious shortcomings in technical knowledge and subject teaching knowledge integrating technology of teachers in higher vocational colleges.

2.2 There is no significant correlation between TPACK level of teachers in higher vocational colleges of arts and education training and other variables, that is, teachers receive more single module training and less comprehensive training

There is a significant correlation between the various elements of teachers' knowledge of integrated technology in (TPACK), as shown in Table 1.

Table 1. Table1 Correlation analysis of each element of TPACK.

Factor	Significance	TK	PK	CK	PCK	TPK	TCK	TPACK
TK	Pearson	1						
	Sig.							
PK	Pearson	.256**	1					
	Sig.	.000						
CK	Pearson	.192**	.375**	1				
	Sig.	.003	.000					
PCK	Pearson	.363**	.452**	.573**	1			
	Sig.	.000	.000	.000				
TPK	Pearson	.463**	.287**	.421**	.537**	1		
	Sig.	.000	.000	.000	.000			
TCK	Pearson	.521**	.012**	.215**	.396***	.472**	1	
	Sig.	.000	.000	.001	.000	.000		
TPACK	Pearson	.556**	.381**	.273**	.511**	.642**	.626**	1
	Sig.	.000	.000	.000	.000	.000	.000	

** The correlation coefficients in the above table are significant at the level of 0.01

Table 2. Table1 Correlation analysis of each element of TPACK.

Factor	significance	Training frequency	Normal major or not	Teaching years
TK	Pearson	-.032	-.221**	-.152**
	Sig.	.423	.000	.006
PK	Pearson	-.067	-.231**	.276**
	Sig.	.186	.000	.000
CK	Pearson	.035	.126*	.251**
	Sig.	.543	.024	.000
PCK	Pearson	-.165**	-.172**	.262**
	Sig.	.003	.002	.000
TPK	Pearson	-.003	-.196**	.173**
	Sig.	.946	.001	.002
TCK	Pearson	-.0625	-.318**	-.034
	Sig.	.287	.000	.042
TPACK	Pearson	.076	-356**	.158**
	Sig.	.108	.000	.005

** The correlation coefficients in the above table are significant at the level of 0.01

In terms of education and training, the frequency of teachers' education and training only has a significant correlation with "Pedagogical-Content" knowledge (PCK), but no significant correlation with the other six factors, which indicates that the training teachers receive is almost related to subject knowledge, and has nothing to do with pedagogical knowledge and technical knowledge. It shows that the content of teachers' training is mainly concentrated on the single "Content knowledge" module or " Pedagogical Knowledge (PK)" module, and the lack of the in-depth integration of "Technological knowledge" and "Content knowledge" by the use of "information technology knowledge", that is, there is a phenomenon that teachers receive more single module training and less comprehensive training.

In terms of whether the undergraduate is a normal major, from the average value, the factors except Content knowledge (CK) of teachers majoring in normal education are higher than those of teachers majoring in non-normal education. It can be seen that, in terms of teaching knowledge, teachers with a bachelor's degree in normal majors are significantly higher than those with a bachelor's degree in non-normal majors. It can be concluded that teachers with bachelor's degree in non-normal majors in higher vocational schools need to supplement pedagogical knowledge.

3 Prominent problems in TPACK training of teachers in higher vocational colleges of arts

3.1. In the education and training of teachers in higher vocational colleges of arts, they pay too much attention to theoretical preaching and neglect the guidance of technical integration training

"China Education Informationization2.0 Action Plan" points out that teachers' innovation ability of Information based teaching is still insufficient, and the deep integration of information technology and subject teaching is not enough, among which" the deep integration of information technology and subject teaching is not enough "is the problem that needs to be solved in the 2.0 stage of information based education .In the TPACK structure of higher vocational teachers, the CK and PCK related to subject knowledge are higher, but the TCK, TK and TPK related to technology are lower .At present, China has invested a huge amount of manpower, material resources and financial resources in the in-service training of teachers. Every year, various kinds of training aimed at improving the professional development of teachers, teachers' information-based teaching ability, and teachers' teaching methods are abundant. To some extent, good training results have been achieved, which effectively promotes the professional growth of teachers. However, it is undeniable that most of the training still adopts the traditional lecturing style, in which the training teachers speak on the stage while the trained teachers listen below. The training has not achieved the best effect due to it, and it is difficult to improve the training interest and effect of teachers from the substantive level [3]. Therefore, it is urgent to change the traditional mode of education and training, transforming the traditional teaching-based training focusing on theory or method into technical practice training focusing on information technology integration of subject knowledge and teaching knowledge.

3.2 In education and training, too much emphasis is placed on the single module of "Content knowledge", "Technological knowledge" and "Pedagogical knowledge", and the training that integrates and deeply fuses the three is lacking

Studies have shown that offering pedagogy and psychology courses is one of the effective ways to strengthen pre-service teachers' integration of technology, teaching method and subject knowledge and improve their information-based teaching ability [4].The level of "Technological Content knowledge" (TCK) and " Technological Pedagogical knowledge" (TPK) related to subject knowledge in the TPACK structure of teachers in higher vocational colleges of arts is low. As can be seen from the above table 2, the frequency of teachers receiving professional training is significantly correlated with Pedagogical Content knowledge (PCK), and has nothing to do with the other six factors. Disciplinary training courses, educational teaching courses and information technology training courses are isolated from each other, lack of internal systematism and integration, so the effect of the integration of the three is not obvious. Therefore, teacher education and training can not only rely on training to improve the level of teachers' subject or professional knowledge. Subject professional knowledge level cannot serve teaching well through the presentation and representation of teaching method knowledge and technical knowledge. Therefore, the integration of Pedagogical Knowledge (PK) and Technological Knowledge (TK) is the key factor to improve teachers' knowledge of Technological Pedagogical Content Knowledge (TPACK).

3.3 The teachers with bachelor's degree in non-normal majors are seriously lacking pedagogical knowledge (PK)

Each element of higher vocational teachers' TPACK knowledge structure has a significant correlation with whether they have received normal education, as shown in Table 2 above. Whether a teacher is in normal major during the undergraduate period reflects, to some extent, whether a teacher has received systematic learning of pedagogy, psychology and other knowledge before employment. Studies have shown that offering pedagogy and psychology courses is one of the effective ways to strengthen pre-service teachers' integration of technology, teaching method and subject knowledge and improve their information-based teaching ability [5].

From the average value, in the TPACK knowledge structure of higher vocational school teachers, the teaching method knowledge (PK) factor of teachers who have received normal professional education is significantly higher than those who have not. This will lead to the lack of teaching method knowledge, pedagogy and psychology knowledge of non-normal majors, which will affect their organic integration and deep integration of teaching method knowledge, information technology knowledge and subject content knowledge.

4 Conclusions and Recommendations

4.1 Conclusions

Through the research and analysis of the above relevant data, we can draw the following conclusions:

(1) The overall TPACK level of teachers in higher vocational colleges is in the middle level, the level of Content knowledge and Pedagogical Knowledge is relatively high, and the Technological Knowledge level is seriously low.

(2) There is a significant positive correlation between TPACK elements, among which TPACK has the highest correlation with TCK and TPK

(3) There is a certain lack of Pedagogical knowledge for non-normal teachers. Whether teachers have received normal education or not is significantly related to each element of teachers' TPACK knowledge structure.

4.2 Recommendations

4.2.1 Focus on technology integrated training on "TPACK"

Relevant foreign studies, such as Chai et al., show that the TK, CK, PK and TPACK levels of pre-service teachers are significantly improved through the learning and training of information technology courses [6].It is suggested that the relevant municipal departments should carry out diversified education and training, and the training content should be changed from the traditional "focusing on the theory or method of training propaganda" to the "emphasizing information technology integration of subject knowledge and teaching knowledge of technical practice" training. For example, establish practical training of information technology application and general courses or professional courses, special training of teaching methods or teaching modes applied to a certain course, "information technology + innovative education and teaching concept + subject content", that is, subject teaching method training integrating technology, and hold practical, effective and operable education training at all levels and of all kinds. To provide teachers with information technology services and how to improve the teaching methods in the course of the subject of a package of operable training programs.

4.2.2 Strengthen the deep integration of "Content knowledge" with " Technological knowledge and Pedagogical knowledge" in education and training

"China's 13th Five-Year Plan for Educational Informatization" points out that IT is necessary to establish and improve the standards of teachers' application ability of information technology, incorporate the cultivation of information-based teaching ability into the training curriculum system of normal university students, include it in the index system for the assessment of school level in colleges and primary and secondary schools, and include teachers' application ability of information technology into the compulsory teacher training hours (credits). Ability improvement should be closely combined with discipline teaching and training, and targeted courses and teaching methods characterized by deep integration of information technology should be carried out [7].The purpose of holding education and teaching training is to improve teachers' education and teaching ability, and the purpose of developing teachers' TPACK level is to realize the deep integration of information technology and subject teaching. It is suggested to strengthen the deep integration of subject content knowledge with technical and pedagogical knowledge in the education and training at municipal and school levels. It is not only to hold some lectures or training to improve teachers' professional or disciplinary knowledge, but also to carry out practical lectures or training that integrate information technology, educational teaching and other knowledge. On the one hand, this can improve teachers' training interest and enthusiasm; on the other hand, it is conducive to improving teachers' ability to deeply integrate modern information technology, innovative educational ideas and subject content knowledge from the overall and operational level, Mishra& Koehler believed that only by gradually developing and improving teachers' TPACK level could the optimal teaching effect be achieved[8], and thus fundamentally improve teachers' professional ability and information-based teaching ability, and promote the high-quality development of vocational education in our city.

4.2.3 Establish a training mechanism of "attending lectures and learning" for new teachers, especially for non-normal majors with undergraduate degrees, for at least half a year or one year

The new teachers in higher vocational colleges are basically from school to school, that is, from graduation to work, or from enterprises to school. For the new teachers, especially those who have not been trained in systematic normal education, their cognition and knowledge of education and teaching are far from meeting the requirements of excellent teachers in higher vocational education, not to mention integrating technical disciplinary knowledge. Throughout other countries, all attach importance to the training of subject education knowledge of integrating technology for new teachers.

For example, the New Zealand government provides one-year learning opportunities for new teachers in order to enable them to better apply information technology and pedagogical knowledge to their teaching, they provide a one-year learning opportunity for new teachers. During this year, teachers have no teaching tasks, but go to high-quality classes to attend lectures, participate in academic organizations and engage in relevant plans to study in other schools, and learn typical practices and teaching experiences. For another example, South Korea, Finland, the United Kingdom and other countries deliver high-quality training courses and typical teaching case videos to new teachers online, and provide teachers with various teaching methods and information technology services to improve the informatization teaching ability of new teachers.

"Attending lectures" is to learn excellent teachers, experienced teachers of education and teaching methods, learning pedagogical knowledge; Teacher "learning" is to learn "technical knowledge, subject content knowledge". It is suggested to establish a long-term training mechanism for teachers to "attending lectures and learning", to carry out regular workshops to improve teachers' teaching ability, and to organize regular activities such as technical exchange and teaching experience sharing, so as to provide a community of practice for teachers to improve their TPACK level at school-level. It is suggested that education administration departments or relevant functional departments of colleges and universities should establish at least half a year or one year of "attending lectures" system for new teachers, which is a dual demand of feasibility and necessity, so as to provide broad space for future development for new teachers, strengthen the ability of vocational education, and provide strong support for training high-quality skilled talents, craftsmen of great powers and skilled craftsmen.

Acknowledgments

This paper is one of the phased achievements of The 14th Five-Year Plan Project of Chongqing Education Science in 2022 "Research on the Improvement of Vocational Teachers' Informatization Teaching Ability Based on the Nine-factor TPACK Model" (Project No. K22ZS336743) and the Science and Technology Research Project of Chongqing Education Commission in 2020 "Research on the Academic Behaviour Analysis and Early Warning Mechanism of Higher Vocational College Students Based on Educational Big Data" (Project No. KJQN202006201)

References

1. Koehler M J, Mishra P. （2005） What happens when teachers design educational technology? The development of technological pedagogical content knowledge. J. Journal of Educational Computing Research, 32:131-152. https://doi.org/10.2190/0EW7-01WB-BKHL-QD
2. Schmidt D A, Baran E. （2010） Technological pedagogical content knowledge (TPACK): The development and validation of an assessment instrument for preservice teachers. J. Journal of Research on Technology in Education, 2:123-149. https://doi.org/ 10.1080/15391523.2009.10782544
3. Liu W.P. (2014) A model of college teacher Training based on Teacher's Professional Development. J. Journal of Hebei University (Philosophy and Social Sciences),2:137-140. https://doi:10.3969/j.issn.1000-6378.2014.02.027
4. Hao Y. Q. (2014) Research on the Cultivation of Preservice Teachers' Teaching Ability Based on Model. D. Master Dissertation of Shaanxi Normal University,1:31
5. Zhang Z. (2016) Research on the Factors Affecting the Development of Pre-service Teachers' TPACK.J. Modern educational technology,1(26):46-50 https://doi.org/10.3969/j.issn.1009-8097.2016.01.007

6. Chai C S, Koh J H L, Tsai C C.(2010) Facilitating pre-service teachers' development of technological, pedagogical, and content knowledge (TPACK).J. Educational Technology and Society, 13:63-73. https://doi.org/10.1080/10476210.2011.622041YV
7.http://www.moe.gov.cn/srcsite/A16/s3342/201606/t20160622_269367.html
8. Mishra. P.. & Koehler. M. J. (2006). Technological Pedagogical Content Knowledge: A Framework for Teacher Knowledge. J. Teachers College Record,108(6):1017–1054 https://doi.org/10.1111/j.1467-9620.2006.00684.x

https://doi.org/10.1051/shsconf/202316702001

Analyse the influence of "cultural and creative + blind box" mode on the commercialization of Chinese traditional culture IP

Yining Luo [1,*]

[1]Asia Australia Business College, Liaoning University

Abstract. In recent years, with the rise of modern culture and the integration of global culture, the development of traditional culture has been a great impact. In order to make more and more people be attracted by traditional culture, the development of commercialization of traditional culture IP has become an inevitable trend and necessary means, and the mining and operation of traditional culture IP has also become an important topic of traditional culture units. As one of the most important commercial elements of cultural brands, Chinese creative products. The success of cultural innovation plays an important role in the commercialization of cultural IP. With the rise of the popular play culture and the popularity of the blind box in recent years, some cultural and travel organizations have followed the trend and started to make use of the dividend of the blind box economy to create the traditional cultural and creative play blind box. Based on the discussion of the marketing concept of cultural and creative blind box, this paper will study the influence of the mode of "cultural and creative + blind box" on the commercialization process of traditional culture by analysing the development status and future prospects of the commercialization of traditional culture IP and the feasibility of the marketing model of "cultural and creative + blind box".

1 Introduction

Since China put forward the reform and opening up policy, the trade between China and Western countries has been increasing, and the Western culture has been pouring in. As China's economy lagged behind the West at that time, many novel things and open minds did not appear in China, so when this unprecedented new culture appeared in the public eye, many people were deeply attracted, especially the young people. Fresh culture and rapid economic growth have opened the eyes of the Chinese people. People have higher and higher incomes, spend more and more on entertainment, and the single social life has gradually become richer. Under such circumstances, the position of traditional culture in people's hearts has been shaken by Western culture, the teapot in people's hands has been gradually replaced by coffee, the music they like has changed from opera to European and American music, and the modern art exhibitions have replaced visiting museums. Some people even only celebrate Western festivals and ignore some traditional festivals and customs. The space for the development of Chinese traditional culture in modern civilization is shrinking day by day, and it is quite common for young people to be enthusiastic about foreign culture [1]. However, in recent years, with the continuous development of China's economy and the continuous promotion of its inter-national status, the country gradually realized the importance of traditional culture to the Chinese nation, and began to increase the publicity and promotion of traditional culture. China has a history of five thousand years, and our ancestors have left many cultural treasures. These treasures are the crystallization of the wisdom of generations and need to be passed on and protected. Many traditional cultural units tried their best to let more and more people to understand and fall in love with traditional culture. However, as the pace of life continues to accelerate, people pursue fast and short-term fast-food culture, and even many people cannot calm down to watch a movie or a book in its entirety. Traditional Chinese culture is often obscure, and it takes a long time for people to have a certain understanding of it through traditional media. In this era of pursuit of speed, the transmission and popularization of traditional culture needs to inject fresh blood. Therefore, a single publicity and promotion cannot make traditional culture stand out in the current cultural environment. In order to improve the influence of traditional cultural IP in the cultural market, it is necessary to combine "text" and "business", and it is particularly important for some representative traditional cultural IP to carry out commercial reform.

2 Current situation of commercialization of traditional cultural IP

IP is the abbreviation of "intellectual property", originally meaning intellectual property, while in the field of culture, IP refers to cultural works with certain popularity and influence, and suitable for secondary

* Corresponding author: 2524789909@qq.com

https://doi.org/10.1051/shsconf/202316702001

adaptation and development on the basis of the original. In the current context of China, cultural IP is not only some literature, film and television works with their own flow, but also some representative national treasures and cultural relics and monuments have become important cultural IP. If these traditional IP want to achieve successful commercialization and enhance cultural influence, it is particularly important for the transformation and development of some high-quality IP. In the long and splendid history of China, the cultural treasures left by our ancestors can be said to be countless. Therefore, Chinese traditional culture has a huge potential market in the process of combining "literature" with "commerce". Under the global e-commerce model, with Chinese traditional culture as the quality content, core sup-port from the source and a large number of solid fans, the product has been connected with a commercial chain under the operation of commercial capital, thus dis-covering huge commercial value [2]. At present, the most successful Chinese traditional culture enabling business belongs to the Palace Museum. The reason why the Palace Museum stands out in the complex and changing cultural market is not only the profound connotation of IP itself, but also the profound understanding of the IP operation of cultural museums. Cultural and creative products of museums are the organic unity of culture and commodities [3]. The so-called cultural creation refers to cultural creativity. On the premise of having a wide audience and a series of cultural themes, cultural enterprises adopt innovative ways to transform culture creatively [4].

As one of the first cultural institutions in China to realize the importance of combining culture with commercialization, the Palace Museum began to increase the mining and operation of cultural IP in 2013, and launched a series of design competitions such as "Bring the Palace Home". These activities have helped the Palace Museum discover many talented designers, and also provided the Palace Museum with new ideas for IP innovation. Many creative cultural and creative products have become popular on the Internet. In addition, documentaries and variety shows on the theme of the Palace Museum have also been appearing in front of the audience, gaining the attention of many young fans. For example, there is a variety show called "New Forbidden City", which invites some famous stars to explore the Forbidden City together with the experts in the Forbidden City and find some elements in the Forbidden City according to the corresponding theme. Finally, students from professional design schools will complete a creative design. This program not only makes the cultural creation of the Forbidden City more famous, but also lets more people know the history of the Forbidden City. This series of measures not only greatly improved the influence of the IP of the Forbidden City, but also improved the realization ability of the IP of the Forbidden City. This year, the sales volume of the cultural innovation of the Forbidden City reached as much as 600 million yuan. It shows that the Palace Museum has made some progress in the innovation of cultural and creative products and the promotion of traditional culture IP.

3 Analysis of the development trend of blind box economy

With the born of "Generation Z", more and more people are crazy about the blind boxes [5]. Blind box refers to the toy box that consumers cannot know the specific product style in advance, which has random properties. One of the stimuli to their buying a blind box is curiosity [6]. Only when the consumer opens the blind box in person can they know exactly what style they have. As the purchasing process is full of unknowns, the stimulus of this uncertainty encourages consumers to make re-peated decisions, which makes people feel addicted for a time, and thus promotes the rapid rise of the blind box economy. Many companies see the prospect and future of this new entertainment consumer goods, have joined the blind box market. Blind box economy has become the most popular in the capital market, attracting a lot of funds and investment. In recent years, in the capital market, many A-share listed companies have begun to layout the blind box market. One of the most popular blind box brands named POP MART generated 8.18 billion yuan in sales in the first half of 2020 [7]. The current size of the blind box economic market is about 15 billion yuan, and it is expected to exceed 30 billion yuan by 2021 [8].

Blind box economy is a typical non-functional consumer product and once consumers have aesthetic fatigue for this kind of goods, it will be a big blow to blind box brands. Therefore, the core of its industrial chain is commercial development around IP, and "blockbuster" products are constantly launched through co-naming popular IP and signing original designers. Although the development of the blind box economy has brought huge market dividends, there are also some problems in the blind box market at the present stage. According to the survey, the blind box consumer group is mainly concentrated in the young people born after 1995 and female players of blind boxes account for 75%, with young people aged 18-29 years old (58%) as the main consumer group [9]. In addition to the use value of the goods themselves, consumers generally pay more attention to the psychological satisfaction and sense of achievement brought by purchasing goods [10]. This kind of customers usually have no rational concept of consumption, the psychology of seeking novelty and comparison will make them addicted to this. If the number of addicts is high, the reputation of the brand launching blind box products will be affected to some extent, and some consumers will have a negative view of them. Therefore, many people believe that for the negative impact of blind boxes, especially on teenagers, parents and schools need to strengthen the education of teenagers, and the government should also strengthen the supervision of the market. In January 2022, the Shanghai Municipal Administration for Market Regulation formulated and issued the Compliance Guidelines for Blind Box Business Activities in Shanghai, marking the first red line in the blind box industry in China. Therefore, in the future, the country will introduce more and more national laws and regulations for the blind box economy, so that the whole blind box market can develop healthily in the legal track.

4 The advantages and disadvantages of "Cultural creativity + blind box" mode

4.1 Advantages

The combination of cultural and creative products with cultural and creative brands has promoted cultural creative IP and cultural connotation to a certain extent. In particular, the blind boxes launched by some cultural and tourism organizations such as museums can not only spread history and culture well, but also bring more development ideas to the cultural tourism and cultural and creative market. With new ways to promote the popularity of traditional IP, so that traditional IP is no longer old-fashioned, attract more young fans around the world, so that traditional culture with new vitality. The marketing model of "cultural and creative + blind box" can greatly increase the sales of traditional cultural and creative industries, and usher in new opportunities for the once ordinary cultural and tourism industries. In 2022, Sanxingdui Museum has launched an "archaeological blind box" cultural and creative products, which is different from ordinary cultural and creative blind boxes. It hides the products in the soil. In the process of opening the blind box, consumers need to personally experience using tools to dig them out, so that people can experience the joy of archaeology, and thus have an interest in cultural relics and culture. Soon after the launch of this series, it was swept away by consumers. It sold 360,000 pieces in just 10 months, bringing tens of millions of dollars in profits, showing the effect of 1+1>2. After the series of blind boxes became popular, it not only brought the topic heat and economic benefits to Sanxingdui Museum, but also made more young people understand and fall in love with the cultural connotation of the museum, thus promoting consumers to make their next consumption. Nowadays, with more than 4 million fans under its official account, Sanxingdui Museum has become one of the most popular museums for tourists. It can be said that cultural and creative blind box is a successful attempt to expand the influence of traditional cultural IP and achieve commercialization, which has a certain positive influence. Cultural and creative blind box plays an important role in both the social influence of traditional cultural units and the commercialization of traditional cultural IP.

4.2 Disadvantages

In the era of "everything can be blind box", although the blind box economy has brought dividends to all walks of life, but due to the chaotic consumption, many negative news also emerge in an endless stream. There are many cases showing that many teenagers are addicted to the blind box because their minds are not mature, and they keep buying the blind box to get the hidden designs they want to own. Some children even have argument with their parents because they are addicted to the blind box and it has affected their studies. After this negative news appeared, the market supervision department increased the attention to the blind box market, once the phenomenon of random purchase will be stopped. However, in essence, the blind box mode is to use people's curiosity and addiction mechanism to stimulate consumers to consume, but with the passing of time, people's curiosity of the blind box will decline and the appeal of the blind box to consumers will also decrease, so the desire to buy the blind box will be reduced. Once people's consumption desire and curiosity for blind boxes decrease, the sales of cultural and creative blind boxes will also be affected. In addition, with the gradual improvement of laws and regulations related to the blind box economy, more and more irrational purchases of blind boxes will gradually decrease. Although this will contribute to the more benign development of the blind box market, it will also reduce the popularity of blind boxes. Another point is that although the blind box economy is still in the growth stage at this stage, the popularity of the blind box among consumers has decreased. Therefore, the growth peak of the blind box economy may be about to reach, and the help to the commercialization of cultural and creative IP may decline.

5 Conclusion

In general, the mode of "cultural and creative + blind box" has successfully brought new development opportunities and new creative ideas for the traditional culture industry. By combining the blind box economy with traditional IP, it speeds up the process of IP commercialization, improves the IP cash ability, and at the same time makes many young people feel the charm of traditional culture in the blind box of cultural and creative industry. People who used to think that traditional culture means a departure from fashion are gradually changing their views and being attracted by the charm of Chinese culture again and again. At the present stage, the blind box economy is still in the stage of rapid development, but with the gradual improvement of the regulatory system of the industry and the slowing down of consumers' curiosity hunting, the growth rate of blind box is bound to decline. At that time, the advantages of "cultural innovation + blind box" may no longer be prominent, and cannot bring long-term effective benefits for traditional IP. Therefore, for traditional cultural IP, the mode of "cultural creation + blind box" is a successful attempt. However, in order for IP to go on the road of commercialization for a longer time, traditional cultural units must make progress as time goes by, develop their strengths and avoid their weaknesses. Only by constantly looking for new business opportunities, combining traditional cultural IP with the new era, continuous innovation and research and development can we further help traditional cultural IP go further on the road of commercialization.

References

1. Yang, R. (2019), On the development status and inheritance of Chinese traditional culture, Building Knowledge Warehouse of China. (01)
2. Bai, D. (2019), Analysis on the development status and reasons of Chinese traditional culture IP, Media Today, 27(02): 26-29
3. Zhang, F. (2016), The development of cultural and creative products in museums under the background of "Internet Plus", Heritage and Conservation Research, 1(02): 22-26
4. Sun, M. (2022), Analysis on the Design of Cultural and Creative Products under the New Culture Economy, Modern Corporate Culture.2022(10):40-42
5. Zhang, Ziwei. (2021). Research on Blind Box and its Marketing Strategy. Advances in Economics, Business and Management Research, 2021(182). 10.2991/aebmr.k.210712.020.
6. Jiang, R. (2022). Analysis of Trend Culture from the Perspective of Social Currency: -- Taking the POP MART as an Example. BCP Business & Management. 2022(18). 183-188.
7. Prospectus of Pop Mart International Group Limited POP MART 2020 Prospectus (2020： 5-7)
8. Report on In-depth market survey and development trend of blind Box Industry in China 2021-2026, [R] Beijing: China Academy of Business, 2021
9. Su, M. (2019), When blind box meets IP consumption China academic journal electronic publishing house (2019:1-3)
10. Liu, J. (2022). Investigation and Research on Consumer Psychology of Blind Box Consumers: Taking "POP MART" Consumers as an Example. Frontiers in Business, Economics and Management. 2022(6). 138-141.

SHS Web of Conferences **167**, 02002 (2023) https://doi.org/10.1051/shsconf/202316702002
CACC 2023

Research on The External Dissemination of Yao Pangwang Festival Culture Based on the Belt and Road Initiative

Zhangyan Li*

Guilin University, Guilin Guangxi 541006 China

Abstract. Yao Panwang festival culture is a characteristic cultural heritage of the Yao ethnic group It is an important part of world excellent traditional culture. The Belt and Road Initiative broadens the channels of cultural exchanges between various countries and regions, makes the exchanges between nations closer and more frequent, and provides more channels for the external dissemination of excellent national culture. This paper starts with the cultural connotation of Panwang festival, excavates the cultural communication value of Panwang festival, explores the external dissemination of Panwang festival culture based on the Belt and Road Initiative, so as to expands the international influence of national culture.

1 Introduction

The Yao Panwang festival culture has a long humanistic history, profound cultural heritage, rich cultural resources, unique national characteristics and is a treasure of excellent national culture. The connotation and value of the festival makes it possible for the Yao Panwang festival culture to spread to the outside world. The Belt and Road initiative has broadened the channels for cultural exchanges between countries and regions, brought closer exchanges between ethnic groups of various countries. And the cultural exchange activities such as cultural festivals, art festivals, film festivals, and television weeks between countries, provides more ways for the external dissemination of excellent national culture. In this context, excavating the unique national cultural resources of the Yao Panwang festival culture and exploring the strategies of external dissemination of the Yao Panwang festival culture based on the Belt and Road initiative are of positive significance for the external communication of national culture.

2 Interpretation of Yao Panwang Festival Culture

2.1 The Connotation of Yao Panwang Festival Culture

Panwang Festival, also known as "Panwang Dance", is a festival for the people of the Yao ethnic group to remember their ancestor Panwang, and it is also the most widespread and grand festival among the Yao ethnic group. In the long historical development, Panwang festival has accumulated a rich cultural interior and strong cultural heritage. During Panwang festival, men, women and children wear their own national costumes, sing "Panwang Song", dance "Long Drum" sacrifice wine and incense, and praise the originator together. Its solemn religious sacrifice ceremonies and songs and dances of various forms and contents can not only reflect the remembrance and worship of ancestors, but also show the spread of secular folk sentiment and the praise of labor, love and life [1].

Feng,H.G. and He,J.Q. believe that the Yao Panwang sacrifice ritual is the main activity of the Panwang Festival, which carries the historical changes, customs, religious beliefs, totem worship, moral concepts, literature and art, music and dance and other cultural phenomena of the Yao ethnic group[2]. Gu,W.X. came to the conclusion that the Yao Panwang Festival is the carrier of Yao culture transmission from the three folk ceremonies of "returning the wish of Panwang", "big song of Panwang" and "dance of Panwang" [3]. Zhou, Q. believes that the Panwang Festival culture shows the rich folk skills of the Yao people, which accumulates the ancient history, culture, customs and habits of the Yao people. It is a cultural phenomenon in which the Yao people consciously inherit their own traditional culture in a specific time and space environment. The Yao people present their common national psychology, national consciousness, aesthetic taste, and unique material and spiritual culture of each ethnic group to the world in different forms [4].

According to the cultural connotation of the Yao Panwang Festival, the author believes that the culture of the Yao Panwang Festival is a very characteristic traditional ethnic culture, which condenses the cultural factors such as the ethnic origin, history culture, agricultural production, religious sacrifices, ethnic dances, traditional skills, national values, ethics, national costumes, and eating habits of the Yao people.

* 115249936@qq.com

2.2 The Value of the Yao Panwang Festival Culture

Panwang Festival has a long history, rich cultural connotation, and unique national characteristics, which not only brings people a variety of fun, but also is conducive to the inheritance and development of national culture, national unity and harmony, and national economic prosperity. These communication values can be summarized as humanistic concept values, aesthetic values, industrial development values, etc. First, the memorial ceremonies, song and dance activities in Panwang festival culture all embody the humanistic concepts of national unity and harmony, friendship, mutual respect, mutual trust and so on. These concepts are in common with the concepts of other countries, and are the spiritual bond for building a harmonious society and a harmonious world. Second, the unique charm of the Panwang song and the strong and courageous long drumming in the Panwang festival culture, the fine and luxurious clothing, a wide variety of silver jewelry, colorful Yao brocade, exquisite embroidery, etc., the music, dance, clothing, silver jewelry, Yao brocade, embroidery, etc. with unique national characteristics vividly show the aesthetic value of Panwang festival culture. What's more, they can be made into cultural brand products with unique national characteristics and characteristics of the times. Coupled with effective industrialization operation strategies, they can obtain better economic and social benefits.

3 The Significance of External Dissemination of Yao Panwang Festival Culture under the Belt and Road Initiative

Panwang festival culture has a long humanistic history, rich cultural heritage and cultural resources, and has great communication value. The Belt and Road Initiative has opened up channels for cultural exchanges and cultural communication among ethnic groups of various countries. In this context, the external communication of Panwang festival culture is actively respond to the national Belt and Road Initiative and enhance national cultural self-confidence; It is the need to improve the level of external communication of Panwang festival culture, expand the international influence of national culture, and also promote the inheritance and development of national culture and the economic prosperity of ethnic regions.

3.1 Responding to the Belt and Road Initiative and Enhancing National Cultural Self-confidence

At present, some endangered folk traditional cultural artworks and art performance forms have been reproduced into exquisite cultural products and sold well in domestic and foreign markets after scientific protection, development and integration; Some national literary works that have been passed down from generation to generation and embody the national spirit and wisdom have been carefully collected and integrated, and have been compiled into books in multiple languages, which have been circulated at home and abroad, and some have even been included in the treasure house of world culture [5]. With the implementation of the Belt and Road initiative, cultural exchanges and communication activities between countries and ethnic groups have become frequent and close. This is a positive response to the Belt and Road initiative to expand the international influence of national culture, and further enhance the self-confidence of national culture.

3.2 Improving the Level of External Dissemination of Panwang Festival Culture and Expanding the International Influence of National Culture

Panwang Festival culture contains cultural heritage of worldwide value, which is unique in China and the world. However, at present, the excavation, research and publicity of Panwang Festival culture are not enough. Previously, there were also some cultural research academic teams or organizations, scholars on the ancient Yao village culture, folk culture, food culture, clothing culture, etc. They conducted a series of research investigations, and came out with some results. These studies mainly focus on the development and protection of culture, inheritance and development, however, compared with the external transmission of traditional elite culture, regional folk culture in the external communication research is quite limited. First of all, there are few relevant academic research results, and there are few in-depth and comprehensive studies. Secondly, the government's external propaganda department mostly disseminates folk cultural elements from the macro level, such as drama, clothing, folk music, painting, etc., which lacks historical and humanistic views, and is not innovative and vivid enough in the form of propaganda, and it is difficult to arouse the resonance and empathy of foreign people. Exploring the strategy of external dissemination of Panwang Festival culture is of positive significance to improve the level of external communication of Panwang Festival culture and expand the international influence of national culture.

3.3 Promoting the Inheritance of National Culture and the Prosperity of Ethnic Regional Economies

Panwang Festival is an outstanding national cultural heritage and an integral part of intangible cultural heritage. Because most Yao villages are located in relatively remote mountainous areas, economic development is relatively lagging. They may be affected by a variety of subjective and objective factors, and the scale of the Panwang Festival in Yao villages in some areas is slowly shrinking; the ceremony is gradually simplified, and the influence is getting smaller and smaller. And the cultural skills inherited from ancestors such as long drumming, memorial ceremonies, etc., due to the death of old artists one after another, there is no successor, and it is on the verge of being lost. In particular, the Yao people in Southeast Asian countries such as Vietnam, Myanmar, Thailand, and Laos who have emigrated overseas are in a difficult situation in terms of festival ceremony inheritance. If the inheritance and communication of the cultural heritage of the Panwang Festival in Yao villages is not strengthened, the cultural heritage of the Yao Panwang Festival will gradually become exhausted or empty. Therefore, exploring the strategy of external communication of Panwang Festival culture is also the need to promote the inheritance and development of national culture and the economic prosperity of ethnic regions.

4 Reflections on the External Dissemination of Yao Panwang Festival Culture Based on the Belt and Road Initiative

The Belt and Road Initiative has provided convenient conditions for political, economic and cultural exchanges between countries and peoples. With the opportunity, it is better to comprehensively and deeply excavate the cultural characteristics of Panwang Festival, create Panwang Festival cultural boutiques, give full play to the multi-party strength of the official and non-governmental sectors, expand communication organizations, and build diversified communication channels with the help of emerging scientific and technological means and new media.

4.1 Excavating the Cultural Resources of Panwang Festival and Creating Fine Cultural Communication Content

As an important part of Chinese national culture, Panwang Festival culture has unique cultural charm and characteristic cultural resources, and was included in the national intangible cultural heritage in 2006. The Panwang Festival culture, which has a far-reaching influence on China and the world, has unique and distinctive national characteristics and a charming traditional art living culture, which has great communication value. It also has research value in ecological aesthetics, ethnology, anthropology, sociology, etc. Panwang Festival cultural masterpieces are the spiritual bond of the nation. They can reflect people's spiritual and cultural demands. Panwang Festival culture is not only the culture of China, but also the culture of the world. As the saying goes, "the more national, the more global." Only by comprehensively and deeply excavating unique national cultural resources, integrating rich cultural creation achievements, creating Panwang Festival cultural boutiques, and industrializing the packaging, publicity and sales of boutique cultural brands can we better promote Panwang Festival culture to the world.

4.2 Giving Full Play to the Power of Official and Non-governmental Organizations and Expanding Communication Organizations

Under the background of the Belt and Road initiative, it is a complex and systematic project to use cultural exchange activities between ethnic groups in various countries to promote the Panwang Festival culture to the world to expand the international influence of national culture. It not only needs the propaganda power of official organizations of the national government, such as Confucius Institute educational institutions, local foreign affairs propaganda departments, and consulates and embassies in foreign countries, but also encourages non-governmental organizations such as enterprises and institutions, universities, research institutions, etc. to participate in publicity, and at the same time, it is necessary to cultivate the awareness of cultural communication subjects of people from all walks of life. Panwang festival culture is not only the culture of the Chinese Yao people, but also the culture of the Yao people around the world, in the process of the spread of Panwang Festival culture, it is wise to fully integrate the superior resources of all parties at home and abroad, strengthen the cultivation of overseas Yao people's cultural propaganda subject awareness, and use their foreign resources to promote the external dissemination of Panwang Festival characteristic national culture.

4.3 Building Diversified Communication Channels with the Help of Emerging Scientific and Technological Means and New Medias

Internet of Things technologies such as Internet +, artificial intelligence, and cloud computing have promoted society into the era of big data. Emerging scientific and technological means and the rapid dissemination function of new media provide convenience and powerful communication channels for the external dissemination of culture. The external communication of Panwang Festival culture needs to seek diversified communication channels, and multi-pronged measures can change the current situation of weak external communication. First of all, it can be combined with the current Internet technology to take the industrialization development path of Internet + sports, Internet + tourism, Internet + art, etc. At the same time, through high-tech, make the content of the Panwang festival that reflects the national culture,, such as the memorial ceremony, song and dance performance, food and clothing, into micro-movies, micro-videos, micro-lessons, brochures, etc., and then use emerging media communication channels, such as official Weibo, WeChat public account, portal website, network forum, etc. to release micro-information and information of Panwang Festival culture, and vividly and intuitively disseminate Panwang Festival culture in an all-round and multi-faceted way. Also, you can use translation software to conduct translation research on Panwang Festival culture in English, Thai, Vietnamese, etc., and use multiple languages to vigorously promote Panwang Festival culture.

4.4 Carrying Out In-depth Cultural Exchanges With the Advantage of the "Belt and Road" Initiative

The implementation of the "Belt and Road" initiative has brought closer exchanges between the peoples of countries, regions and ethnic groups with different historical origins and cultural backgrounds, and has paved channels for the wide dissemination of excellent cultures of various countries. In this context, taking advantage of the opportunities provided by the "Belt and Road" initiative, China actively carries out more extensive cultural and artistic exchanges, academic discussions, cultural forums, talent cooperation and other international cultural exchange activities with countries along the route, so as to strengthen cultural exchanges and mutual learning between different countries , promote the common development of multicultural in the world, promote the Panwang Festival culture to go abroad, integrate its cultural essence into the world culture, and enable mankind to share the fruits of cultural development. First of all, with the help of various international cultural exchange platforms provided by the "Belt and Road" initiative, cultural activities such as the Panwang Festival Cultural Academic Forum and the International Ethnic Culture Seminar will be held, and in-depth cultural exchanges will be carried out with countries along the route, so as to strengthen cultural exchanges and cooperation and mutual learning. Secondly, mobilize the forces of Southeast Asian countries along the "Belt and Road" to integrate the Panwang Festival culture in Southeast Asia in external communication, form a Panwang Festival cultural circle, and establish international organizations and overseas communication institutions to study Panwang Festival culture, strengthen the power of Panwang Festival cultural communication, and thus enhance the international influence of Panwang Festival culture.

5 Conclusion

Panwang Festival is an outstanding national cultural heritage and an integral part of world intangible cultural heritage,with rich cultural heritage and great external communication value. Under the background of Belt and Road initiative, exploring the cultural resources of Panwang Festival , showing the world the culture with Chinese national characteristics, expanding the international influence of national culture, and promoting national cultural self-confidence, it is necessary to give full play to the power of official and non-governmental organizations, to expand communication organizations and build diversified communication channels with the help of emerging scientific and technological means and new media.

Acknowledgment

2022 Guangxi Universities' Young and Middle-aged Teachers' Basic Research Ability Improvement Project (2022KY1559); 2022 School-level Education Science Project "Research on the Integration of Guangxi Ethnic Minority Culture into the English Culture Courses Teaching in Private Universities"

References

1. Deng, X. (2017)"Panwang Festival": Cultural Interpretation of Sacred Traditional Narrative in Contemporary Context: A Case Study of Shanxiang, Hekou Yao Autonomous County, Yunnan.China Nationalities Expo, 11:118-119+244..
2. Feng, H.G, He, J.G. (2010)*Research on the Ritual of the Yao Panwang Sacrifice and the Yao Panwang Festival*. Minzu Publishing House, Beijing.
3. Gu, W.X. (2011)On Panwang Festival and Yao Cultural Inheritance: Cultural Inheritance from the Perspective of Educational Anthropology,2.
4. Zhou, Q.(2013) On the Cultural Function of the Lanjing Yao Panwang Festival.Journal of Chifeng University (Chinese Philosophy and Social Sciences Edition),34:125-126. DOI : 10.13398/j.cnki.issn1673-2596.2013.05.046
5. Mao, H. l., Lu, Y. (2011)The significance and strategy of protecting the intangible cultural heritage of Panwang Festival in Yao villages .Journal of Guangxi University for Nationalities,33:110-114.
6. Zhang,Y.Q.(2014) Research on customs of Panwang festival of Dongshan Yao ethnic group. Guangdong Technical Normal University, Guangdong.
7. Ma,J.H.(2018) The social life of overseas Yao.Art Review.21.
8. Zhong,Q.L.(2007)Development of cultural industry of Panwang Festival of Yao nationality in Guangxi.Huazhang,4.
9. Tang, Y.Q. (2011)On the value of "Panwang Song" of Jianghua Yao ethnic group in Hunan Province[J].Value Engineering, 21:292-293. DOI: 10.14018/j.cnki.cn13-1085/n.2011.21.099
10. Zheng, X. (2013) Discussion on the protection and inheritance of Yao culture. Journal of Hunan Institute of Socialism, 3: 65-68.

Retro Futurism: The Resurgence of Y2K Style in the Fashion Field

Xiaochun Yang [1,*]

[1] Henan Donghua Techical School, Zhengzhou, Henan, China

Abstract: Y2K style was born in the late 1990s at the end of the 20th century. This period was at the turn of the new and old centuries. At that time, people were full of yearning and longing for the future world. Therefore, this aesthetic style with both the sense of future technology and a little retro flavor was integrated into the design field. Y2K style has been popular for a period of time, and began to decline since 2003. Later, new popular style gradually took her place. In recent years, with some design fashion brands integrating the Y2K aesthetic style into clothing design again, fashion icon and celebrities have spread and promoted the Y2K style, and young people in the era of Z have their own unique aesthetic pursuit, which has made the resurgence of Y2K with retro futuristic style from the minority to the public, and gradually formed a new atmosphere in the fashion circle. This article uses qualitative research methods to study the development and evolution of Y2K style as well as its current impact on the fashion field and the reasons for its popularity again.

1.Overview of Y2K Style

1.1 Background and reason of Y2K style

The word "Y2K" comes from the "millennium bug" crisis in computers. In the 1990s, computer science and technology were not developed, and the programming system was still relatively backward, which may lead to a series of serious problems when the computer processed the date after 2000. The original meaning of the word Y2K was an important bug in the processing of dates by early computer programs. At that time, people who were about to enter the 21st century believed that this bug might lead to the paralysis of global infrastructure, and the word Y2K was born. Where Y represents "year", while K represents the Latin prefix "kilo", representing 1000. Y2K is now used to refer to the year 2000, David Eddie abbreviated the "Year Two Kilo" representing "2000" as "Y2K" in June 1995. Therefore, "Y2K" began to spread as a synonym for computer problems in 2000.[1]

At first, "Y2K" refers to the program bug of the computer, but later it refers to a special visual sensory style culture emerging in the western countries from the late 1990s to the early 21st century.[2]At the end of the 1990s, the Internet was in the transitional stage of the transition from the new century to the old century. During this period, the Internet was popularized all over the world. Due to the "millennium bug" crisis, the public was full of unknown fears about the coming new century, but more people were optimistic. They had unlimited imagination and vision for the future, and imagined the future as a utopian country with highly developed technology. Finally, this imaginative concept of the future was extended to the visual sense culture, which gave birth to the "Y2K" aesthetics. "Y2K" style is usually included in "Y2K Aesthetics", referring to all clothing design, product design, music genre and visual effects that meet the "Y2K" aesthetic. In the overview of the encyclopedia website fandom, Y2K style prevailed in popular culture from 1995 to 2004. Named after Y2K Bug, it has a distinctive aesthetic period and integrates fashion, hardware design and technology.[3]

2. The Development Process and Characteristics of Y2K Style

2.1 Y2K style features from the 1990s to the early 21st century

The formation and development period of Y2K aesthetic style was from the 1990s to the beginning of the 21st century. The characteristics of "Y2K" style in this period mainly included the following two types of characteristics: the first type was the cold mysterious style that used colorless and cool colors as the main colors, heavy material texture and tight and smooth lines. The second category is the future technology style using high saturation colors, translucent texture materials and high-tech core technology applications.

The clothing industry has always been the leader of fashion trends. In 1995, at the Thierry Mugler fall couture show, the designer Thierry Mugler incorporated his imagination of the future world into the design. This is also the first appearance of Y2K style in the show in the

* Corresponding author: yangxiaochunzoe@outlook.com

1990s. [3] The shape of the future steel machinery female soldier in the show is also a typical representative of the futuristic fantasy in the Y2K style. The glittering gold and silver metal texture materials bring a strong sense of future science and technology. The shape of the semi-human and semi-robot also seems to confirm the public's infinite fantasy about the high development of future science and technology. The smooth and sharp arc shape contour lines also bring a soft feeling to the cold material texture. It seems that everything is full of utopian fantasy about the future world.

Then in 1999, Alexander McQueen launched the Givenchy fall 1999 fin de siècle series, which shook the classic and elegant style of Givenchy once released. Inspired by the Y2K millennium bug crisis, the model wore the "android couture" clothing and walked on the silver mirror T-stage. In addition to using metal texture and PVC fabric, the clothing also decorated the surface of the clothing with circuit board patterns and high-saturation color LED light emitting tubes as design elements, Alexander McQueen conveyed people's vision and expectation for the future of technology after the millennium through clothing design. In the same year, the hot release of the film "The Matrix" injected new popular elements into the Y2K style. The characters in the film were mostly in achromatic and cool colors as the main colors, and the tight-fitting knitwear, shiny black tight-fitting leather jacket, ultra-long black windbreaker, black sunglasses and electronic digital technology accessories. The combination of these fashionable pieces and design elements shows the rational, calm and high-tech fantasy of the future world. Just four months after the film was released, John Galliano, then the design director of Christian Dior, was influenced by " The Matrix". In the show of Christian Dior fall 1999 couture, we can see the smooth PVC coat, metal coat, leather boots, and black eye makeup like wearing sunglasses. These elements are imitating and expressing the coldness and technology of " The Matrix".

The years 1999 and 2000 were the peak of Y2K style. As the most typical popular symbol of the millennium, Y2K style can not only be seen in fashion shows, films and television works, but also in many European and American singers who released records around the millennium, having been affected by Y2K style in their own modeling, and having applied Y2K aesthetics in their own works. For example, Britney Spears told a fantasy story about the future world in her music video "Oops I Did It Again" in 2000. In the music video, Britney Spears told that a Martian living on Mars fell in love with an astronaut from the earth. Britney Spears, as the heroine, wore tight tights made of bright red latex in the music video. The tight style highlights her soft and beautiful body curve, and her face was painted with metallic silver-gray eye makeup, and the dancers beside her also wore silver metallic leather and PVC translucent clothing to dance in the imagined Martian living environment. The image of Britney Spears in red tights in the film has also become one of her classic shapes. In addition to Britney Spears, TLC also uses a lot of PVC, metal texture materials, strong futuristic makeup, high saturation hair color and eye makeup in the clothing modeling of "No Scrubs". Michael Jackson's "dream" mv uses a lot of future science fiction elements in the scene setting and clothing modeling.

The popularity of Y2K style in western countries has also led to the rapid spread of this trend from the West to the East. At the other end of the world, stars in Asia have also begun to follow suit. For example, singers such as Faye Wong, Coco Lee, and Lee Jung Hyun have also incorporated Y2K aesthetics into their album's overall image design. Y2K style is now popular in the world, which also proves that the public is full of expectations and good imagination for the coming new century.

Not only clothing, movies and music are affected by Y2K style, but also the appearance design of some game consoles, such as PlayStation, Xbox and game UI design, are full of Y2K aesthetics.[4] In terms of architecture, in 1997, the Encounter restaurant on the top floor of the Los Angeles airport theme building, which is known as the city's cultural and historical monument, began to open. The building where the restaurant is located looks like a spaceship that has just landed. The interior of the restaurant is mainly decorated with simple and smooth curves, and at the same time, it is equipped with white and purple mysterious lights. The whole restaurant and building seem to be visitors from outside space, This aesthetic feeling full of futurism just meets the public's imagination of the future world.

From the above examples, it can be seen that from the late 1990s to the beginning of the 21st century, Y2K aesthetics has had an impact on design aesthetics in many fields, especially on clothing design. Y2K style also quickly became popular in the world and became a trendsetter.

2.2 Continuation of Y2K style - McBling style (from 2003 to 2008)

With the advent of the 21st century, due to the bursting of the Internet foam and the fatigue of the public's appreciation of the same aesthetic, Y2K style has gradually declined, and its popularity has been replaced by a new style, which is McBling. McBling is a popular aesthetics from about 2003 to 2008. It inherits some features of Y2K aesthetics. Although McBling style and Y2K aesthetics are often discussed together or mistakenly considered to be the same style, and McBling style has a certain aesthetic impact on the resurgence of Y2K more than a decade later, there are still obvious differences in the initial style characteristics between the two.

The word McBling is a compound word, which is composed of Mc and Bling. Mc is taken from McMansion. McMansion refers to those fake mansions used by the media in the 1980s. These fake mansions are built with cheap building materials and sold at a high price, which is used to satirize the aesthetics of ostentatious and vulgar taste. Bling refers to shiny things, so the term "McBling" is ironic and used to refer to the colorful and complicated clothing style at that time.

The first signs of McBling aesthetics began around 2000-2001. One of the earliest examples of this style is the music video of Madonna's song "Music" in 2000. But the iconic figure that really made the style of McBling rise is Paris Hilton. At that time, the reality show "The Simple Life", which she participated in, was popular. Her clothes in the program caused a wave of imitation among young people. This wave of imitation also marked the end of the Y2K era and the arrival of the McBling era.

Paris Hilton, the representative figure of McBling style, became a household name because of the popularity of the reality show "Simple Life". Although she was not a practitioner in the fashion industry, she led the fashion trend at that time. Although her dress style has been wildly criticized by the media, it has been warmly pursued by the younger generation. The popularity of McBling style was guided by the popular culture at that time. In addition to the core representative of the style, Paris Hilton, the singers Britney Spears, Christina Aguilera and so on were also the leaders of Y2K style. In the campus youth films such as "Cheap Girl", "Legal Beauty" and "Leading the Way", we can also see the clothing style of McBling's sweet hot girl style.

To some extent, the McBling style can be said to be a partial continuation of the Y2K style, which is also the reason why some people later mistakenly believe that the two styles are the same. The McBling style of the 2003-2008 period has the following main characteristics: bold and exaggerated color matching, reflective fabric material, use of rhinestone sequin material, and tight short clothing style. For example, a jacket shorter than the navel, a low-waisted pants or a miniskirt. Compared with the previous Y2 style, the McBling style weakens the cold future sense of technology, and uses a large number of bright and exaggerated colors, especially the use of pink, which makes the McBling style more feminine than the Y2K neutral style.[5]

3."Y2K" Style Returns

In about ten years after 2008, the Y2K style seemed to fade out of people's sight temporarily, as if it had disappeared. However, from 2018 to 2020, with the emergence of the virtual idol of the Metaverse, the impact of the COVID-19 epidemic and the global economic recession, the negative development of the earth has spread among the public, and the "Y2K" style, which carries the future utopia, has returned to the public's view again and quickly swept the world [6], the resurgence of Y2K style seems to have caused a boom. In addition to many design brands and designers incorporating Y2K style into the latest design works, many star trendsetters have also begun to pursue Y2K style.

The most representative example of Y2K style's return is the early autumn series of cooperation between Dior men 2019 and the famous Japanese artist Hajime Sorayama. The nearly perfect giant bionic female robot standing in the show is the representative work of Hajime Sorayama, Sexy Robot. Sexy Robot, a Utopian product, blurs the boundary between the real human body and the robot. Hajime Sorayama displays his imagination of the perfect female body on the cold and emotionless robot. The combination of cold metal materials and the feminine soft curves, and the use of multiple lights with the blooming cherry blossom patterns to create the image of blooming flowers, which is also the right expression of Y2K's fantasy future aesthetics.[7] In addition to the show installation design, which is full of utopian imagination, this season's clothing design is also full of the spiritual core of Y2K aesthetics. The overall black and gray clothing is dotted with pink cherry blossoms. The clothing fabric processed by high-tech metallization technology is glittering with the light of metal texture. The cannage pattern is cut on the surface of the fabric by laser etching technology. The inside of the cold outside contains a trace of beauty and tenderness, which is exactly consistent with the Y2K style filled with Utopian ideas.

In addition to Dior men, many luxury brands have also begun to try emerging cultures, such as the Archelight Sneakers in the spring and summer 2018 series launched by Louis Vuitton, which are inspired by future trends and basketball shoes in the 1990s. It is characterized by the extremely high arch on the sole, the ultra-long tongue, and the high heel. The color of the upper also uses the gray color of metal technology, the high saturation fluorescent color, and the patterns with special design sense. It looks like a laser light from a distance, making the shoes look very virtual technology sense. In Jeremy Scott's autumn and winter 2018 series, there are many metal texture fabrics and transparent materials used. In the clothing silhouette, the jacket and skirt design inspired by the design of circuit boards and spaceships can be seen. The use of these elements can also show that designers are full of yearning for the era of technology. Moreover, Balenciaga set the launch of the 2021 fall series in the virtual game world, and showed the ready-to-wear release of the series to the players through the release of the online game "Afterworld: The Age of Tomorrow". The story background of the game is set in 2031 in the future, and the world after ten years is imagined through fable stories. The designer expresses his imagination of future clothing through gamification. In this game, you can see the flowers in full bloom, the hands that can blow fire and the twisted flying car in the air, all of which seem to make the audience enter a strange world full of fantasy. The dress of the characters in the game is also avantgarde and modern, and the presentation of this way also fully expresses Balenciaga's endless imagination of the future world.

The McBling style, which is considered to be the continuation of Y2K style, is also an important branch of style development in the process of the resurgence of Y2K style. Among them, the most representative clothing brand is Blumarine. Under the leadership of the current design director Nicola Brognano, the clothing brand that has been established for more than 40 years has also changed its style in recent years, from the original intellectual and elegant classic style in the past few years to the current popular sweet and cool Y2K style. Blumarine focuses on exploring and reconstructing the style of pop idol in 2000,[8] Taking the popular idol in the 2000s as the inspiration muse, he drew on the current popular millennium trend and refined the original classic elements of the brand to combine and innovate, creating a new Blumarine. For example, in the 2022 spring and summer series, the overall use of bright high-saturation colors such as pink and yellow, sexy bold designs that expose large areas of body skin, frequent low waist lines and butterfly design elements seem to create a dreamy millennium dream.

In addition to these fashion brands, some stars and fashion icon are also following the trend of the millennium style. In May 2020, Lady Gaga released her new album "Chromtica". From the album cover to the MV modeling, it was full of Y2K style. In the album cover picture, Lady Gaga wore a black leather corset which was fixed in the middle of an oversized metal ring. Her left arm wore a metal armor, and her right leg was a prosthetic made of metal materials. Such a semi-human and semi-mechanical shape also created a strong sense of technology. Besides, Korean women's group BLACKPINK used high-saturation hair color, unique makeup and metal material accessories to create a futuristic visual effect in the preview of the new album "How you like that" in 2020.

The resurgence of Y2K style in the fashion circle is not only a large-scale outbreak, but also a synchronized fashion between the East and the West. In June 2020, the female inspirational competition variety "Sisters Riding the Winds and Breaking the Waves" was popular in China. Although Zhu Jingxi, one of the contestants, failed to make her debut in a group, her unique dress style left a deep impression on the audience. Her unique future and scientific makeup and clothing in the program not only made her stand out among the contestants, but also made Y2K style popular in China. In the program, Zhu Jingxi has not only been interpreting the stage of Y2K style, but also said that she wants to be a Y2K style women's team. In her music works, she also expressed blind optimism and vision for the future, which also coincides with the ideological connotation of Y2K aesthetics. In July 2021, SuperELLE magazine even teamed up with a group of famous Chinese celebrities to shoot fashion blockbusters with the theme of # back to y2k # to pay tribute to the retro futurism style.

Two years later, the aesthetics and style of Y2K are still being imitated by everyone, even with increasing popularity. The word Y2K has been viewed by more than 90 million people on a social networking platform named Xiaohongshu, and the total number of videos with the word Y2K tag on TikTok video platform has reached 630 million. On July 20, 2022, well-known singer and actor Ju Jingyi released a photo of the cowboy Spice Girl on Weibo, which attracted a lot of attention. This set of summer-limited Spice Girl makeup and dressing style, which was positioned as Y2K style, was also imitated by netizens. According to the trend of the number of Y2K notes and "Ju Jingyi" notes provided by ReMai Data, the release of this model has aroused the attention of Y2K style.

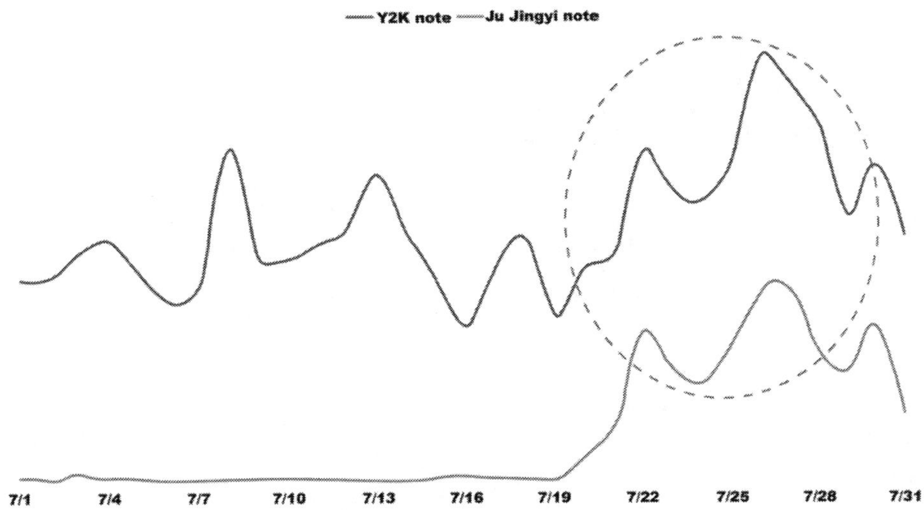

Fig.1. Transmiting tendency

Picture Source: ReMai Data

The term #Ju Jingyi Cowboy Spicy Girl Makeup# currently has 26.65 million views in Xiaohongshu, among which butterfly denim top, blue eye makeup and Y2K Spicy Girl three-strand braid provide rich imitation inspiration for users of Xiaohongshu. Many fashion lovers have uploaded photos of themselves imitating # Ju Jingyi Cowboy Spicy Girl Makeup #, and the combination and promotion of star effect and popular trend can also rapidly expand public awareness of Y2K style.

4.Research on the Reasons for the Resurgence of "Y2K"

Jame Laver, a fashion culture researcher, believes that fashion trends are cyclical. The so-called old is new, and the old can also become a trend. Over a period of time, the change of trend can be repeated repeatedly. In Laver's law, new elements may enter into silence after a cycle. After a period of accumulation, a few elements may be used again.[9]As a style that was popular more than 20 years ago, Y2K style is also in line with this law after being out of fashion for more than 10 years. The philosopher Jacques Derrida put forward "Hauntology" in 1993. Derrida believes that things are differences, infinite differences. It produces and disappears, it's something and it's not something. In fact, the public is constantly exposed to the product of retro futurism vision. Through the act of disappearance, the era has put some things ahead of time and made them something we are destined to encounter repeatedly. The era we live in is not an entity, but a Hauntology formed by the overlapping of the past and the future. The public always looks at the things of retro futurism with a familiar and strange feeling.[10] The resurgence of Y2K style is based on the current review of the past and unlimited vision of the future, which is the typical feature of retro futurism. Elizabeth Guffey believes in the article "Crafting Yesterday's Tomorrows: Retro-Futurism, Steampunk, and Making in the Twenty-First Century" that if futurism is a term that describes our expectations for the future, then retro futurism describes how we remember these visions.[11] The birth of Y2K aesthetics is at the intersection of two centuries. In the past two years, with the emergence of the COVID-19 epidemic and war, which are full of unstable factors, it seems to have returned to the situation where the future is unknown but has infinite beautiful imagination more than 20 years ago. At the same time, with the public's reliance on science and technology and the introduction of new concepts such as the metauniverse and NFT virtual idol, the public is also full of unlimited imagination for the future development of science and technology. It is also normal for the optimistic Y2K aesthetics to become popular again under such circumstances.

5.Conclusion

Y2K style has existed as a short-term fashion style phenomenon and has become popular again after more than ten years. Its return not only integrates new features, but also brings some influence to the fashion field. The Y2K style that returns again is also like a link linking the development of fashion aesthetics in the 20th and 21st centuries.

References

1. Ding Yi. (2000)The origin of Y2K. Knowledge is Power.9
2. Chen Chun, Zhao Ming.(2021).Analysis on the Characteristics of "Y2K" Clothing Style and Its Style Return.West Leather(19),55-56.
3. Jiang Peilin, Liu Lijun.(2022).Audience Analysis of Y2K Youth Subcultural Clothing Aesthetics. West Leather(06),47-49.
4. 2023.https://aesthetics.fandom.com/wiki/Y2K
5. Add a chicken leg,2021.What is Y2K that Korean women's groups love? A video explains it thoroughly. https://www.bilibili.com/video/BV1ZM4y1G7Jq
6. Meng Mei, Ge Lingye. (2022).Research on the Innovative Application of Millennium Culture in Fashion Design--Taking "Y2K" Aesthetics as an Example. Design35(17):11-13. https://doi.org/10.20055/j.cnki.1003-0069.000006.
7. Li Bingjin. (2021).Aesthetic Interpretation of Y2K Style . Art Science and Technology.(18):143-146. https://doi.org/10.3969/j.issn.1004-9436.2021.18.048
8. Han Peiwei, Jin Yongshan. (2022).A Study on the Fashion Characteristics of "Y2K" Aesthetics from the Perspective of Youth Subculture. Grand view of art,(19):130-132.
9. James Laver.(1937).Taste and Fashion: From the French Revolution to the Present Day. George G, Harrap and Company Ltd. London
10. Liu Xiaoying.(2005).Dialogue "Ghost" -- Analysis of Derrida's Hauntology.Theoretical Investigation(06):37-42.https://doi.org/10.16354/j.cnki.23-1013/d.2005.06.010.
11. Elizabeth Guffey.(2015).Crafting Yesterday's Tomorrows: Retro-Futurism, Steampunk, and the Problem of Making in the Twenty-First Century. The Journal of Modern Craft.249-266. https://doi.org/10.2752/174967714X14111311182767

Cultural Heritage Game Design Based on the Collective Memory Reconstruction Model

Wenqian Liu [1], Xiaomei Nie [1,*]

[1] Shenzhen International Graduate School, Tsinghua University, University Town of Shenzhen, Nanshan District, Shenzhen, 518055, China

Abstract: Current methods for constructing collective memory of urban cultural heritage mainly rely on landscaping and museum exhibitions, however, these methods lack engagement and fun, making it difficult for people to form deep memories and a sense of place for the city. This paper proposes an urban cultural heritage game based on the collective memory reconstruction model. Combining location-based service and augmented reality technologies, the game provides a first-person gaming experience with narratives and levels correlated with historical and cultural attractions in the Nantou Ancient City, Shenzhen. Subsequently, a controlled experiment was conducted to verify the effectiveness of the game design. A total of 25 users participated in the experiment. Results show that the proposed design can enhance players' touring experience, cultural perception, and place attachment. This paper explores the theory and application of collective memory reconstruction and provides an empirical reference for cultural heritage game design.

1. Introduction

The ongoing process of urbanization has resulted in the extensive renovation of traditional neighborhoods, leading to a homogenization of urban landscapes and business patterns. Consequently, the revitalization of cultural heritage and collective memory has become increasingly critical in urban development.

The concept of collective memory was first proposed by French sociologist Maurice Halbwachs in 1920. It was defined as a social construct that results from the sharing of past events among members of a particular social group [1]. In recent interdisciplinary research, collective memory is often synonymous with terms like "cultural memory" or "urban memory". To ensure clarity in this paper, collective memory refers to the memory of urban spaces and cultural events shared by members of a particular social group with a common cultural background.

Collective memory can be divided into three aspects: memory subject, memory carrier, and memory mechanism [2]. The subject is people in the city, carriers can be tangible or intangible elements, and the mechanism is the process of interaction between people and place. Traditional approaches to collective memory construction like landscaping or exhibitions require significant resources [3, 4], and browsing museum exhibits alone is insufficient for visitors to develop a deep memory of cultural heritage or build a sense of identity and belonging to the city.

In recent years, cultural heritage games have been proposed as a solution to improve engagement and cultural perception during visits [5]. Gaming emotions are also proven to improve the relationship between people and places [6]. Previous studies have applied location-based service (LBS) and augmented reality (AR) technologies to games [7], but a theoretical framework and improved game design are still needed to fully explore their potential benefits on collective memory.

This paper proposes an urban cultural heritage game based on the collective memory reconstruction model. The game aims to enhance players' touring experience, cultural perception, and sense of place for the city by connecting cultural sites with their background stories. The game was designed and developed using Unity 3D, which provides players with LBS function by Mapbox and AR interaction through the combination of AR Foundation and Azure Spatial Anchor. To validate the game's efficiency, a user test was conducted and the collected data was analyzed. The paper includes sections on the theoretical background, game design process, user test and results, and conclusions with suggestions for future work.

*EMAIL: liu-wq20@mails.tsinghua.edu.cn (Wenqian Liu); nie.xiaomei@sz.tsinghua.edu.cn (Xiaomei Nie)
ORCID: 0009-0009-3014-2222 (Wenqian Liu); 0000-0002-8901-8981 (Xiaomei Nie)

2. Theoretical background and related work

2.1. Collective memory reconstruction

This paper proposes a collective memory reconstruction model, based on the research by Qian et al. [2] and Lak et al. [8], with three key elements: person, place, and person-place interaction:

- Place is the direct carrier of collective memory, providing spatial threads through visual symbols, and conveying cultural information via explanatory media like texts, images, and videos.
- Person-place interaction involves experiencing intangible cultural connotations through tangible carriers/media after entering a space.
- People project their values into space, and then construct collective memory by perceiving space and narrative threads, which benefits the formation of the sense of place.

As shown in the figure, the reconstruction of collective memory relies on landscaping or cultural transmission. However, traditional media used in person-place interaction may not be engaging enough to facilitate in-depth memory construction, and the relationship between collective memory and the emotions generated from person-place interaction remains unexplored.

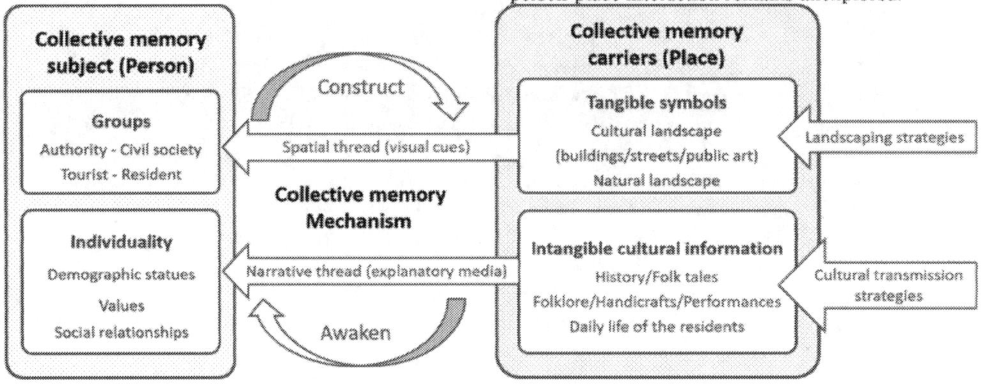

Figure 1: The collective memory reconstruction model

2.2. Cultural heritage game design

Cultural heritage games have been recognized as an effective tool for reconstructing collective memory. Several design frameworks have been proposed to guide the game design stages:

- The "FRACH" framework proposed by Andreoli et al. [9] divides game design into preparation, conception, development, and evaluation.
- The game space selection framework proposed by Sharma et al. [10] classifies mixed reality spaces into four types: individual, isolated, continuous, and ubiquitous.
- The mechanism-dynamics-aesthetics (MDA) framework proposed by Hunicke et al. [11]: Mechanism refers to the rules and operations, dynamics refers to the feedback and state, and aesthetics refers to the emotional experience that is expected to be achieved in the game.

Combining these methods, this paper proposes a framework for designing urban cultural heritage games based on collective memory reconstruction model, which will be elaborated on in the next section.

3. Overview of the game design

3.1. Urban cultural heritage game design based on the collective memory reconstruction model

This paper proposes a design framework for urban cultural heritage game based on the collective memory reconstruction model. The framework distinguishes between tangible and intangible memory carriers, with the former serving as a spatial thread to provide physical environments for people to perceive the city and the latter serving as a narrative thread to provide objects for memory reconstruction. The game serves as a medium that combines these threads to create an immersive experience, enabling players to learn about the history and culture of the place and develop emotional connections to it.

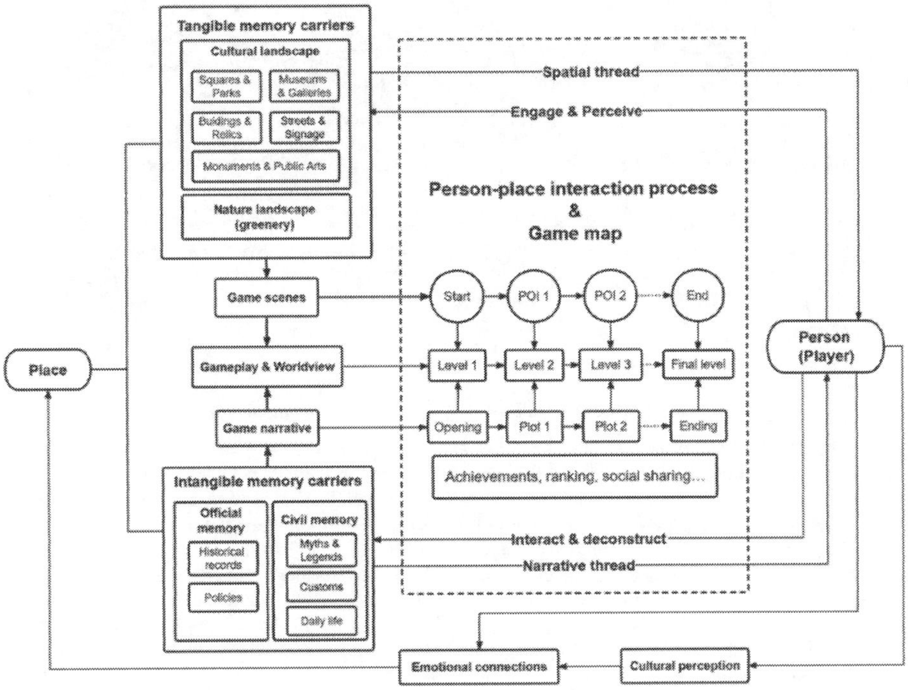

Figure 2: The design framework for urban cultural heritage game based on the collective memory reconstruction model

As the medium for person-place interaction, cultural heritage games must align with both the spatial and narrative threads. The following design methods can be used:

- Game design definition: Designers must select suitable locations and cultural elements for gameplay and worldview. Narrative adventure games, where players explore environments and collect items to solve puzzles, are a common genre.
- Game space selection: The selected locations can be divided into multiple points of interest (POI) to form a game map. The gaming experience depends on environmental conditions, with historic streets offering an isolated experience and cultural parks offering a continuous experience.
- Game narrative design: It consists of characters and stories, which can be drawn from history, folklore, or real people. Linear structures could be used to ensure coherence and integrity.
- Game level design: Mechanisms such as puzzles, battles, and quizzes can be used in each game level, with players rewarded with items or achievements after completion. Rewards should be associated with gameplay, and rankings and social sharing can be added to enrich the experience.
- Game aesthetic design: The visual style should cater to the game's worldview. The interface should be concise and easy to use, and sound effects should be used to stimulate positive emotions.

3.2. Game design and development

3.2.1. Case study on Nantou Ancient City

This paper uses the Nantou Ancient City (hereinafter Nantou) in Shenzhen, China as a case study to evaluate the effectiveness of the design framework. With a history spanning over 1700 years, it served as the economic, political, and cultural center of the coastal area of Lingnan for generations. The city retains its block layout since the Ming Dynasty and historical landmarks such as the South Gate and the Xin'an County Hall [12]. However, it has more stores than cultural attractions, and the existing attractions rely heavily on text or images, making it difficult to leave a lasting impression on visitors. To address these issues, this paper selects Zhongshan South Street and Zhongshan North Street as the primary experience route and highlights the points of interest and their history along the route as the spatial and narrative threads. The POIs are listed below.

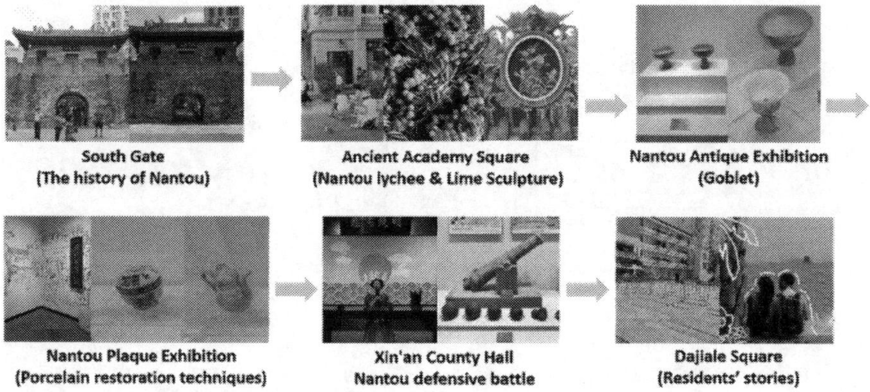

Figure 3: The POIs and history chosen from Nantou Ancient City.

- South Gate: As the representative building of Nantou with over 600 years of history, it serves as the starting point of the route.
- Ancient Academy Square: The lime sculptures around the square are national intangible cultural heritages, and the square is the first open space after entering the town, therefore chosen as the transition POI.
- Nantou Artifact Exhibition: It displays artifacts that represent the city's profound history, and goblet and porcelain restoration techniques are selected as props for the narrative.
- Nantou Plaque Exhibition and Dongguan Hall: These two POIs serve as transitional points.
- Xin'an County Hall: It highlights the political status of the City and represents the history of the defensive battle against the British invaders, therefore chosen as the key POI of the route.
- Dajiale Square: As the largest open space in Nantou where various activities have been held, it serves as the end of the route to showcase the stories of Nantou residents.

This paper presents "Memories of Nantou", an urban cultural heritage game that integrates spatial and narrative elements. The game was developed using C# language and Unity3d game engine and can be played on Android devices. The interactive narrative, developed with FairyGUI, centers around finding cultural memories in Nantou. Players explore the city in a first-person perspective, interacting with non-player characters (NPCs) or viewing introductions in the game's backpack to learn about Nantou's history. The main NPCs, Wen and Xuan, were designed as friendly children, drawing inspiration from Hakka culture in Shenzhen. The game employs Mapbox for location tracking, AR Foundation for interactions, and Azure spatial anchor service to activate corresponding AR scenes. The game's aesthetic design is based on cultural elements in Nantou, with user interface design adhering to Android design guidelines. Each game level is accompanied by background music and characters' voices, providing players with instant visual and sound feedback. The relationship between levels, plots, and interactions is shown below.

3.2.2. Game implementation

Figure 4: The mapping relationship of plots, levels, interactions, and screenshots

4. Methods

4.1. Research Hypotheses

To evaluate the feasibility of the design framework, this paper conducted a controlled experiment comparing the collective memory formed in game-guided and self-guided tours. The study focused on the history and culture of Nantou and evaluated the effectiveness of collective memory reconstruction by measuring players' cultural perceptions. The level of place attachment was also examined, as suggested by Qian et al. [2] and Li et al. [13], collective memory is an important factor in place attachment. The hypotheses proposed in this paper are as follows:

- H1: The game design provides an engaging touring experience.
- H2: The game design enhances players' cultural perception.
- H3: The game design enhances players' level of place attachment.

4.2. Questionnaires

This study employed three questionnaires validated in previous research to assess participants' gaming experience, cultural perception, and place attachment:

- The gaming experience questionnaire was based on Georgiou et al.'s [14] Augmented Reality Immersion (ARI) scale, and comprised 14 items in five-point Likert-type to measure engagement, engrossment, and immersion.
- The cultural perception test included 8 closed questions to measure knowledge from the game.
- The place attachment questionnaire was designed based on Williams et al.'s [15] and Yao's [16] studies, with 9 items in five-point Likert-type to measure place attachment and satisfaction.

4.3. Participants

The experiment involved a sample of 25 participants (13 females and 12 males, with college degrees, aged between 20 and 35). None of the participants had prior knowledge of Shenzhen's history and culture, nor did they reside near Nantou. The average level of knowledge regarding Nantou was rated as "very little" (1.79/5). The participants were randomly divided into two groups: the control group (CG), who were given a self-guided tour of Nantou, and the experimental group (EG), who experienced the "Memory of Nantou" cultural heritage game.

4.4. Procedures

The experiment employed a combination of qualitative and quantitative research methods, including questionnaires, semi-structured interviews, and game data monitoring. The total duration of the experiment was approximately 1.5 hours. Before the experiment, participants were required to complete a background research questionnaire to eliminate any irrelevant factors that may interfere with the study. Both groups participated in a 1-hour tour, with the EG using the cellphones provided by the experimenter to play the game, while the CG visited the attractions independently. After the experiment, participants were asked to complete a place attachment questionnaire, with the EG also required to complete a gaming experience questionnaire (< 5 minutes). After that, participants completed a cultural perception test (< 10 minutes), and then they were interviewed one-on-one to collect their subjective feedback (< 15 minutes). In addition, a data monitoring program was set up in the background of the game, recording participants' gaming time, scores, and page browsing time for in-depth research.

Figure 5: Photos of the experimental group (left) and control group (right) during the experiment

5. Results

The Cronbach.α coefficient for the gaming experience questionnaire was 0.936, and the subscales were 0.936, 0.836, and 0.889, respectively. The Cronbach.α coefficient for the place attachment questionnaire was 0.955, and for the subscales was 0.919 and 0.932 respectively. These values indicate that the questionnaires employed in the study were reliable, and the results obtained from them are suitable for further analysis.

5.1. Gaming experience

According to the table below, players expressed positive opinions regarding the engagement of the game, stating that they were interested and motivated to actively seek cultural elements in their surroundings. Moreover, players were focused on the game, expressing concern about plot development and task completion status. However, immersion was weaker, which may have been affected by the realism of the 3D models. Furthermore, since players experienced the AR game in the real world, they were

https://doi.org/10.1051/shsconf/202316702004

likely to be disturbed by lighting, noise, and other environmental factors, which has been previously mentioned in ARI research [11]. Overall, the results suggest that the proposed game can provide a satisfying touring experience for players.

Table 1 Results of the gaming experience questionnaire

Questions			Mdn	AVG±SD
Engagement	Q1: I like the types of interactions in this game	4.5	23.5	23.6±2.7
	Q2: I think the interactions are novel	4		
	Q3: I wanted to spend time playing the game	4		
	Q4: I wanted to spend time completing the game successfully	4		
	Q5: It was easy for me to play the game	3.5		
	Q6: I didn't have difficulties in game control	3		
Engrossment	Q7: I was curious about how the game would progress	4	16.5	15.9±2.1
	Q8: I often felt suspense by the game	4		
	Q9: If interrupted, I wanted to returning to the game	4		
	Q10: Everyday thoughts and concerns faded out during the game	4		
Questions			**Mdn**	**AVG±SD**
Total immersion	Q11: I felt that the characters/objects were real	4	15.5	15.3±2.1
	Q12: I felt that what I was experiencing were real scenes	4		
	Q13: I didn't have any irrelevant thoughts or external distractions during the game	4		
	Q14: I lost track of time in real world during the game	4		

5.2. Cultural perception

The results in the table below indicate a significant difference in the average score between the EG (AVG=67.7) and the CG (AVG=27.9). The game has a positive impact on the players' cultural perception, particularly in the knowledge that is highly interactive and related to the spots (T3-T5). The CG reported that current attractions failed to leave a lasting impression on them, and some participants did not even notice the introduction video in the museum. The EG reported gaining a more comprehensive understanding of the city under the game's guidance. However, some players mentioned that they paid less attention to the exhibits because they were too focused on completing the game.

Table 2 Results of the cultural perception test

Index	Question type	EG score (n=12)	CG score (n=13)
T1	Native culture learning (collecting lychee game)	91.7	92.3
T2	Native culture learning (lime sculpture introduction)	50.0	15.4
T3	History learning (Dongguan Hall introduction)	58.3	7.7
T4	History learning (first-person shooting game)	75.0	15.4
T5	History learning (South Gate introduction)	83.3	23.1
T6	History learning (finding goblet scraps)	83.3	30.8
T7	Native culture learning (Kintsugi quiz)	33.3	7.7
T8	Native culture learning (Juci quiz)	66.7	30.8
AVG		67.7	27.9

5.3. Place attachment

As shown in the table below, the game promoted players' place attachment and satisfaction. Players changed their attitudes towards the city by discovering its history in the game, and they suggested more interactions for other cultural attractions should be added to the game in the future.

Table 3 Results of the place attachment questionnaire

	Questions	EG (n=12)			CG (n=13)		
			Mdn	AVG±SD		Mdn	AVG±SD
Place attachment	Q1: I'm very attached to Nantou	4	13.0	13.5±2.3	3.31	11.0	11.0±2.0
	Q2: I identify strongly with Nantou	3			2.69		
	Q3: Nantou is the best place to experience Shenzhen's history and culture	3			2.69		
	Q4: No experience in other cultural site can compare to the one in Nantou	3			2.31		
	Questions	EG (n=12)			CG (n=13)		
			Mdn	AVG±SD		Mdn	AVG±SD
Satisfaction	Q5: In general, I'm very satisfied with this tour	4.5	21.5	21.6±2.1	4	18.0	17.3±2.6
	Q6: I was deeply involved and felt pleasant during this tour	4			4		
	Q7: This tour is worthy of my time and effort	4			4		
	Q8: I have a better understanding of Shenzhen's history and culture after this tour	5			3		
	Q9: I'm willing to revisit Nantou in the future	4			3		

5.4. Background monitoring data

As shown in the table below, it is evident that players tend to spend more time on game levels that offer greater contrast and interactivity with the real environment. Furthermore, as time increases, players develop a deeper understanding of the historical and cultural content embedded in these scenes.

Table 4 Results of the background monitoring data

Scenes	Time Mdn (IQR)	Spearman's rho	
		Correlation Coefficient	Sig. (2-tailed)
Guide (South Gate intro)	366.5 (114)	0.616*	0.033
Ruins (Collecting goblet scraps)	118.0 (115)	-	-
Lychee (Lime sculpture & lychee intro)	31.0 (21)	0.482 (lychee) 0.606* (sculpture)	0.112 (lychee) 0.037 (sculpture)
Lychee game (Catching lychees)	330.5 (222)	-	-
Antique (goblet & handicrafts intro)	195.0 (164)	0.583*	0.047
Craftsman (Kintsugi & Juci)	115.5 (36)	0.666* (Juci) 0.307 (Kintsugi)	0.018 (Juci) 0.331 (Kintsugi)
Dongguan Hall (History)	101.0 (9)	0.790**	0.002
Nantou defense battle (Intro)	37.0 (10)	0.731**	0.007
Shooting game (defeat enemies)	144.0 (150)	-	-

6. Discussion

The experiment results support the hypotheses proposed. Firstly, the proposed design provides an engaging touring experience. Players showed particular interest in the defending Nantou game, which they played for the longest time, likely because it presented a real historical scenario and created a strong visual contrast. The AR interaction was also highly praised by players, indicating its potential for use in collective memory research.

Secondly, the proposed design has a positive effect on enhancing players' cultural perception. Interacting with NPCs was noted as a particularly effective way to deepen understanding of the city. However, players were less likely to read the introductions in the backpack, suggesting that knowledge integration into AR interactions may be more effective.

Finally, the proposed design has a positive effect on enhancing players' sense of place. By marking attractions on the map and displaying virtual scenes in the real world, the game effectively highlighted the cultural features of Nantou and garnered positive attitudes among players. Additionally, gaming emotions such as curiosity, surprise, and accomplishment were found to benefit the formation of collective memory.

7. Conclusion

This paper proposes a design framework for an urban cultural heritage game based on the collective memory reconstruction model. An AR game with LBS is designed for Nantou Ancient City based on the framework, and a controlled experiment confirms its positive impact on players' touring experience, cultural perception, and sense of place. The game's interactive dialogues and tasks inspire players to explore the city actively, learn its history and culture, and form a deep collective memory and sense of place. This paper also explores the application of AR interaction in the field of cultural heritage, expanding the role of emerging human-computer interaction technologies.

Future research suggestions include exploring AR game interaction for continuous and ubiquitous spaces with multiple players playing together, as well as adaptive difficulty levels for players of different ages and educational backgrounds, to make the design more widely applicable.

Acknowledgements

This research was supported by Shenzhen Key Laboratory of next generation interactive media innovative technology (No. ZDSYS20210623092001004). The authors would also like to express gratitude to the participants for their valuable contributions to this paper, and special thanks to the resources provided by Nanshan Museum.

References

1. M. Halbwachs, On collective memory, 1st. ed., University of Chicago press, Chicago, IL, 1992.
2. L. Qian, J. Zhang, C. Zheng, P. Liu, J. Zhang, H. Zhang, A review of studies on collective memory from the perspective of geography, Human Geography 30 (2015) 7-12. doi:10.13959/j.issn.1003-2398.2015.06.002.
3. L. Xiao, The Study on City Memory of Conservation and Regeneration of Historical Environment of Xiguan, Guangzhou, Ph.D. thesis, South China University of Technology, Guangzhou, GD, 2018.
4. T. Wen, L. Liu, Activation and Inheritance of Urban Memory and Cultural Performance in Museums: An Analysis of Tourist Comments on a Museum in Guangzhou, Tropical Geography 39 (2019) 267-277. doi:10.13284/j.cnki.rddl.003120.
5. E. Lacka, Assessing the impact of full-fledged location-based augmented reality games on tourism destination visits, Current Issues in Tourism 23 (2020) 345-357. doi: 10.1080/13683500.2018.1514370.
6. T. Oleksy, A. Wnuk, Catch them all and increase your place attachment! The role of location-based augmented reality games in changing people-place relations, Computers in Human Behavior 76 (2017) 3-8. doi:10.1016/j.chb.2017.06.008.
7. A. Lehto, N. Luostarinen, P. Kostia, Augmented reality gaming as a tool for subjectivizing visitor experience at cultural heritage locations—case lights on!, Journal on Computing and Cultural Heritage (JOCCH) 13 (2017) 1-16. doi:10.1145/3415142.
8. A. Lak, P. Hakimian, Collective memory and urban regeneration in urban spaces: Reproducing memories in Baharestan Square, city of Tehran, Iran, City, Culture and Society 18 (2019) 100290. doi:10.1016/j.ccs.2019.100290.
9. R. Andreoli, A. Corolla, A. Faggiano, D. Malandrino, D. Pirozzi, M. Ranaldi, G. Santangelo, V. Scarano, A framework to design, develop, and evaluate immersive and collaborative serious games in cultural heritage, Journal on Computing and Cultural Heritage (JOCCH) 11 (2017) 1-22. doi: 10.1145/3064644.
10. H. N. Sharma, S. A. Alharthi, I. Dolgov, Z. O. Toups, A framework supporting selecting space to make place in spatial mixed reality play, in: Proceedings of the Annual Symposium on Computer-Human Interaction in Play, CHI PLAY '17, ACM Press, New York, NY, 2017, pp. 83-100. doi: 10.1145/3116595.3116612.
11. R. Hunicke, M. LeBlanc, R. Zubek, MDA: A formal approach to game design and game research, in: Proceedings of the AAAI Workshop on Challenges in Game AI, AAAI Press, San Jose, CA, 2004, pp. 1722.
12. Szns.Gov.Cn, Nantou Ancient City Transformation, 2021. http://www.szns.gov.cn/ztzl/nsrdzt/ntgcdb/index.html.
13. Z. F. Li, X. Y. Nie, Mechanism of collective memory of cultural tourism destinations on place attachment of tourists: A case of Wuzhen, Pingyao and Phoenix town, Areal Research and Development 37 (2018) 95-100.
14. Y. Georgiou, E. A. Kyza, The development and validation of the ARI questionnaire: An instrument for measuring immersion in location-based augmented reality settings, International Journal of Human-Computer Studies 98 (2017) 24-37. doi:10.1016/j.ijhcs.2016.09.014.
15. D. R. Williams, J. W. Roggenbuck, Measuring place attachment: Some preliminary results, in: NRPA Symposium on Leisure Research, San Antonio, TX, 1989.
16. D. Yao, K. Zhang, L. Wang, R. Law, M. Zhang, From religious belief to intangible cultural heritage tourism: A case study of Mazu belief, Sustainability 12 (2020) 4229. doi:10.3390/su12104229.

A Study on Visual Innovation of Macau Souvenirs Packaging in the Context of Multicultural Communication

Yue Zheng*, Xianlu Li, Caiyu Zhang, and Xuan Wang

Guangdong Technology College, Zhaoqing 526000, China

Abstract: This paper takes Macau souvenir packaging as the research object and explores the visual innovation of Macau souvenir packaging and its significance in cultural communication from the perspective of multicultural communication. The paper first introduces the background of Macau's multiculturalism and the cultural value of Macau souvenir packaging, followed by an analysis of the impact and driving role of multicultural communication on the visual innovation of Macau souvenir packaging. The visual innovation of Macau souvenir packaging is analyzed from aspects such as patterns, colors, and design elements, using AI algorithms for validation and conducting visual innovation design for souvenir packaging. This paper provides valuable insights and references for the design and development of Macau souvenir packaging, contributing to the promotion of Macau souvenir packaging in cultural communication and the tourism industry.

1 Introduction

Macau is a city that concentrates the cultural traditions and characteristics of both Chinese and Western cultures. As a Special Administrative Region in southern China, Macau's tourism industry is one of the most important economic pillars of the region. With the advancement of the Greater Bay Area construction in China, Macau's tourism industry is playing an increasingly important role and position. Macau not only has a rich historical culture and diverse tourism experiences but is also a unique and charming tourist destination, attracting more and more tourists to travel and consume.

The development of Macau's souvenir food industry has become one of the most representative and advantageous industries in the region due to the growth of the tourism industry. Macau's souvenir foods not only have unique flavors and quality but also serve as an important carrier for the inheritance of Macau's culture, representing its history and culture. However, behind these industries, the packaging design and brand image of souvenirs also play a crucial role[1]. They can not only enhance the recognition and acceptance of Macau tourism and souvenir foods among regional consumers but also promote consumption upgrades and cultural exchanges within the region [2]. Therefore, the connection between Macau souvenirs and the tourism industry is inseparable, and the inheritance and development of Macau souvenirs are an indispensable part of Macau's tourism industry.

2 Overview of Macau's Multiculturalism

Macau is a multicultural city that, due to its history, geography, and politics, has absorbed cultural elements from different regions and ethnic groups, forming a unique and diverse cultural landscape. In Macau, you can see European-style architecture from the Portuguese colonial period, such as Portuguese-style churches, castles, and squares, which reflect the influence of Portuguese culture on Macau. At the same time, you can also see many Chinese-style buildings, such as A-Ma Temple, Lou Kau Mansion, and Na Tcha Temple, which incorporate the essence of both Chinese and Western cultures, showcasing Macau's unique cultural charm.

In addition to architecture, Macau's culinary culture is also diverse. In Macau, you can taste cuisines from Portugal, China, Indonesia, and other regions. For example, Portuguese egg tarts and Portuguese-style chicken rice are famous Macau dishes, while Cantonese cuisine and Macau's unique blend of Portuguese-Chinese dishes also attract many tourists.

Furthermore, Macau boasts a wealth of cultural events and festivals, such as the Macau Grand Prix, the Drunken Dragon and Lion Dance Festival, and the Procession of Our Lady of Fatima. These events showcase the charm of Macau's multiculturalism and attract tourists from all over the world.

* Corresponding author: zhengyue@gdlgxy.edu.cn

3 Mutual Promotion of Macau's Tourism and Food Industries

Macau is a vibrant and culturally appealing tourist destination with numerous historical and cultural attractions and is well-known for its unique souvenir foods [3]. Macau's souvenir food offerings are diverse, including famous Portuguese egg tarts, almond cookies, and beef jerky, among others. These delicacies have attracted countless tourists to come and taste.

At the same time, Macau's tourism industry has also promoted the development of the local souvenir food industry. As more and more tourists flock to Macau, the souvenir food industry has experienced significant growth, constantly innovating and improving. To meet the demands of tourists, local souvenir food manufacturers are continuously developing new varieties and flavors of souvenir foods to attract more tourists to purchase and taste [4]. It can be said that Macau's tourism and souvenir food industries have formed a mutually beneficial interactive relationship. The development of the tourism industry drives the development of the souvenir food industry, while the innovation of the souvenir food industry attracts more tourists to visit Macau. This interactive relationship not only promotes local economic development but also enriches the tourists' travel experience [5].

4 Methodology

Under the backdrop of multicultural communication, Macau souvenirs, as a type of food product that blends unique cultural essence and regional characteristics, are widely appreciated by tourists and local residents alike. In this context, the importance of exploring visual innovation in the design of Macau souvenirs becomes increasingly prominent [6]. Based on this, the present study combines traditional methods with AI algorithms to develop an innovative visual design process, which is shown in Figure 1.

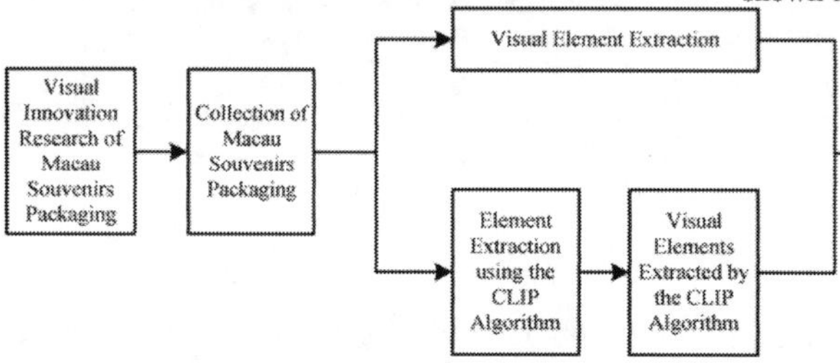

Fig. 1. Flowchart of visual innovation research of Macau souvenir packaging.

First, we collect and study Macau souvenir food packaging, extracting visual elements such as color, texture, shape, pattern, and so on. Then, we upload the collected souvenir food packaging images to an artificial intelligence platform, using image recognition technology to extract content and elements from the images, such as text, graphics, products, etc.

We correlate and compare the visual elements observed by the human eye and the extraction results from artificial intelligence to identify commonalities and differences between them. We then categorize and organize these findings into a collection of keywords. Next, we input these keywords into an AI design platform, employing AI algorithms to create a series of brand-new Macau souvenir food packaging design proposals. Designers then modify, optimize, and innovate based on these AI-generated design proposals to produce the final souvenir food packaging products.

This process can be iteratively and continuously optimized to achieve better design outcomes and market feedback. By utilizing artificial intelligence technology to extract content and elements from images, we can more quickly and accurately analyze and extract key information. This helps designers better understand consumer needs and trends, ultimately improving design efficiency and quality.

5 Diverse Visual Symbols of Macau

The diverse culture of Macau is extensively embodied in the design of souvenir packaging. By integrating Macau's unique architectural features, festivals, events, and folk elements into the packaging, the cultural value and regional brand image of the souvenir products are not only elevated, but consumer attention is also attracted, thereby promoting the dissemination of urban culture. The analysis of packaging with iconic features analyzed in this paper is shown in Table 1.

Table 1. Setting Word's margins.

Souvenir Brands	Produces	Innovation Visual Elements
PASTELARIA YENG KEE	Snoopy Almond Cake	Macau Tower
PASTELARIA YENG KEE	Snoopy Original Caramel Candy	Macau Grand Prix Circuit
KOI KEI BAKERY	Portuguese Cashew Cookie	The Ruins of St. Paul's in Macau
KOI KEI BAKERY	Black Sesame Peanut Crispy Candy	Largo do Senado

KOI KEI BAKERY	Peanut Crispy Candy	Na Tcha Temple
PALAEIO LISBOA	Almond Cookie	Hotel Lisboa Macau
PASTELARIA CHUI HEONG	Almond Cake	Vintage advertisement font

5.1 Macau's Iconic Color

The colors of the Macau Special Administrative Region flag are green and gold. These two colors carry symbolic meanings and market value in business, which can help brands establish stronger market competitiveness and brand culture while attracting more consumers and tourists, contributing to economic development [7]. Simultaneously, using the colors of the Macau flag for product packaging makes the products more prominent and easily recognizable in the market, helping to draw consumers' attention. Moreover, as Macau is a well-known tourist destination and gaming hub, utilizing the colors of the Macau flag can enhance brand awareness and recognition, strengthening the brand's market competitiveness [8], as shown in Figure 2(1).

5.2 Portuguese Charm

Macau is a place with a long history and rich culture, which includes the influence of Portuguese culture. In Macau's packaging design, many visual symbols related to Portuguese culture can be seen, such as Portuguese-style architecture, patterns, and color schemes.

The Portuguese color scheme has an international appeal, and using these colors can better attract the international market, expanding the product's sales channels and scope. Its unique color system can endow products with a distinctive image and characteristics, helping brands establish a unique image and style that sets them apart from competitors. For example, the Macau Portuguese egg tart flavored chocolate launched by the Feitoria do Cacao brand, as shown in Figure 2(2), breaks away from the conventional brownish-yellow impression of egg tart products. Instead, it features blue and white Portuguese-style tiles as the main image, and the internal food also adopts the shape of Portuguese-style tiles. This design better reflects Macau's cultural characteristics, showcasing its rich history and cultural heritage, and helps consumers better understand and recognize Macau.

5.3 Folk Elements

Macau is a part of China, and thus, many Chinese folk elements can be seen in Macau's packaging design, such as traditional Chinese patterns and color schemes featuring Chinese elements. For example, the Macau Koi Kei Bakery peanut candy series packaging, as shown in Figure 2(3), features a comprehensive scene depicting the old-time streetside sale of peanut candies and incorporates the Na Tcha Temple, a popular tourist attraction in Macau. The Macau Lei Loi Bakery almond cake packaging , as shown in Figure 2(4), employs the outline drawing technique, combining the image of the Lei Loi Bakery store with the Ruins of St. Paul's and incorporating the store's specialty food products into the illustration, effectively promoting the brand and attracting customers to purchase [9].

Fig. 2. Macau souvenir packaging cases.

6 Visual innovation using artificial intelligence drawing

To verify the elements extracted from Macau souvenir food packaging as mentioned earlier, this study employs the CLIP (Contrastive Language-Image Pre-training) model for element extraction in Macau souvenir food packaging [10]. CLIP is a multimodal model based on contrastive learning, which is a pre-training method or model based on contrasting text-image pairs. In this model, the Text Encoder part uses a text transformer, while the Image Encoder part employs ResNet50 as the foundation for fine-tuning [11]. This approach enriches the categories recognizable by the model, including Portuguese and Chinese architectural design elements as well as relevant cultural symbols and more.

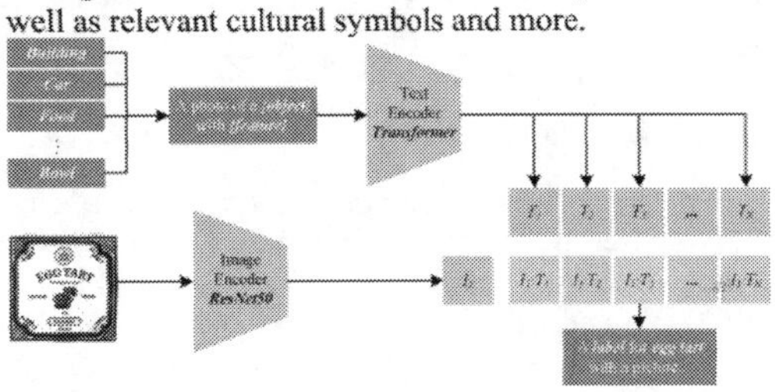

Fig. 3. Flowchart of data for extracting image design elements using CLIP model.

6.1 Design element extraction

After deploying the CLIP model, the output threshold is set to output at least 40 words. The selected souvenir packaging images are extracted sequentially. As shown in Figure 3, the extracted content of the souvenir packaging includes: a label for egg tart with a picture of a bowl of eggs on it and a blue ribbon around the bottom of the label, and a red ribbon around the bottom of the label, Art & Language, official art, Azulejo, an illustration of, modern European ink painting.

The CLIP model can reproduce the content of the image well, even accurately pointing out the use of Portuguese tiles - Azulejo. Subsequently, the extracted elements are refined and the statistical results are as follows: The most frequent vocabulary: Internationalist style, which is mentioned in 56% of the souvenir packaging. This is followed by elements with local characteristics in the packaging, with 44% using Macau architecture, 28% using Portuguese design elements, and 24% using Chinese elements, for details, see Table 2. From this, it can be concluded that Macau souvenir packaging mainly starts from a modern style, with a high proportion of elements such as Macau architecture and Portuguese style. Iconic Macau buildings, such as the Ruins of St. Paul and Macau Tower, are often reused, and iconic buildings and figures are often combined. Most of the packaging is designed according to a formula: food name + food image + Macau elements, which is consistent with the previous analysis.

Table 2. Statistical table of Macau souvenir packaging design elements.

High-Frequency Words	Percentage	Common Combinations	Corresponding Colors
Internationalism Style	56	-	-
Macau Architecture	44	Characters, Multi-Language	-
Modern Illustration	36	-	-
Portuguese Design	28	Tiles, Regular Shapes	Blue and White
Chinese Elements	24	Chinese Architecture, Line Drawing Style	Red

6.2 Design with guidance

Under the guidance of relevant design elements, this study utilizes the Stable Diffusion model as an efficient design inspiration extractor [12]. With the prompt [Cake label with image of moon cake on top, Portuguese style ceramic tile in the middle, art and language, official art, illustrations of modern European ink painting, Internationalist style, Portuguese design elements], a large number of images were generated. Among the generated images, visually innovative ones were selected for detailed depiction. Figure 4(1) is the concept map directly generated by Stable Diffusion, and Figure 4(2) is the result of the author's secondary processing based on Figure 4(1).

The main color scheme of the image is blue and white, and the primary design style comes from Macau's unique Portuguese-style ceramic tiles. These patterns are full of the beauty of Western classical art. The main visual element is Macau's classic souvenir, the almond cake. The overall design embodies Macau's historical culture and local characteristics, allowing tourists to better understand Macau's history and culture while also bringing new momentum to the development of Macau's tourism industry. At the same time, the effectiveness of the method is validated, making it easy for relevant design practitioners to reproduce and design souvenir packaging that conforms to Macau's characteristics and contemporary trends based on this foundation.

Fig. 4. Design under the guidance of AI.

7 Conclusion

In the context of multicultural communication, the study of visual innovation in Macau souvenir food is a field full of challenges and opportunities. By summarizing the design elements of Macau souvenir packaging and incorporating artificial intelligence for validation, this study has successfully proposed an effective method for visual innovation. This method not only enriches the cultural connotation of Macau souvenir food but also enhances its market competitiveness, making a positive contribution to the dissemination of Macau's culture.

However, this research is just a small beginning in the study of visual innovation in Macau souvenir food. In the future, we still need to continue exploring more cultural elements, more design innovations, and more applications of artificial intelligence to achieve a more comprehensive and in-depth development of Macau souvenir food. We hope that future researchers can make more breakthroughs and contributions in this field, allowing Macau souvenir food to become an even more unique and influential cultural brand.

References

1. A. K. Agariya, A. Johari, H. K. Sharma, U. Chandraul, and D. Singh, "The role of packaging in brand communication," International Journal of Scientific & Engineering Research, vol. 3, no. 2, pp. 1-13, 2012.
2. E. S. Wang, "The influence of visual packaging design on perceived food product quality, value, and brand preference," International journal of retail & distribution management, 2013.
3. L. E. Wells, H. Farley, and G. A. Armstrong, "The importance of packaging design for own-label food brands," International Journal of Retail & Distribution Management, vol. 35, no. 9, pp. 677-690, 2007. doi:10.1108/09590550710773237.
4. J. Tang and T. Wang, "Development and develop strategy of Macau creative tourism souvenirs," in 2013 International Conference on Engineering, Management Science and Innovation (ICEMSI), 2013: IEEE, pp. 1-4.

5. W. H. Kong and T.-Z. Chang, "The role of souvenir shopping in a diversified Macau destination portfolio," Journal of Hospitality Marketing & Management, vol. 21, no. 4, pp. 357-373, 2012.
6. A. L. Brody, B. Bugusu, J. H. Han, C. K. Sand, and T. H. McHugh, "Innovative food packaging solutions," Journal of food science, vol. 73, no. 8, pp. 107-116, 2008.
7. B. Mohebbi, "The art of packaging: An investigation into the role of color in packaging, marketing, and branding," International journal of organizational leadership, vol. 3, pp. 92-102, 2014. doi:10.33844/ijol.2014.60248.
8. B. Rundh, "Packaging design: creating competitive advantage with product packaging," British food journal, vol. 111, no. 9, pp. 988-1002, 2009. doi:10.1108/00070700910992880.
9. U. R. Orth and K. Malkewitz, "Holistic package design and consumer brand impressions," Journal of marketing, vol. 72, no. 3, pp. 64-81, 2008. doi:10.1509/jmkg.72.3.64.
10. A. Radford et al., "Learning transferable visual models from natural language supervision," in International conference on machine learning, 2021: PMLR, pp. 8748-8763.
11. Z. Kong et al., "Automated Maxillofacial Segmentation in Panoramic Dental X-Ray Images Using an Efficient Encoder-Decoder Network," in IEEE Access, vol. 8, pp. 207822-207833, 2020. doi: 10.1109/ACCESS.2020.3037677.
12. R. Rombach, A. Blattmann, D. Lorenz, P. Esser, and B. Ommer, "High-resolution image synthesis with latent diffusion models," in Proceedings of the IEEE/CVF Conference on Computer Vision and Pattern Recognition, 2022, pp. 10684-10695.

Speech Signal Processing and Analysis Based on Mongolian long-tune Folk Songs

Sile Hu[1], Shu Liu[2], Shuzhen Ma[2], Axu Hu[2*]

[1] Inner Mongolia Preschool Education College for The Nationalities, Ordos, Inner Mongolia;

[2] Key Laboratory of China's Ethnic Languages and Information Technology of Ministry of Education, Chinese National Information Technology Research Institute, Northwest Minzu University, Lanzhou China

Abstract: This paper mainly analyzes the speech signal of Mongolian long-tune folk songs through the method of speech acoustic analysis. From the perspective of duration, it studies the changes in the length of the passages, sentences and vocabulary of Mongolian long-tune folk songs, discusses the phonetic and acoustic characteristics of Mongolian long-tune folk songs, provides basic data for the protection and digital inheritance of oral culture, and explores the establishment of oral culture database.

1 INTRODUCTION

Both Chinese national vocal music and western bel canto pay attention to the important role of breath in singing. Mongolian long-tune folk songs have the characteristics of wide range, long rhythm, huge structure and diverse singing methods, so it is very important to regulate the breath during lengthening. It can be said that breath is the lifeline of Mongolian long-tune folk songs. No wonder that Mongolian long-tune masters Hazab and Baoyin Deger both attach great importance to the special role of breath in the singing of long-tune folk songs [1]. Hazab believes that the interpretation of breath is mainly realized by breathing, and the most basic feature of Mongolian long-tune is "long". Abundant breath is the basic skill of good "long" interpretation. Aiming at the problem of breath, Hazab put forward the breath theory of long-tune folk song singing - the theory of "three gases connected". He believed that only by correctly using the "qi of saddle bridge", "qi of belt" and "qi of chest buckle", can a good long-tune folk song be performed. From the perspective of methods and means, the current international research on singing acoustics shows a trend of close integration with modern science and technology [2]. From the perspective of the unit nature of researchers, most of them belong to computer, singing performance, language communication, electronic engineering, psychology and other disciplines.

The definition of Mongolian long-tune mainly comes from three aspects [3]: first, the popular concept used by the people; Second, the concept put forward by the older generation of long-tune singing artists; The third is the new concepts and technical terms proposed by music theorists. Among them, on the basis of inheriting the unique singing method of long-tune folk songs, Hazab, an outstanding artist of long-tune folk songs, proposed a set of appropriate concepts and terms, and applied them to his singing and teaching activities, which promoted the development of long-tune folk song singing art towards systematization, standardization and nationalization.

The oral culture of any nation has been gradually formed in the long river of human evolution for millions of years. Therefore, the process of the emergence and development of this "cultural species" is no less than the formation of natural species, and it is the precious wealth of human wisdom. The extinction of natural species should be protected mainly by establishing seed banks and gene banks. However, there is no consensus on the protection of such "cultural species" as Mongolian long-tune in the world. How to effectively rescue and protect our national vocal language and oral culture is an urgent problem we face. In 2005, Mongolian long-tune was selected into the world-class intangible cultural heritage list, and the research on Mongolian long-tune has become a worldwide topic.

At present, the research on long-tune mainly focuses on music theory, artistic expression, historical development, cultural connotation, performance methods, etc., with rich research contents and fields [4]. However, due to the limitation of research technology and means, there is little research on the production principle and pronunciation mechanism of Mongolian long-tune, which is the basis for revealing the physiological mechanism of emotional expression of long-tune [2]. With the help of modern phonetics methods based on medicine, computer and language science, a number of physiological acoustic parameters of pronunciation are extracted, an oral cultural parameter model is established, the relationship between parameters and pronunciation actions is analyzed, and the mechanism of voice production is explored. This provides new inspiration for the research and protection of Mongolian long-tune and

* Corresponding author: 125397932@qq.com

other oral cultures, and also finds new applications for the accumulation of modern phonetics research for many years. It can deepen the theory of speech production and open up a new field of oral culture research and protection.

The research on the noumenon of Mongolian long-tune mainly includes the following categories: the first category is the collection and induction of Mongolian long-tune, the second category is the research on the historical origin and generation context of long-tune, the third category is the research on the noumenon form of long tune music, the fourth category is the research on the singing art of long-tune, the fifth category is the aesthetic research on long-tune, the seventh category is the research on the music culture of long-tune, and the eighth category is the interdisciplinary research on long-tune. These works provide cultural accumulation for the research on the mechanism of long-tune [5]. Therefore, the courseware uses modern instruments to analyze data, and uses the methods of linguistics and other disciplines to explore the physiological performance of the vocal skills of Mongolian long-tune folk songs. This multi-disciplinary and multi-modal research will be a new research trend in the field of long-tune research in recent years.

2 METHODS

The experimental design is to complete the selection, recording and segmentation of corpus in the pre-processing stage, clarify the rules for file naming and preservation, and establish a database. Save the recorded voice files before analysis, classify them according to different themes, and save them in the same folder to form a corpus. Then use the cooledit software to segment the recording results. Divide the whole song into sections, sentences, and vocabulary to prepare for later analysis and research. There are 2 speakers, Ao Erigeji Dema is a female, one of the three national long-tune inheritors in Alxa; Mao Bayier is a male, teacher of the Alxa long-tune folk song inheritance base.

3 ANALYSIS OF CAROL DURATION

The first paragraph of the male and female vocal carols "Fu", "Xu" and "Jiang" was selected, and the duration of the first paragraph of the three Mongolian long-tune was calculated in the praat speech analysis software.

From Table 1, it can be concluded that the first paragraph of "Jiang" in the Mongolian long-tune consists of four sentences, and the female vocal carol is the longest, because each sentence of the Mongolian long-tune has drag tune, nogula and falsetto singing. The order of duration of male voices is the same as that of female voices. However, compared with the female Mongolian long-tune "Jiang", the male Mongolian long-tune "Jiang" is longer, because the male Mongolian long-tune is generally more vigorous and stable.

Table 1. Statistics on the duration of the first segment of male and female carols.

Theme	Long-tune name	Female voice duration	Male voice duration
Carol	Xu	66.63	86.10
	Fu	52.81	71.86
	Jiang	87.47	153.87

3.1 Length analysis of carol sentences

The breath is related to the strength, speed, continuous and relaxation of the song. Therefore, it requires the ability to inhale, "knot", connect, and relax. According to the needs of the song content and the length of the sentence, inhale or exhale. breath. In singing, the breath must form a strong support point in order to control the uniformity of the breath and use the change of strength freely, that is, the abdominal breathing method and the so-called "Dantian Qi method" [3]. From Table 2, in terms of the sentence duration of male and female carols, the three Mongolian long-tune sentences of male voices are all longer than those of female voices. The length of each sentence is consistent with the length of each Mongolian long-tune. When analyzing the length of the passage, it is found that the male Mongolian long-tune is longer than the female Mongolian long-tune, because the breath of the male voice is thicker and calmer than that of the female voice. The sentence length of each carol is very stable, usually only a few seconds apart. The length of the seventh sentence of the female Mongolian long-tune "Xu" is not very different from the length of the previous six sentences, but the seventh sentence of the male Mongolian long-tune "Xu" is the same as There is a big difference in the length of the first six sentences, which shows that the male voice skillfully uses the dragging tone at the end of the last sentence of the long-tune.

Table 2. Comparison of the sentence lengths of male and female vocal carols

Long-tune name	Sentence	Female voice duration	Male voice duration
Fu	first sentence	33.86	44.42
	second sentence	32.28	40.79
Xu	first sentence	6.88	7.39

https://doi.org/10.1051/shsconf/202316702006

Long-tune name	Sentence	Female voice duration	Male voice duration
Fu	first sentence	33.86	44.42
	second sentence	32.28	40.79
	second sentence	6.31	6.75
	third sentence	6.30	6.12
	fourth sentence	6.30	6.77
	fifth sentence	5.26	5.85
	sixth sentence	6.50	7.26
	seventh sentence	12.86	27.52
Jiang	first sentence	17.42	39.88
	second sentence	22.81	38.09
	third sentence	19.37	36.08
	fourth sentence	23.32	36.41

3.2 Analysis of carol vocabulary duration

The word is the most basic grammatical unit in the Mongolian long-tune, and it has its own prosody characteristics, so the stress position of the word plays a very important role in the rhythm of singing [4]. In terms of the analysis of the duration of carol vocabulary, the first paragraphs of the three carols were counted, and a comparison table of duration was listed, as shown in Table 3.

From Table 3, it can be concluded that the male Mongolian long-tune and the female Mongolian long-tune cannot correspond to individual vocabulary, which is due to the fact that the Mongolian long-tune is sometimes improvised, and the lyrics mastered by the two singers are not the same, but the tune is exactly the same. For example, there are four or five words in the first and second sentences of the female long Mongolian tone "Fu", while the first and second sentences of the male Mongolian long-tune "Fu" have only three words. From the statistics of the length of time, it can be concluded that the vocabulary length of the male Mongolian long-tune is longer than that of the female Mongolian long-tune. The length of the Mongolian long-tune vocabulary of male and female voices can basically correspond, but the seventh word of female voice is only 3.49s, while the seventh word of male voice is 20.36s, because the seventh word of male voice is the end of the second sentence, so use the arrival time to Nogula is longer. When the vocal cords vibrate due to the impact of breath during singing, the Adam's apple can be seen vibrating from the outside of the singer's throat. Many of the beautiful "little nogula" (little folds) in Wuritingduo are related to the throbbing of the Adam's apple. "Xiao Zhe" is a variety of pleasing and pleasing folds created and expressed by the movement skills of various organs. "Zhe" means bending and bending, and the Mongolian long-tune refers to the sound pattern of bending into a curved line like a subtle wave, that is, folded sound [5]

4 ANALYSIS OF THE DURATION OF MONGOLIAN LONG-TUNE FOLK SONGS WITH DIFFERENT THEMES

Based on the above methods, this paper selects wedding songs and historical songs to conduct acoustic analysis of three Mongolian long-tune with different themes, and obtains the differences in the acoustic characteristics of Mongolian long-tune with different themes from Table 4.

For each subject matter of male voices, the use of the Mongolian long-tune is longer than that of female voices, which is closely related to the breath of the speaker. The long-tune duration of the three themes is from long to short: carols, wedding songs, and historical songs.

Table 3. Comparison of the vocabulary duration of male and female vocal carols

Long-tune name	Sentences	vocabulary	Female voice duration	Male voice duration
Fu	first sentence	first word	8.48	10.57
		second word	10.13	12.19
		third word	7.48	20.45
		fourth word	7.81	8.09
	second sentence	fifth word	8.74	10.00
		sixth word	8.04	10.08
		seventh word	3.49	20.26
		eighth word	2.98	3.55
		ninth word	8.98	10.44
Xu	first sentence	first word	3.13	3.25
		second word	3.30	4.10

Long-tune name	Sentences	vocabulary	Female voice duration	Male voice duration
Fu	first sentence	first word	8.48	10.57
		second word	10.13	12.19
		third word	7.48	20.45
		fourth word	7.81	8.09
	second sentence	fifth word	8.74	10.00
		sixth word	8.04	10.08
		seventh word	3.49	20.26
		eighth word	2.98	3.55
		ninth word	8.98	10.44
	second sentence	third word	1.59	1.63
		fourth word	1.67	1.68
		fifth word	2.90	3.32
	third sentence	sixth word	1.58	1.58
		seventh word	1.60	1.48
		eighth word	0.73	0.69
		ninth word	2.22	2.12
	fourth sentence	tenth word	1.57	1.58
		eleventh word	1.64	1.63
		twelfth word	3.33	3.38
	fifth sentence	thirteenth word	0.89	0.67
		fourteenth word	0.76	0.66
		fifteenth word	3.44	4.34
	sixth sentence	sixteenth word	1.71	1.61
		seventeenth word	1.70	1.59
		eighteenth word	3.22	3.92
	seventh sentence	nineteenth word	8.32	17.72
		twentieth word	2.12	9.28
		Twenty-first word	2.36	
Jiang	first sentence	first word	2.65	3.81
		second word	7.46	20.47
		third word	5.07	7.19
		fourth word	3.48	7.49
	second sentence	fifth word	4.35	6.52
		sixth word	6.70	9.95
		seventh word	3.76	7.21
		eighth word	7.21	12.92
	third sentence	ninth word	2.07	2.89
		tenth word	9.75	17.02
	fourth sentence	eleventh word	5.01	7.33
		twelfth word	3.78	7.86

Table 4. Statistics on the duration of the first segment of male and female carols

Theme	Quantity	Masterpiece
Wedding song	3	Handsome purplish red horse
Carol	3	The rich and vast Alxa
Historical song	3	Altay Hange

https://doi.org/10.1051/shsconf/202316702006

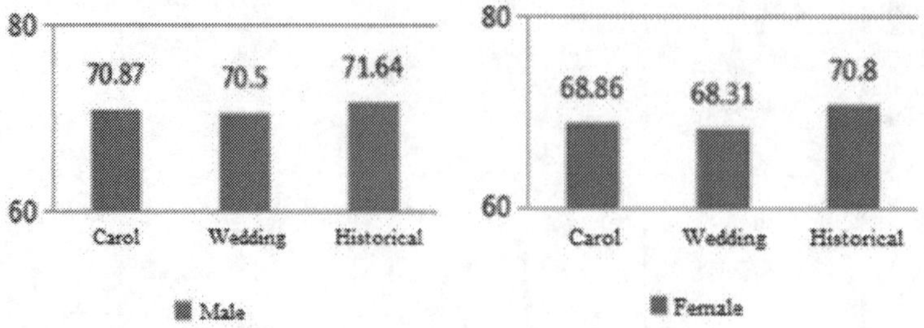

Figure 1. Comparison of the average duration of male and female voice Mongolian long-tune.

5 CONCLUSION

In terms of the length of Mongolian long-tune, the use of Mongolian long-tune for each subject matter of male voices is longer than that of female voices. The three Alxa long-tune with different themes are from long to short: carols, wedding songs, and historical songs. Because the carol is more solemn and solemn, and pays great attention to nogula and falsetto singing, there is a corresponding dragging situation after each sentence. Compared with the other two themes, carols are relatively calm, condensed, slow and long. When singing the carol, both singers were solemn and full of emotion. In terms of the duration of special singing methods, among the three Alxa long-tune with different themes, the one with the largest number of special singing methods and the longest duration is carol.

ACKNOWLEDGMENT

This work was supported by Gansu Provincial Department of Education Innovation Fund for Higher Education Institutions (No.2022A-181), the Gansu Provincial Humanities and social Science special project (No.22ZZ100), the National Social Science Foundation of China (No.Z2210901).

References

1. Hu He, Chen Jiatai, Zheng Yuling. Mongolian prosodic feature acoustic parameter database. Inner Mongolia University Press,2001(1)
2. Wang Bei, Yang Yufang, Lv Shinan. Standard Chinese sentences stressed syllable pitch changing mode study. Acta Acustica, 2002(3).
3. Qingge Ertai. Mongolian grammar. Inner Mongolia people's publishing house,1991(1).
4. Halazhen. Ordos wedding big songs. Inner Mongolia people's publishing house,2011(1).
5. Yu Shanying. Different types of singer resonance peak features singing tone and the mechanism of formation research. Vocal music study,2010.4

The Contemporary Application of Confucius' Being Carefree[1]

Enhui Xing*

Inner Mongolia University of Science and Technology (Baotou, Inner Mongolia Autonomous Region,China 014010,

Abstract: As can be observed from contemporary life, emotions like feeling distressed, anxious or stressful lasts through every stage in life. If these states of mind are not properly managed and regulated, they can lead to anxiety and other emotions, which greatly affect the well-being at all ages, thus spilling over from the individual to family well-being and then even to social harmony. According to research findings in psychology, negotiable views on fate are positively correlated with the perception of social adversity, the escape from psychological distress, and the attainment of happiness[2]. This paper argues that Confucius' thoughts, words and deeds all embody the concept of negotiable fate. In addition, the excellent Chinese traditional culture also contains the wisdom of being carefree, helping people lead a life free from shame or fear due to their sticking to the moral standard and teachings. Confucius' state of "no worries" is undoubtedly a contemporary remedy for confusion and anxiety.

1 Introduction

For the regulation of the psychological suffering, the teachings of Confucius, as part of the outstanding traditional Chinese culture, can provide contemporary references and utility. It can be reflected on the following two aspects: Confucius' philosophy on fate and Confucius' philosophy on being carefree(following one's heart while not exceeding the rules thus achieving the mindset of "no worries"). This paper aims to probe into these two aspects and look for the teachings and philosophy acting as guideline and remedy for contemporary psychological dilemma.

2 The concept of negotiable fate

Drawing on two related studies, the concept of negotiable fate can play an important role in the regulation of psychological suffering and the attainment of happiness. Specifically, Liu Huiying and Hao Xiangsen's study, "The Effects of Perceived Social Adversity, Psychological Distress and Negotiable Destiny on Suicide Risk: A Moderated Mediation Model" demonstrates that perceived social adversity was significantly and positively associated with emotional problems such as anxiety and depression, and negatively associated with feelings of well-being and control. Negotiable fate perceptions can help people gain a sense of control, as well as assign meaning to unexpected outcomes and persist in pursuing goals in undesirable situations. It can enhance an individual's sense of self-efficacy, sense of meaning in life and subjective well-being, maintain psychological fitness, and thus reduces the risk of self suicide[3]. In Chang's research, a scale was investigated and analyzed to conclude that negotiable fate promotes college students' subjective well-being and that the experience of meaning in life plays a part in mediating it[4].

3 Confucius's philosophy on fate

In Chen Manqi's article, it is mentioned that "negotiable fate refers to the belief that individuals can negotiate their own fate by exercising their subjective initiative within the limited resources available[5]. In this sense, Confucius' philosophical wisdom embodies the negotiable view of fate. The concept of negotiable fate is reflected in Confucius' words and deeds. If we can meditate and savor the negotiable view of fate in Confucius' teachings, it will be of significant help to enhance subjective well-being, subside anxiety, and maintain psychological health, family happiness, as well as social harmony.

3.1 To know the destiny given by heaven.

The Analects says, "Without knowing the fate, one cannot be a gentleman", which means You cannot be a gentleman without knowing the regulation of Heaven and the responsibility people are born to take. This is because the regulation of Heaven is the law of historical development. It also refers to the responsibilities and duties that individuals are born with in this world. Without comprehending the regulation of heaven or the given responsibility, one cannot know the trend of the times. As Mr. Yang Bojun paraphrases in his *Translation and Commentary on the Analects of Confucius:* Confucius was not a fatalist, but he also puts emphasis on knowing the fate[6].

*357176331@qq.com

Confucius' view of knowing the fate is still relevant and helpful in contemporary times. From the perspective of nature, all things in this world have their own way and their own laws. From the perspective of human being and society, family and the environment around are both given from birth, without the possibility of being chosen, and each person has his or her own duty and responsibility. On the basis of knowing our destiny and accepting it, Confucius had a higher level of understanding and pursuit, which was not to compromise and not to give up, and to live in the present.

3.2 Trying our best.

Knowing the destiny does not mean to do nothing or not trying hard. Rather, it means perseverance with a clear objective in mind and trying hard to make the best on the basis of the fate given. This is just another example illustrating negotiable fate. He did not compromise, did not give up, and lived in the present. In *The Analects*, it is recorded that Confucius once said, "When building a mountain out of soil, people may choose to stop and give up, although it's almost done. When filling up the low-lying land, people may choose to stick to it, although it's just the beginning of a tough course." Confucius uses the piling of mountains and the filling up of low-lying land to teach his students about the merit of perseverance. According to Mr. Yang Bojun's comprehension, It would be one's determination that helps in striving forward and finally succeed[7]. The perseverance of Confucius is evident in this example. Another example is the description of Confucius' experience and mentality in Mr. Ma Yong's book *Three Thousand Years of Chinese Confucianism*: "From the day Confucius returned to Lu until the day of his death, he was not stuck in the state of feeling depressed by his political disillusionment. Instead, he devoted himself to compiling ancient literature and training his disciples with the diligence of 'not knowing that old age was approaching', and the result was not only the formation of a basic theoretical framework and talent pool for the development of Confucianism, but also the setting up of a fairly complete set of wisdom resources for the future development of Chinese society"[8].

From what has been presented above, we may clearly feel that Confucius' words and actions well illustrate his spirit and philosophy of accepting fate and keep perseverance. But perseverance does not imply that there is only one path through which people can step toward their aims. Just as the following part illustrates.

3.3 Confucius expressed his understanding on a real gentleman by saying that "Gentlemen are not limited to their expertise" and chose to make changes according to the development of the situation in reality.

Coming back to *The Analects*, "When building a mountain out of soil, people may choose to stop and give up, although it's almost done. When filling up the low-lying land, people may still choose to stick to it, although it's just the beginning of a tough course." In his translation of *The Analects*, Mr. Yang Bojun adds to his understanding of this sentence that the chapter can also be explained in this way: "It is like piling up a mountain of earth, when almost done, if I (should) stop, I will. If (it should) be kept forward, then I will go forward[9]."According to Mr. Yang Bojun's understanding, we can consider that the principle of stopping and continuing here is based on "only righteousness". With the passage of time, changing in space or shifts in circumstances, one makes flexible and appropriate choices. But this does not mean a wavering of will or a lack of firmness on principle. On the contrary, as can be seen from Confucius' view of loyalty and forgiveness, such flexibility is based on a firm adherence to principle. It is a negotiable philosophy toward life, and it helps in accepting and adapting to the conditions. This is what we are still proud of and benefit from today. It is perhaps in this state of acceptance and adaptation that helps Confucius achieve the mentality of "no worries".

4 Confucius' philosophy on being carefree: following one's heart while not exceeding the rules

Confucius' attitude toward the given state of life can be fully interpreted as an exercise of subjective initiative in a negotiable view of destiny. As the article *Fate attribution, acquisition of sense of control and the functions of psychological adjustment in Confucian culture* suggests, the Confucian concept of negotiable fate makes it easier for individuals to make flexible attributions, increase their sense of control, and thus cope with major life events and realize psychological adjustment[10].The philosophy and method of "dialectical view on destiny" can be still applied in contemporary times. While functioning as a mental regulation, it also helps in finding out the inner voice and develop our subjective initiative. By constantly being motivated in a given living conditions and bravely facing the constant changes, the individual seeks meaning while at the same time contributing to the development and progress of society through the realization of their own values.

Holding the above-mentioned view of Confucius on fate, in the face of uncontrollable destiny, one adheres to principles and relentless pursuit in order to realize one's mission. While At the same time, this is only part of Confucius' philosophy. Confucius also pursued a carefree state of "following one's heart and not exceeding one's rules".How could we achieve the state of carefree, or at least making improvements towards that mindset?

Confucius gives us a hint in *The Analects*:

4.1 To enjoy and thus being away from worries

4.1.1 Acceptance of all that has been given to life with joy

It is written in *The Analects* that Zigong once said, "What is it like to be poor but without flattery and rich but without pride?" Confucius replied, "It is acceptable, but still not as good as those who are poor but happy, and rich but well-mannered." it is the philosophy guiding us to be happy with what we are and what we are given, whether being poor or rich. When talking about Yanhui, Confucius comments, "The virtuous Hui is also a man who lives really simple life but does not change his happiness." Even in the midst of poverty and hardship, Yan Hui was still happy and eager to learn. Perhaps it was because he was happy with himself, and he had the inner peace of mind and could then enjoy the present moment and pursue its meaning.

Such contentment and concentration on the present saves us from the pain of restlessly seeking but not being able to get. What is there to worry about when you are in the company of joy? This is also true in contemporary circumstances: not to complain, and not to worry too much. Be still happy when you are poor, and be always polite although you are rich. Is this not a negotiable view of destiny? Such a state of mind and action is also acting on the change of destiny through the exercise of subjective initiative. This is just the contemporary use of Confucius' wisdom.

4.1.2 Savoring happiness from engaging in the present:

Material pleasures are always limited and fleeting, but the pleasure from spiritual pursuits is worth savoring. Confucius was happy to devote himself to what he loved, and he admired those of his disciples of the same kind. In the Analects, Confucius says of himself, "He is a man who has become indignant and forgotten his food, and who is happy to forget his worries, not knowing that old age is approaching." His pursuit of learning and concentration makes him even forget to eat and forget to worry. The pleasure and joy that comes from these spiritual pursuits is the source of happiness. Through the relentless refinement of what he loves, and the joy of doing it, one can naturally achieve the realization of self-worth, which is also a source of happiness, as well as a source of meaning and fulfillment.

Being engaged in the pursuit of one's loved career and having no spare time and energy to be distracted, thus enjoying the ultimate happiness which does not depend on anything outside. Focusing on the present moment brings a sense of happiness and being real, and naturally, there will be no place at all for anxiety or helplessness. Such concentration also undoubtedly leads to self-improvement, which will help to grasp new opportunities and thus realize negotiation with destiny. This is precisely what Confucius' "happiness without anxiety" teaches us in our lives today.

4.2 Self-cultivation helps bring us away from worries.

As Mr. Huang Renyu says in *Confucius and Mencius*, "Confucius, as recounted in *the Analects*, has a sense of relaxation..." And such a relaxed state and realm of Confucius stems from Confucius' wisdom and philosophy. We savor the words and actions of Confucius in order to try to get closer to such inner quietness and openness[11].

4.2.1 A gentleman seeks help from himself rather than blaming the outside world or fighting for rights.

The Analects states, "A gentleman demands of himself, whereas a normal person demands of others." In fact, it is the constant demand on oneself that enables self-cultivation, progress and improvement. Instead of thinking that others are bad, it may be a good way to gradually correct ourselves. To learn and make improvements, we are supposed to begin by correcting and cultivating ourselves. In *The Book of Rites*, there is a saying : "When a gentleman misses a shot, he should seek his own fault. You are not supposed to blame the target, rather, your own archery skills are to be improved. Actually rather than blaming, we should understand it as finding the internal cause and starting to act on it.

Meanwhile, by looking for flaw from ourselves, rather than blaming others, we are already free from worries and resentment toward others. As The Analects states, "If you are strict to yourself and rarely blame others, you will be far from resentment." If one can look to oneself and ask for more from themselves and less from the outside, one will be away from worries, resentment, and grievances. The Rites also has statements about being free from worry and resentment: "If one can correct himself often and does not ask from others, there will be no resentment." This kind of peace and free from resentment may be the reward of inner peace and freedom from worry. If you do not feel guilty, why should you worry or fear?

The perception of desire in Chinese traditional culture, especially Confucianism, is still of great relevance to our times. While the lack of purpose is obviously not conducive to development and progress, if it comes to the other end, the endless pursuit and desire for fame and fortune can lead to serious psychological imbalance and misfortune. This is why "not be pushy" is the key. Not seeking, not forcing, not complaining and not blaming, can help contemporary people get rid of anxiety, focus on the present moment and get out of their psychological difficulties. The power of "not be pushy" can bring inner peace and calmness, and reduce the pain of seeking but still not being able to obtain. While this may also help in concentrating on the present and thus making achievements. Conversely, when one chooses to take risks in the constant pursuit desires, it proves to have undesirable consequences in the long run.

4.2.2 Moderation and adherence to principles bring the mindset of carefree.

To be moderate means to be neither redundant nor deficient. Mr. Yang Bojun said, "Being moderate is the highest moral standard of Confucius"[12].According to the Analects, Confucius once said, "To be moderate is a true virtue, and it is the highest standard! The people have not had it for a long time." However, being moderate is not a state swaying from side to side without adherence to principles. Mr Wang Meng has the following interpretation of being moderate: "In modern life, the word 'moderate' even makes people think of being a smoothy. The reason might be that the state of being moderate is one that is neither radical nor manic, and not to the extreme as well...But it is also the firm adherence to principle[13].

As has been mentioned above, being moderate is the adherence to principles that are just right. So how can adherence to principle make one free from worries? As *The Analects* states, "Wealth and nobility are the desires of men; if they are not obtained in the right way, I will not accept it. Poverty and lowliness are the abominations of men; if we do not get rid of them in a righteous way, I would prefer not change the present situation." So to become rich and be an official is what people want, but if it is not obtained in the right way, it is also unacceptable to a gentleman. By the same token, a gentleman will not get out of a situation like poverty or lowliness, which everyone avoids, if he does not get out of it in the right way. Here the principle is benevolence and righteousness. So a gentleman loves his money and takes it in the right way. This is how inner peace and freedom comes- from adhering to principles.

This is also true today. By being moderate and sticking to principles, there will be much less distress or moral condemnation in the mind, So we can understand that if we stick to our principles, then we have no worries. It is just a decision made by ourselves. With such moral self-consciousness and subjective initiative, it is natural to be frank and feel at ease. Conversely, if one loses the blessing of benevolence and morality, it is naturally detrimental to both the physical and mental well-being and happiness.

5 Conclusion

Confucius' practice of negotiable fate and the carefree state of mind remains a contemporary treasure, guiding us in our contemporary lives to the joy of being free from worry by being free from guilt. It can also be an effective solution to the contemporary experience of pain from daily frustration. This philosophy is conducive to the cognitive development of the individual and to the experience of freedom from worries, which in turn enhances one's sense of well-being and facilitates harmonious coexistence. In contemporary society, such perseverance and pursuit, acceptance and empowerment can reduce the social pressure on the individual, and at the same time relieves personal suffering. On the other hand, we will be able to devote to a cause that interests us, fulfilling our mission and at the same time making contribution to the society.

References

1. Fund: Inner Mongolia Autonomous Region Higher Education Society 2021-2022 Key Project of English Teaching Reform in Inner Mongolia Autonomous Region Universities (Project No. SZ2021001-A)
2. Liu, Hui-Ying, Hao, Xiang-Sen. The effects of perceived social adversity, psychological distress and negotiable fate on suicide risk: a moderated mediation model [J]. Chinese Journal of Clinical Psychology Vol. 30 No. 4 2022:954-958.
3. Liu, Hui-Ying, Hao, Xiang-Sen. The effects of perceived social adversity, psychological distress, and negotiable fate perceptions on suicide risk: a moderated mediation model [J]. Chinese Journal of Clinical Psychology Vol. 30 No. 4 2022:954-958; pp. 954.
4. Chang, Paul R., Bai, Baoyu, and Zhong, N. The influence of negotiable fate on subjective well-being: the mediating role of the sense of meaning in life [J].Chinese Journal of Clinical Psychology Vol.25 No.4 2017 p.724.
5. Chen Manqi. The relationship between negotiable fate and future expectations - the mediating role of emotional experience and the moderating role of collectivist perceptions of social reality [J]. Journal of Fujian Normal University (Philosophy and Social Sciences Edition) No. 2, 2016 (General, No. 197) Journal of Fujian Normal University (Philosophy and Social Sciences Edition) General, No. 197 p. 167
6. Yang Bojun. The Analects of Confucius: A Translation and Commentary [M]. Beijing: China Book Bureau, 2021. p. 17
7. Yang Bojun. The Analects of Confucius: A Translation and Commentary [M]. Beijing: China Book Bureau, 2021. p. 133
8.Ma Yong. Three Thousand Years of Confucianism in China [M]. Guiyang: Confucius Hall Book Bureau. 2022. p. 35.
9.Yang Bojun. The Analects of Confucius: A Translation and Commentary [M]. Beijing: China Book Bureau, 2021. p. 133
10.[Lou Bing. Fate attribution, acquisition of sense of control and the functions of psychological adjustment in Confucian culture [D]. 2019. p. 4.
11.Yang Bojun. Translation and commentary on the Analects of Confucius [M]. Beijing: China Book Bureau, 2021. p. 236.
12. Yang Bojun. Translation and commentary on the Analects of Confucius [M]. Beijing: China Book Bureau, 2021. p. 92
13.Wang Meng. A Brief Interprtation of the Analects [M]. Beijing: Beijing United Publishing Company, 2019. p.4.
14. Wang Meng. A Brief Interprtation of the Analects [M]. Beijing: Beijing United Publishing Company, 2019. p. 242."

A Comparative Study of Chinese and Japanese Folk Culture

LingLi [1,a], CuiNa [2,b*]

[1] Shaanxi University of Chinese Medicine, Xianyang, Shaanxi, China

[2] Shaanxi University of Chinese Medicine, Xianyang, Shaanxi, China

Abstract. China and Japan are close neighbors who belong to the same East Asian Chinese character culture system, and have many similarities in language and folk culture. This thesis takes folk culture as a research perspective, and compares the similarities and differences between Chinese and Japanese cultures in terms of food culture and taboo culture respectively. Firstly, the article explains the definition scope and classification criteria of folk culture; secondly, the two points of food culture and taboo culture are selected for comparison from material folklore and spiritual folklore. Food culture focuses on table manners, cooking standards and eating habits, and analyzes the reasons for the formation of food culture from three aspects: economic development, geographical environment and social culture. Finally, the scope of taboo culture is large, so this paper selects two representative differences in numerical taboos and color taboos for discussion, showing the preference of Chinese and Japanese folklore. Finally, a short summary of the significance of folk culture for cross-cultural communication is presented.

1. Introduction

China and Japan have had close cultural ties since ancient times, and both countries belong to the Confucianism cultural circle and the Chinese character culture circle. When President Xi Jinping met with the Japanese delegation to China, he said that the foundation of Sino-Japanese friendship lies in the people, and it is necessary to enhance civil exchanges. When meeting with Japanese Prime Minister Fumio Kishida, he proposed that the two sides should continue to take advantage of the unique advantages of geographical proximity and humanistic ties to carry out exchanges. Therefore, this paper compares the folk cultures of China and Japan as comparative countries. Additionally, the exchange of folk culture is not only a practical need for national development, but also a basic need for cross-cultural communication, which is conducive to reducing friction and conflict between the two sides of communication. Not only that, culture is the unique spiritual symbol of a country and nation, and the teaching of culture class is essential in Chinese international education. At the same time, folk culture, as the cornerstone of the national cultural system, is good for foreign students to know and understand Chinese culture and heritage more rationally and objectively, and to feel the charm of Chinese traditional culture. For teachers, the teaching of folk culture is conducive to the complete and accurate promotion and dissemination of the excellent Chinese national culture, and truly realizes the practical significance of teaching Chinese as a foreign language.

2. Definition and classification of folk culture

The concept of folk culture in China has appeared as early as the Western Zhou Dynasty, which is described in *The Book of Rites*: "Therefore, the ruler of the people, the chapter of good to show the people's customs"; *the Book of History*: "Chu folk customs, prefer to ride in a short car"; *the Book of Han*: "Changing folkways, transforming folk customs". In 1846, the British scholar Thomas formally introduced the concept of "folk" on the basis of Saxon, and combined "folk" (people, folk) with "lore" (knowledge, learning). Which is not only a summary of folk customs, but also refers to the knowledge of the study of this phenomenon. [1]

Folklore has developed rapidly in recent times, and scholars from different countries have different perceptions of folklore. Yanagida Kuni, the founder of Japanese folklore, divided folklore into three major categories: tangible culture, language arts and mindfulness phenomena. In addition, by combining linguistics with the knowledge structure of folklore, "olk culture is used as an entry point to under-stand and grasp the language of a country." [2] Folklorist Zhong Jingwen points out that folklore is the living culture created, enjoyed and inherited by the masses of people, which is a collective, inherited and patterned folk culture, and its promotion and transmission are based on oral transmission, behavioral demonstration and psychological influence. He divides folk culture into four categories: material folklore, social

[a] dorialingling@163.com
[*] Corresponding author: [b] 479231169@qq.com

folklore, spiritual folklore, and linguistic folklore. Material folklore includes production folklore, food folklore, clothing folklore, housing folklore and architectural folklore, which occupy an important position in the material and spiritual life of each nationality and can satisfy people's physical needs as well as higher-level needs, including security, belonging, self-esteem and self-fulfillment. Spiritual folklore is a kind of folklore about ideology developed from material and institutional culture. It consists of beliefs, taboos and rituals, and is a psychological experience generated by people in the process of understanding and transforming nature and society. Once this experience becomes a collective psychological habit and is passed down from generation to generation in the form of specific behaviors, it be-comes one of spiritual folklore. Since folk culture involves a lot of contents, this paper selects two branches of food and taboos for detailed discussion to compare the differences between Chinese and Japanese folk cultures. [3]

3. Chinese and Japanese folk food culture

3.1. Manifestation of food culture differences

3.1.1. Table manners.

Table manners are the essence of food culture, and familiarity with table manners is a necessary quality for a mature cross-cultural communicator. In terms of chopsticks, Japanese chopsticks are much shorter than Chinese chopsticks. Because of the meal sharing system, each person has his or her own dish to eat without helping others. Chopsticks are placed horizontally before eating, and the heads of chopsticks are pointed because Japanese people love to eat fish, so it is easier to separate the fish from the spines by using chopsticks with pointed heads. [5] Also, the owners of chopsticks in Japanese households are fixed and cannot be mixed. Ryozo Sugimoto once mentioned, "The usage of Japanese tableware shows that Japan is a country of individualism." In China, chopsticks are placed vertically and long chop-sticks are used to hold dishes between each other, representing hospitality. And in families chopsticks are communal and do not need to be distinguished. [6] When ordering a meal, Japanese people usually finish all the food on the table as a courtesy, which is also the source of Japanese frugality. The Japanese also allow "wolf and tiger" eating [7] to show their enjoyment of food and praise to the chef. The Chinese, on the other hand, do not want their guests to finish their meals, as it is considered rude to run out of food. At the Chinese family dinner table, "Eat without saying anything, sleep without speaking." has been the practice of parents in raising their children.

3.1.2. Cooking standards.

In Chinese food culture, the standard of cuisine is to pursue the full range of Aroma, color and taste, and it is believed that food needs to be cooked carefully to truly realize its value, so more than 20 different cooking methods such as "frying, pan-frying, deep-frying and steaming" have been produced. People in different regions have also developed local food characteristics in different environments, resulting in the eight major cuisines in China. In *the Analects of Confucius*, the saying "The more delicate the grain is, the better, and the thinner the fish is cut, the better." is a high summary of our food culture. It means to describe a very careful diet. In Japanese food concept, cooking will destroy the original nutritional value and flavor of food, so there are five ways of Japanese cuisine: raw, boiled, grilled, fried, and steamed. Among them, the most widely used method is "boiling", which forms one of the characteristics of Japanese food: lightness. [4] In order to pursue the original taste of food, Japanese people also use limited seasonings, mostly soy sauce, and according to the geographic location of Japan, it extends in an arc from the northeast to the southwest, forming two major culinary systems, Kanto cuisine and Kansai cuisine, both of which prefer soy sauce as the main ingredient.

3.1.3. Eating habits.

The Chinese have long been influenced by the five elements of yin and yang and Traditional Chinese Medicine, and believe that the five internal organs and six internal organs of the human body are interconnected. [8]Therefore, our body need to be supplemented with warm food, i.e., "eating the five grains does not make you sick". The staple food in the traditional Chinese diet is rice and wheat, and the side dishes are mainly vegetables and meat. Through production, the staple food also consists of steamed buns and noodles. Japan likes raw food and cold food, and its diet structure and content is mainly fast. The main food is mainly rice, such as sushi and rice balls, and the side dishes are mainly vegetables and fish, with less meat. The most famous sashimi in Japanese cuisine highlights the raw and cold nature of the Japanese diet, and its most common ingredient is tuna. [9]

3.2. Reasons for the difference in food culture

3.2.1. Economic development.

Food culture can, to some extent, reflect the level of development of a country's productivity, with a material basis for the existence of food culture. Beijing Ape Man began to fire roasting stone baking in history, as the source of Chinese cooked food culture; the cavemen caught fish and shrimp; the Hemudu People planted grains; the Han Dynasty introduced cooking methods; the Tang, Song and Ming and Qing dynasties food culture flourished. Chinese food culture has lasted for more than 1.7 million years and continued to mature with the development of the ancient economy. Japan was less civilized in the primitive society period, and was in the tribal stage in the 3rd century A.D. In the 7th century A.D., Chinese food culture was introduced by Japanese envoys and foreign students, and its food culture began to develop gradually. In addition, the Japanese Emperor Tenmu issued a "meat ban" in 676

A.D. [10], so the range of ingredients was somewhat restricted.

3.2.2. Geographical environment.

Folklore has strong selectivity and adaptability to the natural geographic environment. There are differences in geography and environment, and naturally there are differences in food and folklore. China's vast territory and diverse climate are suitable for the growth of different plants and animals, making the variety of source ingredients diverse. Chinese people take advantage of the excellent conditions, "rely on mountains to eat mountains, rely on water to eat water", according to local conditions, there are different types of local cuisine, for example: Sichuan cui-sine, Cantonese cuisine, Huaiyang cuisine, etc.. Japan is an island country, consisting of more than 4,000 islands, so it has abundant fishery resources and relies on water to eat, and its diet is rich in seafood. The scarcity of resources has led to a strong sense of crisis and the formation of a custom that favors lightness and freshness.

3.2.3. Social culture.

Social culture is also an important reason for the difference in diet. In traditional Chinese culture, emphasis is placed on the five elements of yin and yang, the doctrine of Chinese medicine and health, and attention is paid to the "harmony" in cooking. During the Spring and Autumn Period, Yan Ying took cooking as an example: "Harmony is like making a meat soup, cooking fish and meat with water, fire, vinegar, sauce, salt and plum, and cooking them over a wood fire. The cook mixes the flavors so that the various flavors are just right; if the flavors are not enough, the seasonings are added, and if the flavors are too heavy, the seasonings are reduced. The gentleman eats this kind of meat soup, used to calm his mind" Because of the diversity of the social environment, the Chinese food culture is determined by "harmony". Whether stir-fried, fried, steamed, or boiled, the ultimate goal is to bring out the best in Aroma, color and taste. To the Heian Period in Japan, Japanese food culture was gradually formed on the basis of Chinese food civilization. After the Meiji Restoration, Western culture was introduced to Japan, which improved the overall quality of Japanese food culture, combining the beauty of both Eastern and Western cultures. Later on, the Japanese revered the habit of frugality and rarely expanded on food culture. Therefore the development of the cuisine was limited.

3.3. Taboo Culture

3.3.1. Definition of taboo.

Taboo, known as "tabu" in the fields of ethnography, folklore and anthropology. The specific meaning of "tabu" is the prohibition and restraint of thought and action in order to prevent the incurring of blame and calamity through thought and action. The scholar Wan Jianzhong, in his book *Taboos and Chinese Culture*, once mentioned that there are three interconnected key points in the understanding of the basic characteristics of taboos, namely, that they belong to negative behavioral norms, that is, they are irresistible to the punishment of national beliefs and taboos at the psycho-social level [11]. This paper focuses on the differences between Chinese and Japanese folk cultures with a focus on numerical taboos and color taboos.

3.3.2. Number taboos.

In China, "four" is an inauspicious number because of its harmonic sound of "death"; "seven" is also disfavored because of its harmonic sound of "qi". "Six" and "eight" are both very popular because "six" is a sign of success and "eight" is a sign of wealth. On special days, the Chinese usually choose even numbers to avoid suffering from widowhood. The cultural connotation of "two" is richer, and is regarded by Han Chinese as the meaning of pairs, while Manchu and Korean people taboo the funeral on even days to avoid bad things in pairs; Guangzhou people taboo "three", because its harmonic sound is "scattered ", the locals will replace three points with two sixty points.

In Japan, the appearance of "four" and "nine", will make the public very sensitive because of the resonance of "death" and "suffering". Therefore, there is no number four, nine, fourteen, twenty four in Japanese hospitals. And there is no number four in parking lots and hotels, but there is a number nine room in prisons, not a number four room, in order to punish prisoners for their pain but not for their death. When giving gifts, Japanese people prefer odd numbers, especially "three" and "seven". The legend of the seven gods of good fortune has been passed down in Japanese folklore, and its origins come from various religious sects, including Shintoism, Buddhism, and Taoism. The number "three" is both an auspicious Christian number and a number considered by Pythagoras to have reached a new unity beyond one and two. In *the Japanese Chronicle of Ancient Matters* and *the Japanese Book of Records*, it is written that the three generations of Amaterasu Omikami's royal grandson came to Japan with three divine weapons, the most frequent of which is number "three".

3.3.3. Color taboos.

Chinese people like red, red symbolizes joy, good luck, folk marriages revered red, the bride wears red wedding clothes. Red can drive away evil spirits and attract good fortune, and in the year of birth, wear more red clothes to avoid disaster. Red symbolizes the development of smooth, "red people" "red list" "red luck" and so on all indicate positive development. Chinese people do not like white, white represents the color of evil funeral, after the death of a loved one to do "White Ceremony" for it, wearing mourning. White symbolizes failure and shallow knowledge. There are white flags for losing wars, "white papers" in exams, and white faces in operas. Nowadays, due to the influence of Western culture, white elements are becoming popular in China.

The Japanese like white, which is light and elegant, symbolizing purity. White appears most frequently in Masamune Atsuo's *Manyeki*. The aesthetic sense of ancient Japan centered on snow, moon and flowers. In the traditional Japanese wedding ceremony, the bride must wear a white dress and the guests must wear a white tie. White is a symbol of good luck, and it is written in *The Chronicle of Japanese Books* that Emperor Takatoku and Emperor Gwangin renamed their years "Hakusei" and "Hakugame". [12] In Japanese, the word "bare or loyal" is generally used to represent the color red, which represents a mysterious color and has more of a negative meaning. In traditional Japanese songs and dances, white represents decency, red represents villainy, and the opposite is true for Chinese face painting. In summary, China reveres the color red, which symbolizes positive meaning. White represents a more negative meaning. The Japanese attach more importance to white, symbolizing purity and innocence, while red mainly represents mystical ritual colors and festivity, but to a lesser extent than in China.

4. Conclusions

Folk culture is an important part of a country's culture, and different countries have formed their own unique folk culture. The accelerating process of globalization requires us to be aware of the cultural similarities and differences of other countries, to respect the differences and understand their folk culture in the process of foreign communication. This paper selects two East Asian countries, China and Japan, and explores them mainly in terms of food and taboos. Through the comparison, we have a clear understanding of the basic characteristics of the folk cultures of the two countries, which provides some reference for people's daily interactions in cross-cultural communication situations. In teaching Chinese as a foreign language, Chinese teachers can better grasp the strength of cultural export, respect the differences and avoid cultural conflicts.

Acknowledgments

Project Name: Research on the Foreign Translation of Chinese Medicine Texts in the Perspective of Cultural Confidence (Project No.:2019M038).

References

1. Luo Bo. (2007) Interpreting Chinese culture from the history of the evolution of the concept of vulgarity to folklore [J]. Journal of Guangxi Socialist College. (01):53-55.
2. Ke Ling. (2006) Thinking about Folk Culture in Teaching Chinese as a Foreign Language [J]. Journal of Yunnan Normal University. (04):5-11.
3. Ruan Jing. (2012) On the Application of Folk Culture in Teaching Chinese as a Foreign Language [J]. Shandong Social Science, No.205 (09):43-47.
4. Jiang Y. (2022) The differences and causes of Chinese and Japanese food culture--a review of Chinese Food Culture [J]. Journal of Food Safety and Quality Inspection, 13(20):6787.
5. Wang Yiming, Lv Ruquan. (2017) The differences between Chinese and Japanese food culture from the perspective of table manners [J]. Food Science, No.378(10):74-75.
6. Chen Zhen. (2018)An examination on table manners in China and Japan [J]. Scenic spots, No.354 (11):194+230.
7. Xu Shunying. (2009) An introduction to Japanese table manners [J]. Chinese Out-of-School Education,(S5):191+216.
8. Gao Cong. (2018) Cultural differences between China and Japan from the perspective of eating habits [J]. Northern Light, No.252 (10):71-73.
9. Zhang, O.,Yu, A. N.. (2017) Experimental investigation of the differences between Chinese and Japanese food culture from the subtleties [J]. Food Science, (22):135-136.
10. Hou Qiaohong. (2011)An analysis of the differences between Chinese and Japanese food customs and the reasons for their formation [J]. Journal of Chifeng College (Chinese version of philosophy and social science), 32(10):113-116.
11. Wang Man. (2011) A comparative study of taboo culture in Chinese and Japanese folklore [D]. Heilongjiang University.
12. Wang Shengbo. (2017) A comparison of the cultural symbolism of Chinese and Japanese color words "red" and "white"[J]. Journal of Qiqihar University (Philosophy and Social Science Edition), No.239(01):136-139.

Application of green plant elements in office space

Wenchao Wang[1,a], Jiaqi Chi[2,b], Siqi Niu[3,c]

[1] School of Design and Art, Shenyang University of Architecture, Qingdao, Shandong, China

[2] School of Design and Art, Shenyang University of Architecture, Shenyang, Liaoning, China

[3] School of Design and Art, Shenyang University of Architecture, Huludao, Liaoning, China

Abstract: "green plant elements" visualization of modern office space design is the "green plant space" design development process. The transformation from "green plant element" to "green plant space" is the main expression of interior design in recent years. Explore the green plant elements in modern office space innovation. Green plants in the building has good ecological performance, light weight, easy to shape, has become an indispensable material in modern architecture and interior decoration materials. With the help of natural green plants, green relaxation, fresh and other elements, to create a new modern office environment.

1 Preface

According to the shape of green plant elements, the sense of gradation can affect indoor or even outdoor environmental relations, and its combination with spatial environment and layout will affect indoor and outdoor space design. Green plant elements can divide the enclosed environment, but also the open environment can be separated, making the whole space more hierarchical.

With the development of the times, people pay more attention to the topic of health and environmental protection. The high pace and frequency of modern life make people have less energy to take the extra time out to feel the charm of nature and enjoy the feeling of green plants. With the improvement of people's living standards, people's demands for daily work are also increasing gradually. People no longer limit themselves to the standardized and unified office space, but introduce the elements of the outside world into the interior, so that people can feel the breath of the outside world and even see and touch the green plants in the nature while working.

2 Summary of office space

2.1 Emergence of various types of office space

The root cause of the current crisis of office space is that the office environment is single, the visual experience is dull, the lack of humanistic care. In many traditional spaces, people crouch in a narrow space, separated by a simple low partition, the space is simple and single, and there is no corresponding rest, leisure, entertainment, dining areas, resulting in employees in such a working environment can not be physically and mentally satisfied.

The Larkin Building, designed by Wright in 1904, was revolutionary in the history of the workplace. This is the first skyscraper in history. (Figure 1)

Office from 1940 to 1950 advocates open, uninterrupted office space. (Figure 2)

Bob Propst's cubicle office, which he launched in 1967, has long been criticized for using modular and lattice-shaped design to achieve effective and sound management. (Figure 3)

Entering the 21st century, the open and cooperative office has left the cubicle, broken through the hierarchical structure between the boss and ordinary employees, and created an equal, cooperative and innovative working atmosphere. (Figure 4)

Fig.1 Larkin Tower

[a]3368959209@qq.com, [b]122746267@qq.com, [c]nsq0920@126.com

Fig.2 1960s Landscape Office

Fig.3 The interior of the Larkin Building

Fig.4 Landscape Office of the 1960s

2.2 Classification of office space

2.2.1 Development of office space abroad

Early office spaces developed from residential spaces and later evolved into deliberative office spaces. By the end of the 19th century, due to the Industrial Revolution in the West, people pursued high efficiency and high profits, and office space began to appear independently. Since the beginning of the 21st century, work has gradually become the focus of people's lives, people begin to spend more time in office, the development of office space to keep pace with the pace of society, began to appear a variety of office models to stimulate functional space in the form of shared space [1].

With the development of the three industrial revolutions, the city continues to develop, various types of design styles continue to appear, a variety of office equipment constantly updated, so that office space is also constantly improving. From the beginning of the living room office space to today's intelligent, human nature, human and cultural office space, office space gradually mature.

There are seven stages of office space abroad: white-collar factory in the early 20th century, open office space in the 1950s, landscape office space in the 1960s, experimental office space in the 1970s, electronic office space in the 1980s and virtual office space in the 1990s [2].(Figure 5, Figure 6)

Fig.5 Union Carbonized Building

Fig.6 Office Building, Bishop Center, Netherlands

2.2.2 Development of domestic office space

Compared with foreign countries, the economic development of our country in the early stage is slow, and the development of office space is relatively slow. Our country's office space design is generally based on the interior office design of foreign countries.

The office space of our country can be divided into three stages: before the reform and opening -up, the economic development of our country was slow, all kinds of economic development were backward, only the government and relevant units have the office space, can only meet the most basic office needs, all kinds of office space in the 1980s. -- With the rapid progress brought about by the reform and opening up, China's urbanization and industrialization have developed rapidly, and people's ideas and cognition have been greatly updated. Various kinds of information have poured into people's lives. People begin to pursue the needs of office space in all aspects, not limited to the satisfaction of basic functions. Landscape, open and humanistic office forms are all applied to office space [3].(Figure 7, Figure 8)

Fig.7 World Trade Center Shenzhen

Fig8 Guangzhou World Trade Center

2.3 Status of office space

At present, the office environment in our country is weak in office cognition, and the research on office space in our country is relatively late compared with that in foreign countries. Many offices in our country are still limited to the office areas with cells, and do not pay attention to the division of office environment. In order to improve the office rhythm and efficiency, the material and spiritual needs of employees are ignored [4].

Most offices in our country are still attached to fixed cells, such as manager's room, reception room and staff area. The division of office space is too complicated, which breaks down the fluidity, continuity and communication of the office.

The lack of humanistic spirit in the office. At present, the design of the office is limited by the development of social economy and culture. At present, there is no distinct design with its own style.

2.4 Green plant concept

Green plants are non-unicellular plants containing chlorophyll, which can absorb carbon dioxide and water and at the same time convert to sugar and release carbon dioxide and water as a renewable raw ecological materials. Green plants light weight, easy processing, low price and easy molding, has become an indispensable part of building materials and interior decoration materials.

And the use of green plants in the indoor environment can play a regulatory role in beautifying the environment, purifying the air and regulating indoor water balance. Green plants are recyclable renewable resources. Using green plants as design materials can make people feel completely back to nature, leaving no architectural pollution, which is in line with green sustainable design principle.

Today, the speed of social progress is proportional to the environmental pollution, and the consumption of resources is going up in a straight line, so we should use recycled materials in building and interior space.

2.4.1 Development of Green Plants in Foreign Countries

In foreign countries, the use of green plants in space design has become a hot topic. Designers have not only carried green plants directly to use, but according to the needs of space design, give green plants more appropriate shape, through combination, pruning to achieve the desired effect, and the use of plants with the local environment, the real integration of architecture and nature.

For example, the GLEM office of MP design uses bamboo as the main structure connecting ceiling and ground in the interior. The design of this office just uses the green plant elements. Even if the original space changes will not produce material waste, the original building left behind by the bamboo material can also be decorated or re-use. There are many foreign green plants will be outdoor elements into the office space, through the design means and space division, the real realization of the "harmony between man and nature." (Figure 9)

Fig.9 Stereo Greening of Singapore "Garden City"

2.4.2 Development of domestic green plants

At home, the application of green plants in the space is not as good as abroad, there is a quite understanding of green plants can only be planted as a cultivation plant in the space, or even directly transfer green plants, the effect is not satisfactory.

Currently in the domestic green plant elements into the building and indoor examples or very few, most will be green plants as a three-dimensional green use, are directly green plants as the exterior decoration. Green plant elements, as renewable materials, produce less harmful substances than decorative materials with the same properties in the process of construction. With the continuous progress of science and technology and the continuous development of new technologies, Green plant elements have brought more advanced design forms and means for the development of interior design and architectural development [5].

3 Application of green plant elements in modern office space

3.1 Application of green plant elements in modern office space

With the development of the times, the pattern of the office space has developed from the simplification and systematization to the diversification, and the diversification of the office space has also promoted the staff to work more and more onerous, in the modern office, the work rhythm is speeding up the atmosphere, the staff's work request is also more and more strict, the work efficiency must enhance, under such environment, should pay attention to people's physical and mental health. The environment of office space is relatively closed, the use of green plants can adjust the air quality in the office space. When the green plant elements are introduced into the office, people will feel close to nature and yearn for nature, so that people's tension and pressure are reduced. In today's workplace, people's mental pressure is increasing, more green plants to create a quiet office atmosphere and reserved planting space, employees can plant their own, enjoy the process of greening conservation and management, to create a quiet space within their own, office space needs green plants with such a variety of functions as "air purifier" [6].

3.2 Influence of Green Plant Elements on Working Atmosphere

3.2.1 Psychological impact

Now the rapid development of society, social atmosphere impetuous make people's mentality is gradually impetuous, office environment atmosphere can directly affect people's mood and working state. Comfortable, harmonious, natural, fresh working environment, can make people happy, positive, improve work efficiency, on the contrary, dirty, irritable, cold working environment makes people depressed, inefficient. Frankly speaking, office space is to serve people, to provide people with a comfortable, stable office environment, in the office space, the center of space, but also the main space, the use of people is essential. In the process of evolution, people still retain the original idea of yearning for nature and expecting to return to nature. In the design of this office space, the introduction of green plant elements can make people mentally comforted and satisfy. In the office space, the introduction of green plants has a positive impact on people's mental health, makes people feel full of nature, innocence, simplicity and sunshine, imperceptibly affects people's psychology and mentality, and has a positive impact on people's work and life.

In the design, we can use a lot of green plants on the walls and in the center of the space, so that people can feel the encirclement of the green plants at any time and in any place.

3.2.2 Physical influence

The adaptability of people in the environment is very strong, and the impact of the environment on people is also very strong, people and the environment is interdependent and symbiotic, people work in the indoor air does not circulate, the degree of air turbidity will become more serious, people can not get access to fresh oxygen, work efficiency will be greatly reduced, the introduction of green plant elements in the indoor, under the photosynthesis of green plants, absorb the carbon dioxide inside the office space and then convert it into oxygen, for indoor air environment regulation has a great role, people always exposed to fresh oxygen, will greatly improve work efficiency. In today's office environment, people use computers and mobile phones in a large proportion, and the introduction of green plant elements in the indoor office environment, in people's visual fatigue, can effectively alleviate fatigue, has a great role in people's health.

3.3 Cultural extension of green plant elements in indoor office space

In the design of space, each element has its corresponding meaning. In the design of space, we should not only consider the rationality and comfort of space, but also the meaning of space expression and cultural heritage. In the continuous development of society, many elements are given different meanings.

Since ancient times, the elements of green plants have been endowed with many meanings by us. For example, bamboo has been endowed with perseverance, the gentleman has a high moral integrity, and lotus has been endowed with self-restraint, not drifting with the tide, and so on. The use of green plant elements in office space can also be used as a reminder to warn people to stick to themselves and keep their original minds undisturbed by the surroundings in today's social life and work..

3.4 Relationship between green plant elements in indoor office space and outdoor environment

Green plant elements play an important role in the interior and exterior. It is a hinge connecting the interior and exterior. It can be a unique scene in a corner of the interior or a wall in the exterior. The use of green plant elements in the space can not only enhance the vitality of the interior, break the original pattern of rigidity, indifference, and can make indoor and outdoor contact, so that the relationship between the two spaces more closely, so that indoor people feel outdoor scenery and sense of belonging [7].

No matter which kind of indoor space is complementary to the surrounding outdoor space environment, indoor space without outdoor space can not be called indoor space, outdoor space without indoor space can not be called outdoor environment, the two are always combined, and the use of green plant elements is to make the relationship between the two closer, so that the two blend, in the indoor space if green plant elements can not only make the indoor space more comfortable, natural can also make the two more closely integrated, truly achieve "harmony between man and nature."

4 Conclusion

Through the green plants in the design to create a unique understanding of the design, a more close to the green design, the unique elements of China into the design, so that today's design is not only reference to Western architecture, more Chinese characteristics, the introduction of green plants in the building elements, the building and indoor and outdoor space and green plants elements better integrated together, but also better integrated with people and nature.

References

1. Tan ZP 2020.16-17 Modular Design of Joint Office Space [D].Donghua University.
2. Committe on Mathematical Ci Hai.(2002)Mathematical Ci Hai.Shanxi Education Press
3. Lim. 2015.04. On the Application of Green Plants in Office Space and Its Development Trend. Journal of Liaoning Institute of Economic Management and Cadres.Liaoning Institute of Economic Technology.
4. Ni XS. 2017-06-01 On Reading Guidance Strategies in History and Social Teaching. Selected Small Authors;
5. Wong BW, Wei YX. 2015-06-01 "On the Psychological Impact of Indoor Space". "Scientific Chinese";
6. Caori Patricia Takeuchi.Design and Construction of Housing with Structural Elemen
7. ts of Bamboo(Guadua)Laminated.Universidad Nacional de Colombia.Trans Tech ， 2012
8. Chmielecki,Michale.Sustainability and Design Trends in Hardwood,Cork and Bamboo.

 https://doi.org/10.1051/shsconf/202316702010

Research on Smart Home Interior Design in Youth Living Space

Jiaqi Chi [1a], Mingcen Zhao [1,b,*]

[1]School of Design and Art, Shenyang Architecture University

Abstract: Contemporary smart technology has come to an unprecedented height and has become a clear trend for the future development of society, especially for the living space of young people has a particularly important impact. The ability of young people to accept new things is higher than others, and they are willing to try new things, which makes them able to play their role of smart home products more perfectly. The youth group's concern for the update of smart home and the frequency of using smart home make the main sales target of smart home is the youth group. We need to further explore and study the needs and psychological activities of young people for smart home.

1 Preface

This paper explains the current situation and development trend of smart home applications, outlines the characteristics of smart home, explores the needs of youth groups for living space and the needs of smart home, and briefly summarizes the innovative design strategies of smart home in youth living space using the conclusions of these needs to promote the renewal and development of smart home.

2 Smart Home Application Status

The integration of smart home systems into the indoor environment makes people feel convenient and relaxed while using the space. In the daily life, smart home can provide people with convenient ways to do things, so that people can arrange their affairs in a scientific, reasonable and efficient way, making full use of time and resources.

2.1. More Data-driven Life

In the development process of smart home, it is necessary to design according to the living habits and preferences of the user population to efficiently and scientifically manage the products, so as to achieve the effective use of scientific intelligence. Smart home products can display multi-scene mode operation functions and implement personalized services. Smart home can also help people to improve the utilization of time, so as to get rid of the tedious things in the work.[1]

2.2. Safer Living

The emergence of smart home has raised the security system of living space to an unprecedented height, with its sophisticated anti-theft system and highly accurate outdoor monitoring system can further guarantee the safety of our daily life. As well as some control systems in the smart home to achieve full automatic switching technology, record some safety hazards at home, reduce the accident rate to a minimum standard in advance, to ensure that no major accidents occur at home.

[a]122746267@qq.com
Corresponding author: [b,*]1241522702@qq.com

3 Smart Home Development Trend

Smart home is a product of the contemporary era of big data, in this era of information data explosion, the rapid development of smart home has gradually become one of the focus of attention.[2] Smart home is a product of people in the process of improving the quality of life, in the pursuit of perfection in the process of high people's aesthetics and whether the smart home and the living space is highly integrated with the issue of in-depth discussion. In the future, the position of smart home in the interior design of living space is unshakeable.

3.1. Certain Degree of Foresight

The cycle time for modern interior design to release new products is getting shorter and shorter, and more new things are rapidly integrated into people's home life, and users' styles and habits of interior design are changing frequently. Smart home architecture design integrates sustainable design concepts into daily architectural design, so that the architectural design can meet the current development status and make the interior design inherit development in the long run.

3.2. Has the Effect of Enhancing the Quality of Housing

Intelligent home system humanized to enhance the quality of modern life, people's high requirements for life, high convenience intelligent home can meet them. Modern new technology and new materials in the intelligent home system, and has a very broad market, its fundamental improvement of home life environment by the general public highly recognized.

3.3. With Some Personalization

Personalized demand is a unique demand formed under the action of many different factors such as different psychology, thinking and living habits of each person, personalized demand reflects the level of human appreciation and aesthetic level, the polygon of smart home can reasonably and appropriately meet the personalized demand as much as possible. According to personal preferences for lighting, control and appearance adjustment and modification. Smart home further shows people's aesthetic interest and attitude towards life.

4 Youth Group Usage Needs

Most of the contemporary young people's requirements for a good living space are hard requirements, mainly in terms of convenience and comfort, and its functions need to meet leisure and entertainment, rest and sleep, work and study, dining, washing and drying. [3]Moreover, most of the contemporary young people are working outside and living alone. The special characteristics of these lifestyles and some unique living habits make the demand for living space of these people not limited to basic functions. According to the above introduction of smart home combined with interior design theory, we briefly summarize the needs of young people for living space. From various angles, three major demands are derived.

4.1. Multi-dimensional Requirements

The multidimensional theory of interior space aims to establish the relationship between the occupants themselves and the living environment, and on this basis, to achieve the satisfaction of the occupants' sensory experience, emotional needs, psychological effects, cognitive thinking and physical relationship to the living space, in order to achieve the coordination and balance of the living needs.

Sleeping space is a necessary functional space for young people's living space, due to the limited economic ability of young people, resulting in the limited overall area of living space, the need to meet their necessary sleeping space in the same living space and at the same time to meet, for example, recreation, work and study space. Therefore, the division of the whole space is particularly important, and the space needs to be reasonably divided and more finely adjusted. From multiple angles to meet the multi-dimensional needs of young people, reduce unnecessary home decoration, only home implantation space, the most rational space, the most coordinated.

In addition to meeting the multi-dimensional needs of young people living alone through the reasonable arrangement of space and the use of smart homes, the visual experience can also affect their living experience. Color design in space involves expertise in psychology, color science, physics and other disciplines, and colors with different attributes have different effects on human vision and psychology. The choice of color in a space is very important and has a serious impact on the visual comfort of young people, the quality of sleep, the overall atmosphere of the space and so on. The use of blue environment can have a certain effect on slowing down the heart rate, which in turn makes young people stay calm and more relaxed, so they can have a better rest time. [4]If the color of the indoor space is not convenient to adjust and change, you can use intelligent lighting equipment, according to the space color needs, to create the atmosphere, to meet the multi-dimensional needs of the youth group for color.

4.2. Interactivity Requirements

Interaction design was named in the 1980s by Bill Mogridge, the father of modern notebooks, who said "design is the interactive products that support people's daily work and life", and the importance of interaction design for a product can be clearly seen from Bill Mogridge's expression, especially for smart home products, where interaction design with people is particularly important. [5]In the youth group, the system data shows that most young people have a very strong demand for product interactivity. The reason for this is twofold: one is that the mental pressure and work pressure of the contemporary youth groups lead to the growth of negative emotions, which need to be vented in the living space; the other is that the youth groups mostly live alone, and inevitably lonely and isolated in the process, the interactivity of smart home can provide the most basic communication and interaction for these youth groups living alone. The interactive design of smart home is based on the behavior, senses and psychology of young people, striving for a more realistic interactive experience.

Nowadays, a variety of brands of smart home products control through voice commands, freely adjusting lights, music, air conditioning and a series of other products that require manual operation in traditional living spaces, strengthening the interaction experience of young people with the space in the living space. Smart control systems such as Tmall Genie and Xiao Ai classmates (Figure 1) can use wireless networks and Bluetooth connections to realize the intelligent ecosystem, turning the entire indoor living space into a fully interactive space, fully automated and comfortable to enjoy the space.

Fig. 1 Xiaomi AI Speaker

4.3. Belonging Needs

Young people have been studying, studying, and applying for jobs, and they cherish and work very hard for their jobs. It is expected that the creation of the indoor environment will enable the young people to return home after a busy day to unload their fatigue, remove the tedious operation process, effectively increase the feeling of living in the living space, and upgrade the happiness and sense of belonging of the young people back home.

For example, in the kitchen function space, through the rationalization of its flow, the time and steps of operation can be reduced and the cooking efficiency can be improved, so that more and more young people living alone are interested in cooking, gaining the happiness of enjoying food with their own hands, and developing a sense of belonging to the living space. The design of the functional space of the kitchen should take into account the size of the space, the interspersed form of different operation modules and the range of activities of the users in the space, and make a smooth and reasonable functional layout. [6-7]Young people are accustomed to maintaining a good figure, and some people have a fitness hobby, and do not require too much food, but only need to handle things simply, so a simple and clean smart kitchen is essential. In the process of using the kitchen space to obtain a sense of pleasure, the pursuit of the added value of the product.

5 Innovative Design Strategies for Smart Homes in Youth Living Spaces

Through the analysis and summary of the needs of the youth group described above, combined with the study of interior design of living space, we now outline some innovative design strategies of smart home in the youth living space, from the sleeping space, the need to include the sleeping space, living room space, kitchen space, bathroom four aspects to explain.

5.1. Sleeping Space

The most important thing in the whole living space is the sleeping space, one can work in a better condition only if one has enough sleep. Christopher Alexander mentioned in "The Language of Architectural Patterns" that one should not simply put a single bed in an empty room called a dormitory, but one wants to put a bed in a deep space, that is, one wants to be surrounded by something. [8]Therefore, the placement of beds for youth groups must be careful. Create a safe and private space, can be relieved of the day's exertions during sleep time. To use the bed visit, combined with the rendering of light, the entire sleeping space should be created a warm, calm space, the use of sound-controlled light technology to embed the light in the wall seam, both beautiful and practical. The adjustable light can be changed reasonably and conveniently at the time of different needs. And the use of light to create a sense of the bed enclosed within, to ensure the quality of sleep. Reasonable use of intelligent means to create a comfortable home environment to relieve the body and mind.

5.2. Living Room Space

The living room space needs to be for leisure and entertainment, work and study, sports and fitness, etc. The youth group needs such a multi-functional space to relax and soothe their bodies and minds on weekends. In the limited space needs to rationalize the function, the use of intelligent LCD TV wall-mounted in the background wall, lightweight sofa seats can effectively ensure the versatility of the space. Because there is not a lot of time for indoor cleaning, intelligent sweeping home is essential. Depending on the person, the space can be personalized according to the frequency of various people using the space for various functions. If you are a student and workaholic, reduce the entertainment space, expand the study and work area, and place intelligent lifting tables and chairs to improve the structure of the space; if you are a self-publisher and fitness expert, you need a lot of space for activities, and an open sports area is essential (Figure 2). The variability of the space is incomparable to other controls, and the smart home ecosystem can be implanted into the space, using wireless screen casting and voice control to control any home in the space without having to stand up. [9]Smart curtains and clotheslines, using the Internet of Things system, can be switched on and off at regular intervals, so that everything in the house is ready before the occupants return home, waiting for their return.

Fig. 2 Smart Living Room Activity Area

5.3. Kitchen space

Kitchen space needs to maximize the function in the smallest space, but also to ensure that its owner in the kitchen to play enough space must make changes in the operating table. The use of the "I" and "L" forms to connect the storage area, rinse area, operation area and cooking area can make the entire space more fluid to ensure the convenience of the owner's operation (Figure 3). Young people of today are always concerned about their own health, market fitness to keep fit, so they are also very concerned about the control of diet, the main requirement for things is simple cooking, and do not need a lot of oil and salt frying, need to be air fryers, ovens and other quick heating smart products, and the use of smart products to calculate the intake of calories to keep fit while saving a lot of time. In daily life also need to take into account that no one at home for a long time, the need for intelligent alarm system connected to the cell phone, out of the home can also protect the home whether the gas is turned off and some other kitchen hazards.

Fig. 3 Smart Cooking System

5.4. Restrooms

The bathroom takes up a relatively small proportion of the entire living space, but it shows the aesthetic style of a home and the quality of life. Its available smart home in this out also highlights its multi-functional place. The use of smart mirrors is a common phenomenon in the living space of contemporary youth groups, which adopts the Android system and uses ZigBee, wireless network and Bluetooth connection to realize the interaction between people and mirrors, [10]surfacing real-time news, physical condition, weather of the day and other information on the mirror, so that youth groups can learn the information of the day during the morning wash (Figure 4). Through the smart home, the fragmented time can be used efficiently to improve the life and work efficiency of young people. In addition, the smart mirror panel can be changed by its owner to realize personalized scene mode.

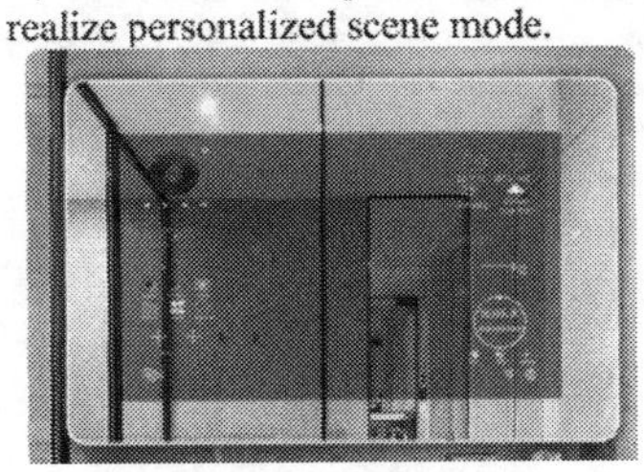

Fig. 4 Smart Mirror

6 Conclusion

The pressure of contemporary young people's life is unprecedented, and the pressure and frustration and fatigue they experience every day are also difficult to relieve, so the comfort and happiness of living space is especially important. With the intervention of smart home, the traditional living space is updated to make it more comfortable and happy. People are constantly pursuing a high quality of life, in which the essential stage is to combine with modern high-tech means, big data, artificial intelligence technology and so on. Modern designers also need to pay attention to the smart home, the smart home better into the living space, to a comprehensive understanding of the smart home, balance the relationship between people and the smart home, in the human-centered design principles to achieve a better design.

References

1. Deng Huixian. The use of Smart Home in Interior Design in the Era of Big Data[J]. Design.2021.34(15):26-28.
2. Liu Y. Research on 5G Network Technology for Improving 4G Network Performance [J]. Digital World.2020.No.174(04):24.
3. Jiang Chang. Research on the Optimization Design of Micro-living Space Based on Young People's Living Behavior[D]. Hebei University of Technology.2022.DOI:10.27105/d.cnki.ghbgu.2022.000434.
4. Liu Rui. A Brief Discussion on the Effect of Color on Human Psychology and Health[J]. Journal of Luohe Vocational and Technical College.2016.15(04):64-66.
5. Song Wanqing. Research on Intelligent Living Space Design for Young People Living Alone[D]. Dalian University of Technology.2021.DOI:10.26991/d.cnki.gdllu.2021.000848.
6. Ge Yanping. Research on the Application of Humanized Interior Design in Modern Interior Design: the Rational Layout of Interior Space in Home Decoration as An Example[J]. Ju She.2021(31):28-30.
7. Wang Yantao, Innovative Research on the Form of Interior Space Design of Youth Apartment[J]. Building Materials and Decoration.2020.No.600(03):85-86.
8. Wang Yipeng, Wang Jie. Three ways of combining architecture and language, historical tracing and theoretical analysis[J]. Southern Architecture.2022.No.207(01):54-60.
9. Li Xinxiang. Research on Intelligent Product Design Based on Scenarios [D]. Shandong University of Construction.2022.DOI:10.27273/d.cnki.gsajc.2022.000079.
10. Li Rong,Zhang Wei,Wu Lifa,Tang Yunfei & Xie Xinguang.(2023).ZPA: A Smart Home Privacy Analysis System Based on ZigBee Encrypted Traffic. Wireless Communications and Mobile Computing. doi:10.1155/2023/6731783.

SHS Web of Conferences **167**, 02011 (2023) https://doi.org/10.1051/shsconf/202316702011
CACC 2023

Analysis of the application of light-transmitting concrete in interior design

Wenchao Wang[1,a], Jiaqi Chi[2,b], Siqi Niu[3,c]

[1] School of Design and Art, Shenyang University of Architecture, Qingdao, Shandong, China

[2] School of Design and Art, Shenyang University of Architecture, Shenyang, Liaoning, China

[3] School of Design and Art, Shenyang University of Architecture, Huludao, Liaoning, China

Abstract: As a new type of green building material, light concrete breaks the concept of traditional concrete, which combines optical fiber, resin and As a new type of green building material, light concrete breaks the concept of traditional concrete, which combines optical fiber, resin and concrete, and can produce multi-form and multi-style products through the change of light, but does not affect its load-bearing. Due to its unique decoration, it is widely used in architecture, landscape, interior decoration, furniture and other major fields. Outdoor light can enter the interior, and indoor privacy will not be conveyed to the outdoors, the change of light and shadow gives people a new feeling, its light transmittance and interior design style light transmittance and interior design style reflect each other, different time periods show different rendering power, the combination of lighting and architecture gives the space a new appeal and meaning.

1. Introduction

Concrete is still one of the most commonly used basic building materials in today's construction. While traditional concrete gives a monotonous, dull and dark feeling in use, the application of light-transmitting concrete changes the usual brutality and heaviness of concrete and introduces light into the interior in another way, and can be used as a decorative material to bring a new light and delicate aesthetic to the interior design environment, giving life and dynamism to the building.

2. Concept and characteristics of light-transmitting concrete

Translucent concrete is a new type of material, also known as transparent cement, in which a large number of optical fibres, a light-guiding material, are inserted into fine concrete, which is then polished and maintained after forming and eventually used in architecture and interior construction. Light-transmitting concrete has a large number of light-conducting fibres inserted vertically into the concrete, allowing light to pass from one end of the light-conducting body to the other, thus achieving a light-transmitting effect.[1] The colour effect can also be obtained by inserting different coloured optical fibres into the concrete according to the design requirements. When the density of the optical fibres is less than 3% of the total density of the concrete material, the mechanical properties are not affected and it has the same strength as traditional concrete, which can be used as a structure and as a load-bearing wall in construction.

Light-transmitting concrete has excellent characteristics such as light weight, light transmission, air permeability, fire resistance, durability, light and noise, and ease of forming.

Transparent concrete, due to its production process, requires the insertion of an appropriate number of light-conducting fibres into the concrete. As a result, the density of light-transmitting concrete is reduced by around 20% compared to conventional concrete for the same volume[2] Due to the addition of light-conducting fibres, the concrete shows different effects under different light conditions, especially during the day and night, with the cooperation of lights, and due to the production characteristics of the material, light-transmitting and opaque, it can well protect privacy and present different artistic and aesthetic features; transparent concrete, due to its When the ambient air is humid, the moisture in the air is absorbed by the concrete's base voids. When the ambient air is dry, the water vapour in the void of the concrete material can be released through the material until the dryness and humidity in the air is balanced; the material of translucent concrete is mainly made of fine concrete, and its fire resistance and wear resistance are comparable to ordinary concrete, so this material can be used in public spaces as a load-bearing and decorative material; as the light-conducting material used in translucent concrete is refractive to light, it can effectively filter out the stronger light. It can effectively filter out strong light and does not produce too much light agitation in the process of use;

[a]3368959209@qq.com, [b]122746267@qq.com, [c]nsq0920@126.com

the translucent concrete is cast in one piece and can be moulded and designed with translucent patterns in advance according to demand, and can be used for decoration after the surface has been decorated.

3. Light-transmitting concrete in interior design

The style of interior design has gradually changed over time, with people's preferences changing from the original sophisticated, colourful collision to the simple and uncomplicated style of today.

The rugged, industrial style of the original concrete made people use it only for the foundation of the building, often wrapped in aesthetic materials, but nowadays, whether it is indoor or outdoor, or walls, floors, partitions, columns and other structural building foundations are always inseparable from the use of concrete, and the emergence of light-transmitting concrete has undoubtedly provided people with a new choice.

3.1 Walls

The wall is the interface that we come into contact with the most within our sight in the interior space, so the decorative design of the wall plays a vital role in people's visual perception. Today's walls are mostly painted or decorated with wallpaper beautification, with direct partial use of concrete rough walls, slightly decorated only to serve the overall design style. Today's walls are mostly used for advertising decoration, map displays, route guidance and sight attraction.

The use of translucent concrete can largely replace the use of electronic screens and ordinary advertising, reducing the waste of resources. The use of translucent concrete, with the use of natural light and artificial light at night, can present a different effect and create a different atmosphere, giving people a new feeling and bringing new vitality and life to the interior space (Figure 1).

Fig. 1 Example of a light-transmitting concrete wall

3.2 Partitions

The partition has a dividing effect in the division of the space area, but in use it does not completely divide the space like a whole wall, but is connected in the separation, with a certain continuity. In use, partitions require constant separation and generally have the need for light and air permeability, while there are no strong requirements for sound insulation or blocking the view.[3] There are no strong requirements for sound insulation or sight lines. The material characteristics and presentation of translucent concrete are ideal for use as partitioning, as it is light-transmitting and opaque, and has the effect of both separating space and providing aesthetic decoration. The use of light-transmitting concrete as partitions in indoor spaces allows light to be mapped through the concrete during the day under natural light, and at night the light from one end of the room to the other, changing the heavy, dull feel of previous concrete partitions and giving a sense of lightness and beauty (Figure 2). The use of translucent concrete as a partition in the bathroom not only solves the problem of waterproofing of the material, but also allows light from outside the bathroom to pass through the translucent concrete and shine vaguely inside the bathroom, which not only saves resources but also brings a different atmosphere to the bathroom (Fig. 3).

Fig. 2 Living room back wall partition

https://doi.org/10.1051/shsconf/202316702011

Fig. 3 Bedroom bathroom partition

3.3 Grounds

The ground is the general term for the ground and floor level of a building. It is mainly used in the building to separate space, to strengthen and protect the structural layer, to meet people's use requirements, as well as sound insulation, heat preservation, waterproofing, moisture-proof and heat insulation, etc., and occupies a large proportion of the human line of sight, so it must meet: (1) the requirements of sturdiness and durability, should not be easy to break, the surface is flat, no dust, etc. (2) Safety, the ground should be non-slip, corrosion resistant, good electrical insulation, etc. (3) comfort, the ground to meet the requirements of decorative[4] , should have a certain colour, pattern, texture, elasticity and sound insulation.

The use of translucent concrete as a ground material indoors not only maintains the characteristics of the original ground, but also adds to the aesthetic appeal (Fig. 4). The use of translucent concrete as a ground material outdoors unifies the park in the landscape and illuminates the ground at night, as well as serving as a walking guide for passers-by on the ground, with the whole style appearing exceptionally plain.

Fig. 4 Outdoor light-transmitting concrete floor

3.4 Top surface

Today's designs are increasingly pursued without mains lighting and the design of interiors is becoming more and more simple, while translucent concrete can partially replace the use of roof lights due to the good light transmission of the light-conducting fibres. With its built-in light-conducting fibres, the top pattern of the concrete can be arranged to achieve the desired effect according to the design idea.

The use of translucent concrete as a roof in the interior can reduce the application of indoor light sources, reducing the waste of resources and achieving the purpose of green design from the other side. In indoor use, it is often placed in places where the lighting requirements are not very clear, such as toilets, showers and private passageways, where the lights are turned on when in use, illuminating with a soft light that does not hurt people's vision and gives a feeling of understated luxury (Fig. 5).

Fig. 5 Light-transmitting concrete top surface

4. Concluding Remarks

With its excellent performance, transparent concrete can be used as a green building material in the design of interiors and exteriors in today's increasingly polluted environment, reducing the use of traditional decorative materials and thus reducing the waste of resources. Light-transmitting concrete combines the strength and hardness of traditional concrete with good light transmission and light conduction, and can be widely used for facades, walls, floors and roofs of buildings and landscapes, as well as for interior and exterior hard and soft furnishings. In today's green design environment, with the advancement of technology, light-transmitting concrete as a green building material will be widely used in various designs one by one.

Reference

1. Li Jia, Li Jiaxin. Analysis of light-transmitting concrete in architectural space[J]. Sichuan Cement,2022(11):8-10.
2. Huang Na, Shi Tieshao, Shi Yu, Li Chunlin, Hu Yuanman. Advances in ecological and social functions of green infrastructure[J]. Journal of Ecology,2021,41(20):7946-7954.
3. QI Kexin, LI Kezhong, WANG Haidong, ZHANG Zhongfeng, TAN Yujie. The application of light-transmitting concrete in outdoor space[J]. Furniture and Interior Decoration,2021(05):104-107.DOI:10.16771/j.cn43-1247/ts.2021.05.019.
4. Zhou Guangde. Research on ecological colorful negative ion interior dry powder coating[J]. China Building Materials Science and Technology,2008(01):17-21.
5. Liu Xin, Li Xinxiang, Hao Zhe. Numerical simulation study of large span highway tunnel excavation[J]. Journal of Shenyang University,2009,21(03):104-107.

Analysis the character of J.S.Bach's Sinfonia in A major (BWV 798) – based on the dynamic and research on related acoustic parameters

Zixuan He[1,*]

[1]Lian Jiang City Experimental School, Department of Music, Piano Study, China

Abstract. Various annotations of the character of J.S.Bach's clavier music in piano perform seem to be under discussion by the pianists and musicologist. However, the extent of dynamic and the way of dynamic changes that had been considered appear scarce. In the essay below, the extent of the dynamic and the way of dynamic changes of J.S.Bach's Sinfonia in A major (BWV 798) in piano perform is under analysis and discuss.

1 Introduction

Since the Baroque clavier to contemporary piano, the Great changes have taken place in the production materials and sound generation principles. In the past hundred years, the character of J.S.Bach's clavier music in piano perform had been analysis-ed and discussed. Alongside this growing dispute there has been much debate over how to define the extent of the dynamic and the way of dynamic changes of J.S.Bach's clavier music on piano performance. However, due to the limitations of the times, there is no recording of the performance in the Baroque period, and the contemporary piano has long changed, researchers can only infer the extent of the dynamic and the way of dynamic changes through the dynamic annotation of the score and the literature records, which gives musicians and enthusiasts a space to explore and seek knowledge.

At the moment, most pianists and musical researchers are disagree with restore or imitate the Baroque clavier tone colour, but they are quite approve of retain the Baroque perform style. As the aspect of dynamics of Baroque clavier music in piano perform, Although many pianists and musical researchers believe that it is necessary to "crescendo and diminuendo" conditionally on the basis of retaining the Baroque music performance style, it should not be much dramatic. However, judging from the various documents read at present, each experts and researchers have different standards and interpretation methods for this viewpoint; Some aspects are even more subjective and generally, without any distinct, objective and specific suggestion.

Therefore, in addition to combining relevant literature and scores, this research attempts to analyse the audio of the same music segment in J.S.Bach's Sinfonia in A major (BWV 798) by different performers by computer technology, quantify and compare the relevant parameters at the level of performance dynamics by means of waveform diagram (dynamic curve), and then draw a more intuitive conclusion.

2 Methods

Compare with the former researches and essays' methods, which shows their methods worth to learn, but are still one-sided and subjective. For example, Chen en en mentioned that she did not agree to imitate the tone colour of Baroque clavier with modern pianos, but she believes that "it is still very necessary to consider the overall dynamics characteristics of the ancient piano music in the Baroque period when performing." [1] Although, She had compared different pianists and researchers' suggestions and came to the conclusion that dynamics and tone colour contrast should be appropriately applied according to specific musical factors, which lacks specific examples, and she still not gives a more specific method of 'how to consider and perform after considered.' [1]

After 9 years, Tan Yun believes that "the Baroque clavier works should be treated with a diversified perspective" [2], because the musical instruments are constantly developing, it is necessary to treat the piano performance with a developmental perspective, and also pay attention to the aesthetic value of the baroque style. He had compared the different pianists' performance of the same Baroque pieces, but he only explained the results of the comparison through his personal description, which also led to the inevitable subjective color of his results. Finally, he also had not given specific practical suggestions [2].

According to the advantages and disadvantages of the above research methods, this research intends to further inherit the previous research methods and use the following three research methods:

* Corresponding author: 506205963@qq.com

2.1 Compare the different scores' dynamic marks

This research had collected 8 different versions of scores of J.S.Bach's Sinfonia in A major (BWV 798). Among them, two of them are original manuscripts in the early and mid-18th century [3,4], three are printed versions from the middle and late 19th century to the early 20th century [5-7], and latter three are domestic and foreign popular versions published in the 21st century [8-10], which are span nearly three centuries. By comparing the editions of these three periods, we can roughly understand the similarities and differences in the interpretation of dynamics in J.S.Bach's Sinfonia in A major (BWV 798) in different periods.

2.2 Compare the different researches' analyses and discussions

This research try to compare different literature to to find out the similarities and differences about the extent of the dynamic and the way of dynamic changes of J.S.Bach's Sinfonia in A major (BWV 798), and try to summarize the commonalities of their thoughts.

2.3 Analyzing the performance audio of different pianists by computer technology

Up till the present moment, the research has selected five performers: Tatiana Nikolayeva (1924-1993) [11], Glenn Gould (1932-1982) [12], András Schiff (1953-present) [13], Zhu Xiao-Mei(1949-present) [14], and Yunchan Lim (2004-present) [15]. These performers are active from the 20th century to 21st century, with a time span of nearly hundred years, and their active places are all over Europe, America and Asia. They have a high recognition in the music world in Performing Bach's clavier works, which has a certain comparative and reference value.

3 Results

Through the analysis of the above three methods, this research obtained the following results.

3.1 The results of compare scores

This research had compared the editions of Holograph manuscript and H.G.M. Darnköhler's manuscript scores [3,4], Carl Czerny [5,6] and Ferruccio Busoni's editions of print score [7], and three 21th century China and International editions: Wiener Urtext Edition, Henle Verlag, People's music Publishing House [8-10]. The extent of dynamic and the way of dynamic changes of J.S.Bach's Sinfonia in A major (BWV 798) in these eight editions' scores will be shown in the following table:

Table 1. Different Scores of J.S.Bach's Sinfonia in A major (BWV 798)

Editions	Extent of dynamic	Way of dynamic changes (and times)
Holograph manuscript (1720) [3]	No any dynamics marks	No any expression
H.G.M. Darnköhler (1745) [4]	No any dynamics marks	No any expression
Carl Czerny (1840) [5]	p-f-fz	crescendo (8) diminuendo (5)
Carl Czerny (1903) [6]	p-mf-f	crescendo (15) diminuendo (9)
Ferruccio Busoni (1891) [7]	p-mf-f-fz-ff	crescendo (13) diminuendo (3)
Wiener Urtext Edition (2013) [8]	No any dynamics marks	No any expression
Henle Verlag (2014) [9]	No any dynamics marks	No any expression
People's music Publishing House (2000) [10]	p-mf-f	crescendo (8) diminuendo (4)

3.2 The results of compare literature

With the revival of Baroque clavier music, the trend of "authentic performance" gradually rises from mid and late 20th century in Europe, which has a dispute with "modern performance style" from 19 century. For the reason of "authentic performance" had encouraged pianists to imitate the tone colour of Baroque clavier when they perform Baroque clavier music on piano, which included that limit or even prohibit the undulation movement of dynamics change.

For the moment, this research had not sought out any literature had discussed the dynamic of J.S.Bach's Sinfonia in A major (BWV 798). However, the dynamic of Bach's clavier works had been mentioned and discussed frequently. Although there is a great controversy between "authentic performance" and "modern performance style", most of pianists and researchers are still consist "crescendo and diminuendo" are necessarily, they believe that the esthetic values of music its-elves are more important than the tone colour of Baroque instruments.

For instance, Alfred Brendel (1931-present), when he was interviewed by David Dubal, he had mentioned that "Musical instruments are not the only way to revival of Bach's music" [16]. This viewpoint was supported by Richard Goode (1943-presents), he emphasized that the best way to interpret Bach's clavier music is to understand the music itself from the heart, rather than the superficial factors [16].

If former two pianists just explained their views of use Baroque music instruments to perform Bach's music, Rosalyn Tureck (1914-2003) had went further and elaborated her views on the dynamic changes of Bach's clavier music works. She disagreed with rely on Baroque music instruments to understand and use dynamics on piano performance of Baroque clavier musics, she believed that "the different of dynamics is the product of Baroque music's structures" and "the difference in dynamics must be as accurate as the music structure." [17]. She claimed that "crescendo and diminuendo are also exist in the piano performance of Bach's clavier music, but it only plays a minor role" [17], because "the dynamic change should be reflected on the first note of new tone colour, the habit of continuing crescendo and diminuendo are not suitable for Bach's music" [17].

3.3 The results of compare pianists perform

This research had tried to collect some pianists' performance audios and analysis them by using computer platform to export audio waveform. There are five pianists had been chosen: Tatiana·Nikolayeva (1924-1993) [11], Gleen Gould (1932-1982) [12], András·Schiff (1953-) [13], Zhu Xiao-Mei(1949-) [14], and Yunchan Lim (2004-) [15]. These pianists were active throughout the 20th century to 21st century, with a time span of nearly 100 years, and their active places were all over Europe, America and Asia. They were highly recognized in performing Bach's clavier music in the music world, which is of certain comparison and reference value.

The 9th bar to 12th bar of J.S.Bach's Sinfonia in A major (BWV 798, Fig.1 in blue shadow) has been chosen to analysis the audio of these six pianists piano performing, because of the dynamic changes in these three bars are quite luxuriant, which can be helpful to find out the differences of the way of dynamic changes between these six pianists performance.

Fig.1. The 9th bar to 12th bar of J.S.Bach' s Sinfonia in A major (BWV 798), cited in Czerny, C. (1840) No.XII, p.42 [5].

After using computer platform, the waveform of pianists perform BWV798's 9th bar to 12th bar were showed in following chart.

4 Analysis and compare

In the table 2 had listed the waveform of five pianists perfomed the 9th bar to 12th bar of J.S.Bach's Sinfonia in A major. Though the analysis of the table 2, this research will compare and discuss the performance of these five pianists, to find out their opinions and deductions about the dynamic changes.

Table 2. Waveform of five pianists perfomed the 9th bar to 12th bar of J.S.Bach's Sinfonia in A major

Pianists	Waveform
Tatiana Nikolayeva (1977) [11]	
Gleen Gould (1955) [12]	
András Schiff (2011) [13]	
Zhu Xiao-Mei (2015) [14]	
Yunchan Lim (2020) [15]	

4.1 Tempo

From the viewpoint of the tempo, it is quite hard to show up from the waveform. However, after contrast their performing audio, it is not difficult can find out that tempo who performed fastest in these five pianists is Gleen Gould, he cost 59 seconds played the whole piece [12]. The second one is Yunchan Lim, who played whole piece and costed 1minute and 14 seconds [15]. Zhu Xiao-Mei and András Schiff's tempo seems at the same level [13,14], which keep between 1 minute and 19 with 21 seconds. Tatiana Nikolayeva's performing is the slowest one, which costed 1minute and 31 seconds [11].

4.2 The tend of dynamic changes

If only from the condition of the waveform demonstrated, these five pianists have different opinions and the ways of performing about the tend of dynamic changes of these three bars: Gleen Gould [12] and András Schiff [13] are support use less undulating dynamic contrast in these three bars' perform; Tatiana Nikolayeva [11], Zhu Xiao-Mei [14] and Yunchan Lim [15] are tend to use more undulating dynamic contrast when play these three bars.

Combining audio and table 2, it can be seen that Gleen Gould [12] and András Schiff [13]'s performing are not very obvious in dynamic change, these three bars' seems maintian in piano and quiet, which seems opposite with Gleen Gould [12]'s previous viewpoint in former paragraph and sounds like it conforms to the characteristics of harpsichord: without "crescendo and diminuendo" .

Compare with Gleen Gould [12] and András Schiff [13], Tatiana Nikolayeva [11], Zhu Xiao-Mei [14] and Yunchan Lim [15] performance are different: the change of tone colours and dynamics are more variable, especially Tatiana Nikolayeva's performance, which can be heard three levels of dynamics changes, Fig.2 had showed the first level: the dynamic changes from c2 to g1 in the top part and shows a diminuendo trend

Fig.2. The three levels of dynamic changes: C--B--A--G

The second level is the dynamic changes of melodies undulate: the short melodies consisting of eight and sixteen notes and appear in the top part and inner part, which were showed in Fig.3, completely consistent with the trend of melody.

Fig.3. The three levels of dynamic changes: the dynamic changes of melodies undulate

The third level is the dynamic changes appear in the bass and showed in Fig.4: gradual diminuendo based on melodic motives.

Fig.4. The three levels of dynamic changes: the bass

According to the waveform, Zhu Xiao-Mei [14] also had some dynamic undulate in her performance. different with Tatiana Nikolayeva [11], she prefers to use diminuendo at the beginning and use crescendo after the beginning of these tree bars (showed in Fig. 5):

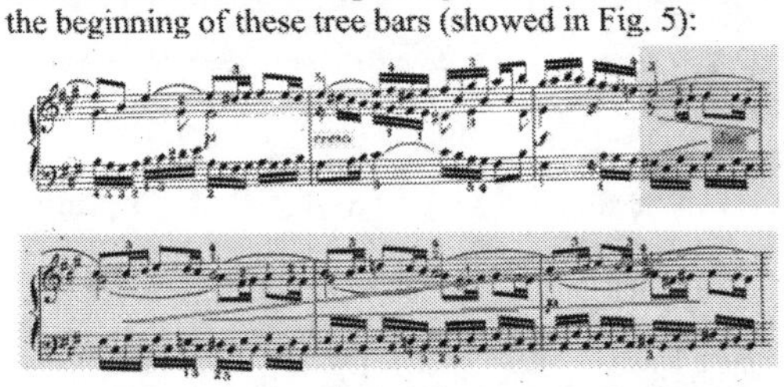

Fig.5. The dynamic changes in Zhu Xiaomei's perform

As for the Yunchan Lim [15], it is quite hard to recognize the trend of dynamic changes in his perform just by analysis the waveform. However, when combine with the audio, the dynamic changes can be clear that the long notes in top part are louder and more prominent, which was similar with Tatiana Nikolayeva [11] and Zhu Xiao-Mei [14]'s perform. he had not do some obviously crescendo or diminuendo during these three bars.

4.3 Extent of dynamic

The extent of dynamic changes can be hard to identified only by waveform and audio, which makes this research have to combine the waveform, audio and the dynamic marks on different scores together to analysis the extent of dynamic changes of these five pianists had used in there perform.

In former paragraph, this research had compare the dynamic marks of 8 different editions scores and had listed their marks of dynamic changes, it is obviously can find that most of them shows the dynamic changes are between "p" and "f". in 9th to 12th bars, the marks of dynamic changes are between "p" and "mf".

For the analysis of the waveform, it just can recognize the trend of dynamic changes, and cannot define the extent of the dynamics. However, when analysis audio, it can be clear to find that the extent of dynamic changes are basically in line with the marks on the scores.

5 Contrast and discuss

After analysis and compare the waveform, audio and scores, this research will continue to contrast and discuss the dynamic changes when play J.S.Bach's Sinfonia in A major (BWV 798) on modern piano.

5.1 The trend of dynamic changes: Is crescendo and diminuendo can be use in Bach's music when play on the modern piano?

By the analysis and the compare of the waveform, it shows that the result are still consistent with the current state of research and performance: there exists comparatively disputes about Whether there should be obvious dynamic contrast or the trend of crescendo and diminuendo. This research had just selected 5 pianists and three of them chose to use crescendo and diminuendo in their performance, which may reflect the that nearly half of pianists may support do not use dynamic changes when play Bach's music on mordern piano, but another half of pianists believe that they can use crescendo and diminuendo in their performance.

However, even in this research have three pianists had chosen to use crescendo and diminuendo in their performance, but the methods and ways of crescendo and diminuendo are still different. Although Yunchan Lim [15], Tatiana Nikolayeva [11] and Zhu Xiao-Mei [14] were all choose to play the long notes in the top part louder to prominent the trend of melody texture, but

Yunchan Lim [15] had not do some crescendo or diminuendo during these three bars. Even Zhu Xiao-Mei [14] and Tatiana Nikolayeva [11] both had crescendo or diminuendo in these three bars, but the trends of crescendo and diminuendo are also not same.

From the above paragraphs, it can be seen that even still disputed, maybe the trends are different, but crescendo and diminuendo can be used in modern piano perform Bach's music.

5.2 The extent of dynamic changes

According to the above paragraphs, it has listed and analysis that the scores between 19th to early 20th century had written the dynamic marks, most of them extent between "p" and "f", some of them can reach "fz" even to "ff". However, according to the tone colour of Baroque time clavier, with the technology at that time, it was seems quite difficult to play timbres' dynamic that could reach "fz" or even "ff"(both harpsichord and clavichord), and also clavichord may can play dynamic "p", but harpsichord is seems to be hard to play dynamic more quiet.

Meanwhile, according to above five pianists perform audio, no matter what kind of contrast they use in their performance, the dynamic extent are also seems to be controlled from "p" to "f" and no any notes sounds too loud.

Of course, there are also some disputes. Li xue mei had mentioned the extent of dynamic in her research, she believes that considered the tone colour and character of harpsichord and clavichord, the dynamic of Baroque clavier music are "between pp to mf" [18]. Nevertheless, according to the history about the invention of fortepiano, which was called "gravicembalo col piano e forte", which combine the advantages of harpsichord and clavichord: can do some crescendo and diminuendo, and the name also can shows the extent of dynamic: piano and forte. So choose the extent of dynamic changes from p to mf maybe more appropriate.

5.3 How to use the trend of dynamic changes in Bach's music: when play J.S.Bach's Sinfonia in A major (BWV 798) on modern piano.

This research had compare the literature in above paragraphs, From the above paragraphs had mentioned, even some musical researchers and pianists are agree with restore or imitate the Baroque clavier tone colour, it seems there are still many musical researchers and pianists are agree with Under the premise of retain the Baroque perform style to do some dynamic changes. For example, Rosalyn Tureck (1914-2003) had mentioned that "To put Bach's new understanding into practice, we need to create a new set of playing skills" to "make Bach's music live on the piano in its proper way, and free it from the piano treatment in the 19th century" [16], She claimed that we should not be biased towards any kind of musical instrument and should not be superstitious about any kind of timbre, and she believes that different performer can make different hand notes to the original score (subjective understanding) [16].

Rosalyn Tureck's opinion had been supported by many other researchers and pianists, such as Richard Good and Alfred Brendel, which had been mentioned in the above paragraphs, especially Richard Good, he said that "People can't have a pair of 18th century ears in the present", "The ideal state is to live in the music from the heart, not from the surface, but through your own feelings." [16]. Which means that how to use dynamic changes in Bach's music on modern piano performing not only needs to feel the music by performer themselves, but also they need to analysis the music and even understand the music by the analysis.

From above paragraphs had discussed that the crescendo and diminuendo can be use in Bach's music when perform on modern piano, but it had not discuss how to use crescendo and diminuendo appropriately in Bach's music. So from this paragraph, this research will discuss how to use the the trend of dynamic changes when play J.S.Bach's Sinfonia in A major (BWV 798) on modern piano by contrast literature, scores and pianists perform audios.

5.3.1 Highlight the long notes

Highlight the long notes in the parts also had been used in many performance, such as Tatiana Nikolayev [11] and Yunchan Lim [15], which were discussed above. Leng Jia [19] had mentioned that when play Bach's polyphonic music on harpsichord, "Especially use the double keyboard harpsichord, can make the different parts melodies layers contrast sharply", and she believes that it is an "advantage", because she claimed that "at that time (Bach's time) play Bach's clavier music (on harpsichords) do not need to much more body movements and basically depends on fingers' touching", but now when we "use modern piano play Bach's music at present not only needs fingers touching but also needs to use body movement to control dynamic and tone colour" [19]. It seems that if the performers want to maintain the "advantage" of "make the different parts melodies layers contrast sharply", they have to adopt some measures.

So, highlight the long notes in the melody parts maybe can be an effective measure to make the different parts melodies layers contrast sharply, not only in order to reflect the direction of the melody line better, but also maintain the performance way in Baroque style.

5.3.2 According to the trend of melodies to choose the trend of dynamic changes

Bach's music works are typically Baroque polyphonic music, and the Sinfonia in A major (BWV 798) also a three-voice polyphonic music, which means this piece was combined with three different melodies. These melodies' notes have different trends. Chen en en had concluded that It is necessary to appropriately use the dynamics and the contrast between light and dark timbre according to specific musical factors [1].

These kinds of situations are also appear in these six pianists performance. According to the previous analysis, which has mentioned that Tatiana Nikolayeva had used three different ways in three voice, which had been shown In the previous analysis, and the most obvious characteristics is every voice dynamic changes are according to the tendency of melodies' rise or fall: when the melodies' note shows a rise tendency, she will use crescendo to perform. If the melodies' note shows a fall tendency, and she will use diminuendo in her perform.

Zhu Xiao-Mei [14]'s perform method are different from Tatiana Nikolayeva [11], in the previous analysis, she prefer to use crescendo in 10th-12th bar. However, according to the audio sound, although not quite obvious, it still can recognize that when she play these three bar have "bright and dark" tone colour contrast: when she play the top voice, the melody will sound "bright" and louder. Conversely, when she play the baritone voice, the melody will sound "dark" and overcast.

6 Conclusion

This research had analysed the dynamic of J.S.Bach's Sinfonia in A major (BWV 798) by compare dynamic marks in different scores and discussion in different literature, contrast and discuss six pianists performing by analysis their performing audio sound and waveform, and finally concluded that:

Firstly, when play Bach's period (Baroque) music work on piano, it seems can use "crescendo and diminuendo" limited, the extent of dynamic can be better to control from p to f.

Secondly, it seems that can adopt to highlight the long notes in melodies to help the different voices melodies contrast more obviously, which may helpful for keep the character of Baroque music style as much as possible

Finally, choose appropriate dynamic changes in perform by analysis melodies' trend may can be useful, not only to contrast dynamic, but also may helpful to precipitate the contrast of tone colour "bright and dark" .

References

1. Chen, E.E. (2010) Tradition and Selfhood: Annotation and deduction the works of late Baroque's clavichord on the piano. In: East Cin Baroque keyboard music: retro and modern. Exhina Normal University. Shanghai. pp. 1- 44.
2. Tan, Y. (2019) The aesthetic dimension plorations in Music., 144(03): 129 -138.
3. Holograph manuscript (1720-1723). Sinfonia 12. In: Public Domain. Staatsbibliothek zu Berlin (D-B): Mus.ms. Bach P 610, Berlin. pp. 54 -55
4. H. G. M. Darnköhler. (1745-1755).Sinfonia 12. In: H. G. M. Darnköhler. (Copyist.), Introduction in German with Clavierbüchlein text. Toulouse Municipal Library (F-TLm): Res. Mus. B 561, Toulouse. p.29.
5. Czerny, C. (1840) No.XII. In: C. F. Peters. (Eds.), Compositions pour le Pianoforte. Public domain. Edition Peters, Leipzig. pp.42- 43.
6. Czerny, C. (1903) XII. In: Fischer, C. (Eds.), Joh. Seb. Bach thirty two and three part Inventions for piano. Carl Fischer's Music Library. New York. pp. 56-57.
7. Busoni, F. (1891) 12. In: Busoni, F. (Eds.), J.S.Bach Klvier werke Busoni-Ausgabc v Dreistimmige Inventionen (BWV 787-801). Breitkopf und Härtel. Leipzig. pp. 32-33.
8. Leisinger, U., Jonas, O. (2013) Sinfonia 12 (BWV 798). In: Leisinger, U., Jonas, O. (Eds.), J.S.Bach Inventions and Sinfonias (2 and 3part Inventions). Wiener Urtext Edition. Wiener, pp. 56-57.
9. Scheideler, U., Schneidt, M. (2014) Sinfonia 12. In: Scheideler, U., Schneidt, M. (Eds.), J.S.Bach Inventionen und Sinfonien. Henle Verlag. Germany. pp. 58-59.
10. He, S.Y. (2000) A major. In: He, S.Y. (Eds.), Bach's Collection of Inventions and Sinfonias. People's music Publishing House. Beijing. pp. 54-55.
11. Nikolayeva, T (1977) 12. In A Major, BWV 798 [Recording]. In Tatiana Nikolayeva J.S.Bach-Inventions Nos.1-15 BWV 772-786 Sinfonias Nos.1-15 BWV 787-801 (CD 501055). USSR: Mezhdunarodnaya Kniga.
12. Gould, G (1955) No.12 in A Major [Recording]. In Gleen Gould - 1955 CBC Broadcast: Sinfonias (3-part inventions) by J.S.Bach. Canada: CBC Broadcast
13. onepiece5973 (2011) András Schiff plays BACH: Sinfonia BWV 787-801. https://youtu.be/Er0jHX6qJl0
14. Zhu, X-M. (2015) Sinfonia No.12 in A Major, BWV 798 [Recording]. In J.S. Bach: Inventions and Sinfonias – Zhu Xiao-Mei–VINYL & CD (CD ACC30350). Leipzig, Germany: ACCENTUS Music GmbH.
15. KUMHO ART HALL (2020) J.S.Bach 15 Three-Part Inventions, "Sinfonias", for Keyboard, BWV787-801/임윤찬. https://youtu.be/AMo-XoSkPxc
16. David, D. (2001) Reflections from the keyboard: the world of the concert pianist. BaiJia Publishing House, Shanghai.
17. Tureck, R. (2009) An Introduction to the performance of Bach. People's Music Publishing House, Beijing.
18. Li, X.M. (2000) The analysis of the dynamic of Baroque's keyboard music. In: HUANGZHONG (Journal of Wuhan Music Conservatory). Wuhan. pp. 63-69.
19. Leng, J. (2002) The analysis of J.S.Bach's Sinfonia. In: Journal of Xinghai Music Conservatory. Guangzhou. pp. 62-65.

A new path of visual communication from the perspective of the metaverse —— Take Liu Yexi's Internet Communication Spectacle as an Example

Dingtai Li[1,*], Tingwei Liang[1]

[1] Huazhong Agricultural University, Department of Advertising and Communication, Hongshan, Wuhan, Hubei Province, China

Abstract. In 2021, the "Metaverse" concept became a big hit. As an exceptional media concept, the metaverse is closely related to the human sensory system, especially visual communication. Through the method of literature research and case analysis, combined with the recent publicity craze of "Liu Yexi" and from the perspective of the combination of the metaverse and short video, this paper argues that the richness and abstraction of the concept of metaverse require short video as a new form of communication. On this basis, the practical experience of "Liu Yexi" visual communication is analyzed from narrative logic, innovation leap, and visual spectacle. Then it is pointed out that the capitalization trend of artificial intelligence is the potential risk the metaverse may face in the long-term development.

1 Question raised: the theoretical logic of visual communication in the metaverse perspective

The concept of "metaverse" first originated from the cyber world constructed by American fantasy writer Stephen Stephenson in the science fiction novel Avalanche in the 1990s. In the novel, Stephenson sees the imaginary metaverse as a kind of open virtual space parallel to the physical space: "a computer-generated world, which the computer depicts on his eyepiece and sends sound into his headphones."[1] Since the concept's inception, the metaverse has been inextricably linked with the human sensory system as a unique medium conception.

On October 28, 2021, Facebook President Mark Zuckerberg announced at an annual conference focusing on augmented reality (AR) and virtual reality (VR) that Facebook would change its name to "Meta" (prefix of metaverse), marking the international social media giant in the field of metaverse opened a primary strategic layout, a time in the world detonated on the topic of "metaverse" heat.

Based on people's visual experience, the metaverse is a transcendence of the natural world and its sensory experience. It not only deepens the connotation of visual communication and broadens people's cognitive boundary of vision but also causes a considerable change in the future media field, arousing attention and hot discussion from all walks of life.

From the current academic research process, scholars in relevant fields at home and abroad are highly concerned about and actively respond to the related concepts and fundamental propositions of the metaverse, and the "metaverse" and its extension gradually show a trend of migration from the social hot spot to the academic hot spot. As early as the 1960s, British anthropologist Greggery Bateson proposed the concept of meta-communication. Based on his comprehensive research in anthropology, linguistics, psychology, information theory, cybernetics, and other multidisciplinary fields, he believed that communication is far from simple information in the sense of the text. "Human linguistic communication can and always operates on many opposing levels of abstraction"[2] and then defines the two aspects of meta-communication.[3] Professor Yu Guoming of Beijing Normal University believes that the birth of the "metaverse" marks the beginning of the future form of Internet evolution.[4] The emergence of the metaverse can be understood as the process of humans constantly breaking through the boundary of the natural world "outwards" and moving towards greater freedom with the upgrading and iteration of media. On this basis, it also puts forward the prospect of the "inward" development of media in the future and the construction of a "mind world" by human beings themselves. He also believes that the core of the rise of the metaverse concept lies in that the return of communication power is the fundamental logic of media evolution in the digital age, and human beings are the most primitive and core media.[5] Professor Hu Yong from Peking University proposed, based on rethinking the concept of meta-media, that, like almost all meta-media, metaverse, as a comprehensive medium, reduces the "former mainstream

* Corresponding author: lidingtai@yeah.net

media" that occupied the dominant position in human society to secondary media and subordinate media.[6] Moreover, from four aspects of context-based consensus communication, voice-based speech communication, avatar-based reflexive communication, and environment-based multi-task communication, this paper discusses the new turn that the metaverse will bring to future communication.

Based on this, this paper takes the metaverse as the theoretical perspective, combines the literature research and case analysis methods, and carries out a literature search for "metaverse" and related keywords through CNKI, Web of Science (WOS), and other databases. Based on clarifying the fundamental theoretical propositions of the metaverse and future visual communication, and further combining with the recently concerned " Liu Yexi " event, this paper analyzes how such a wonder of Internet communication provides a new thinking path for future visual communication in the metaverse perspective.

2 Concept interpretation: Liu Yexi, under the trend of the metaverse

On October 31, 2021, the TikTok account "Liu Yexi" released a concept video lasting 2 minutes and 11 seconds. Liu, a virtual character with Chinese-style makeup, appeared in public for the first time.

At the beginning of the video, the camera zooms in and a woman (Liu) dressed in a Chinese-style red dress is dressing up in front of a dressing table. The first thing people see is her back. At this point, the camera suddenly panned to a crowd of ordinary people behind the woman looking to take photos. As Liu turns around, a young boy steps forward from the noisy crowd and asks the mystery woman in front of him, on behalf of the public, questions such as "Your makeup is ugly" and "Are you human after all?".

As of December 30, 2021, in less than two months, Liu Yexi 's TikTok account has released five videos with similar styles and distinct themes, with a cumulative number of 8.06 million fans and 18.3 million likes, among which the first video has received nearly 3.6 million likes and occupied many hot topics in the hot search list of Douyin continuously within one week—generating extremely high Internet popularity.

Liang Zikang, the head of the team behind Liu Yexi and the CEO of Chuang One Technology, said in an interview that Liu Yexi, wearing eyeliner, eyeshadow, lipstick, long hair and waist, dressed in ancient style and elegant, both actual and mysterious national style image, is a virtual beauty KOL born after three years of technical accumulation of the team and carrying the concept of the universe.[7] The behind-the-scenes team gave Liu Yexi the identity of a "virtual beauty master who can catch monsters" and can interact with people in the real world, such as can perceive people's language, communicating with people, body contact, and then judging human behavior with the cognitive rules and civilization of their world. In the virtual world built by Liu, human beings coexist with themselves in two parallel and independent dimensions of time and space. As video viewers, most users can be attracted by the story of the interaction between the two worlds presented in the video. This constitutes a virtual space -- the metaverse, which can not only project the real world but also be independent of the natural world, with a great sense of immersion and experience.

2.1 Short video form fits the primary trend of future visual communication

In the current communication environment of redundant information, people live in an era of mass information congestion and visual fatigue. "Our eyeballs have never been so busy and tired as today. On the one hand, visual needs and visual desires are constantly climbing. The desire to see has never been so strong as today; On the other hand, the highly visual and mediatization of contemporary culture has provided us with more possibilities and higher quality and more attractive images for viewing."[8] In this context, as an innovative media with rapid development in the era of mobile Internet, the short video is characterized by its more common technical basis, lower production threshold, more diversified scene adaptation, more direct and rapid emotional arousal, more accessible user cost, and easier to transform social capital.[9] In the future, for an extended period, it will maintain the vigorous vitality from "volcanic eruption" to "long stream of water". According to the 48th Statistical Report on Internet Development in China released by China Internet Network Information Center (CNNIC) in August 2021, as of June 2021, the number of short video users in China was 888 million, an increase of 14.4 million compared with December 2020, accounting for 87.8% of the total netizens. Under the hot tuyere of the current metaverse, the transmission form of short video provides the primary development conditions and sufficient imagination space for future visual communication.

Liu Yexi's explosive popularity reflects some commonalities of short video communication forms and presents characteristics different from traditional short videos. Most short video creators will choose the content push strategy of regularly updating and accelerating topics' popularity in the short video platform focusing on traffic and rapid iteration. However, different from the traditional Internet celebrity IP model that constantly updates around the same world view, "Liu Yexi is another set of gameplays", users can hardly see Liu Yexi's continuous and regular updates, even as a " metaverse " character Liu Yexi will not appear in his video alone. She is more like a mother universe and may appear in someone else's child universe.[10] For example, on November 27 this year, Liu Yexi released a video linkage with "Yucatan", another virtual IP owned by One Technology, connecting two characters in the metaverse through short video narration; This video makes "Huihui Zhou" in the real world and Liu Yexi in the metaverse form a correlation in the plot. Using the core concept of beauty, the video narrative conveys the "unknown full picture, no comment", which involves social morality, projects practical issues, and arouses a high degree of discussion among netizens.

2.2 The richness and abstractness of the connotation of "metaverse" call for a new form of communication

The relationship between technology and the virtual world has long attracted attention from all walks of life. As early as 2017, foreign scholars proposed from the perspective of computer technology development that there is no universally accepted definition of "virtual world" in academia and industry. The progress of network technology, the development of mobile Internet, and the rise of distributed computing prompt people to reflect on whether the current theoretical and technical framework can support the "virtual world" in the concept.[11] This year, the concept of " metaverse " suddenly became popular. The richness and abstractness of its connotation once again prompted people to think about the intrinsic relationship between the metaverse, media, and forms of communication.

The evolution and development of the metaverse cannot be separated from the transformation of the media. The earliest metaverse appeared in the literature (science fiction Avalanche), and the development of technology promotes the emergence of new media. The rise and innovation of electronic games have enriched the forms of presentation of the metaverse, such as the popular Sandbox game in recent years, which is an entire virtual world.[12] What will be the following form of communication in the metaverse from text to literature, from computer to video games? In the case of Liu Yexi 's communication, from the very beginning, the creative team not only created a single virtual IP like Liu Yexi but also gave it a unique thought, emotion, and behavior pattern that is different from human society, and thus constructed a highly developed virtual world that blends with the natural world and is independent of each other. In this world, people in real life socialize and interact with a surreal digital surrogate named Liu Yexi, completing the plot narration of short videos in the process of interaction and ultimately giving rise to a virtual civilization that is both mapped to and independent of the real world.[12] From text to literature, from computer to video games, the emergence of Liu Yexi provides a new possibility to express the rich and abstract connotation of the "metaverse" in short videos.

2.3 The application of advanced technology and the frontiers of theory make the visual communication of the "metaverse" possible

Visual technology is an essential carrier of communication activities. "It is some tools and practices that can produce or reproduce visual information for human eyes".[13] The landing process from concept to reality of the metaverse is also a process of the constant evolution of communication technology. With the continuous development and maturity of new communication technologies such as AR, VR, 5G, and cloud computing, the metaverse also has a greater possibility to gradually develop towards the direction of reality, platform, and even industrialization through the iterative upgrading of immersion, participation, and sustainability.

The popularization of mobile terminal equipment, the real-time signal transmission, and the precision and intelligence of algorithm recommendation together reduce the production and transmission cost of image and video information, big data, cloud computing, 5G, VR/AR, and other technologies drive the information transmission to "visualization + mobile screen media + immersive experience".[14] According to the production team behind the scenes, Liu Yexi is a 2.5-dimensional image design between the two-dimensional animation image and the three-dimensional real-world character image. The final landing molding is done through high-precision original painting design, highly realistic 3D modeling, the import of a high-level virtual human engine, high-precision animation capture, and highly exquisite frame modification rendering. In this context, Liu Yexi's explosive popularity proves the necessity and feasibility of developing visual technology to explain the metaverse concept.

3. Phenomenon analysis: the practical experience of "Liu Yexi" visual communication

3.1 The narrative logic of "virtual and real symbiosis" in media space

French sociologist Jean Baudrillard believes that "surrealism shows a higher stage, eliminating the opposition between real and imagined. What is not real no longer remains in the dream or the transcendent. It is in the hallucinatory analog of reality."[15] This kind of "surreal" media space, in the context of the rapid development of artificial intelligence technologies such as AR and VR, dramatically improves the degree of virtualization of the social environment and clears up the boundary between physical reality space and virtual digital space.[14] After the characteristics of "virtual and real symbiosis" in the metaverse world enter the media space constructed by short videos, a very innovative narrative logic is represented in the communication case of Liu Yexi.

In the video update on December 12, Liu holds an oil-paper umbrella as she enters a narrow and dark grocery store, where a TV set plays a missing person notice for an 8-year-old girl. In a short sequence, Liu chases her store clerk sister into a dim factory room filled with electronics, where countless computers display brain graphics that transmit data. The 8-year-old girl is lying on a bed with a device on her head.

In just 4 minutes, the picture of cyber style and the texture of the movie version, through the continuous reversal and exciting story plot, presents a "science fiction short video" about justice, family affection, and the battle between good and evil, vividly displays the narrative logic of "symbiosis between virtual and real", which not only brings people future-oriented science fiction imagination, It also conveys the universal media motifs and values of human society.

3.2 The innovation leap of the existing visual communication logic

In most cases, innovation is not a groundbreaking breakthrough from "0" to "1" and from nothing but a new leap based on the rearrangement and combination of existing or old elements and critical inheritance.

Liu Yexi's craze relies on short videos as the essential communication medium. Its narrative framework, story logic, transitions, and special effects mostly rely on the existing relatively mature technical chassis and the Douyin platform's interactive characteristics. Through Liu Yexi's official account, users actively interact with each other, making the comment section of videos become the second communication field. While generating the heat of topic discussion, it also continuously enriched the "virtual person design" of Liu Yexi. This kind of IP to create "routine" and marketing ideas in the current communication environment is not strange. However, based on learning from various forms, methods, and logic of current visual communication, Liu Yexi's communication boom integrates characteristics and critical development and cleverly finds a communication road that fits his development stage and responds to the boom of the "metaverse" era.

3.3 Spectacle of visual consumption

In the late 1990s, American scholar Douglas Kellner, based on inherits the theory of "landscape society", proposed the theory of "media spectacle" -- media spectacle is defined as those media cultural phenomena that can reflect social values, guide modern life, and have conflicts and drama at the same time.[16] In recent years, domestic scholars have extended and expanded the concept of "network image spectacle" based on "media spectacle" and investigated the human scene, cultural form, and thinking logic presented by network image in the process of production and communication that can stimulate and satisfy human curiosity from the aspects of production technique, text content, and appearance.[17]

At present, the consumption operation of visual culture promotes the prevalence of the "eyeball economy". The more "new, strange, and strange" visual pictures, the easier it is to get consumers' favor and promote cultural consumption behavior. Looking at Liu Yexi's communication case, the production team first constructed narrative wonders using decentralization and intertextuality, connecting the "metaverse" space parallel to the real world with the daily life of human beings, brewing the transition of the story; Through the expression of superficial moments, impact pictures and familiar emotions, the audio-visual spectacle is constructed, and then the unique interactive properties of short video platforms break the one-way transmission of traditional images, establish interactive wonders, and finally ignite the topic heat by the concept of "metaverse" and its unique mode of communication, forming a spectacle of public opinion. Suppose the discussion of "metaverse" from all walks of life was mainly at the conceptual level before. In that case, the media spectacle formed by Liu Yexi makes the concept of metaverse "fly into ordinary people's homes" and form a correlation with the daily life, media contact, and cultural consumption of every ordinary person.

4. Critical thinking: capitalizing concern for future visual communication

Currently, Liu Yexi is in a golden development period from concept to reality and rapid rise. In a foreseeable period, a series of subsequent works released by Liu Yexi still have great potential to induce a new wave of communication and construct broader media wonders. However, it cannot be ignored that Liu's "great popularity" also brings some worries for us to re-examine future visual communication.

From a practical point of view, Liu's initial positioning as a virtual beauty blogger is one of many in the industry. When Chinese virtual icon Ling Ling brought Gucci lipstick, consumers questioned the authenticity and credibility of her experience. When virtual IP is combined with commercial consumption, the characters in the virtual world cannot truly experience and consume the product, which is essentially contrary to the value expectation of consumers for a long time to "bring the goods and host the broadcast".

In the long run, the technology capitalization trend is a more alarming phenomenon. The biggest investor behind Liu is China Win Capital, an investment giant with over 100 billion yuan in assets. The three years of technology accumulation that Liu has experienced also shows that advanced visual communication technology and its application must be connected to the mass operation of capital behind it. Some scholars have long expressed concerns about the intellectualization of capital. The capital intervention will make artificial intelligence appear as compulsory progress and become the memory to absorb and extract human spiritual labor.[18] From this sense, no matter the producer, transmitter, or viewer of Liu Nighxi, there seems to be a risk of becoming a "labor machine" under the pressure of capital. When Liu Yexi gains a large wave of capital and traffic through the open new media platform, will Douyin, as a content production platform for "grassroots" creators, welcome a batch of capital players from now on? In this context, where will ordinary content producers and consumers go from here? The jury is still out.

5. Conclusions

Based on the research background of future-oriented visual communication from the perspective of the metaverse, combined with the wonders of network image caused by Liu Yexi, this paper holds that the communication form of the short video is in line with the primary trend of future visual communication under the concept of "metaverse". It analyzes Liu's practical experience in visual communication from three aspects: narrative logic, innovation leap, and visual spectacle. Finally, from the perspective of future development, the capitalization trend of artificial intelligence is the ethical controversy and moral hazard that the metaverse visual communication represented by Liu Yexi may face in the future.

However, as the concept of "metaverse" is still in its infancy in the current academic research at home and abroad, the available literature and empirical research are relatively weak, which also leads to a weak link in the theoretical logic of this paper. The theoretical understanding and practical experience of the "metaverse" must be further summarized.

References

1. Neil Stephenson, Guo Ze translated. (2009) Avalanche. Sichuan Science and Technology Press, Chengdu.
2. Bateson, G. (1973) Steps to an ecology of mind: Collected essays in anthropology, psychiatry, evolution, and epistemology. Paladin,150.173.
3. Ruesch, J., Bateson, G. (1951) Communication: The social matrix of psychiatry. W.W.Norton & Co.
4. Yu Guoming. (2021) Evolution Logic of Future Media: Iteration, recombination and dimensionalization of "Human Connection" -- The Future from "scene age" to " metaverse" to "mind World". Press Circles, 10:54-60. https://doi.org/10.15897/j.cnki.cn51-1046/g2.20211228.002
5. Yu Guoming, Geng Xiaomeng. (2022) Why "metaverse" : The Future ecological picture of media Society. Journal of Xinjiang normal university (philosophy and social science edition), 03. https://doi.org/10.14100/j.cnki.65-1039/g4.20211119.002
6. Hu Yong, Liu Chunyi. (2022) Metaverse as media: reevaluating gains and losses of communication. Press Circles, 01:85-99. https://doi.org/10.15897/j.cnki.cn51-1046/g2.20211228.002
7. Hedgehog Commune. (2021) Who is Liu Yexi? https://www.thepaper.cn/newsDetail_forward_15256969
8. Zhou Xian. (2008) The Turn of Visual Culture. Peking University Press, Beijing.
9. Peng L. (2019) Short Video: "Transgene" and Rebreeding of video productivity. Media, 01:34-43. https://doi.org/10.15897/j.cnki.cn51-1046/g2.2019.01.005
10. Liu Ying. (2021) An interview with Liang Zikang, CEO of CreativeOne Technology: Liu Yexi is the forerunner, and the trend of "meta-universe" cannot be resisted. Modern Advertising, 21:34-35.
11. Kim J.L. Nevelsteen. (2017) Virtual world, defined from a technological perspective and applied to video games, mixed reality, and the Metaverse. Computer Animation and Virtual Worlds,05. https://doi.org/10.1002/cav.1752
12. Fang Lingzhi, Shen Huangnan. (2022) The concept of technology and the change of civilization: metaverse research. Industrial economic review, 01:5-19. https://doi.org/10.19313/j.cnki.cn10-1223/f.20211206.001
13. Jacques Lacan, Wu Qiong translated. (2005) The wonder of visual culture. China Renmin University Press, Beijing.
14. Chen Shihua, Wang Lei. (2020) The Logic and Approach of Visual Communication of "Chinese Dream". Journal of Nanchang University (Humanities and Social Sciences Edition), 51(06):94-102. https://doi.org/10.13764/j.cnki.ncds.2020.06.011
15. Mark Poster. (1988) Jean Baudrillard Selected Writings. Stanford University Press, San Francisco.
16. Douglas Kellner, Media Spectacle, Shi Anbin translated. (2003) Perspectives on Contemporary American Society and Culture. Tsinghua University Press, Beijing.
17. Wang Changxiao, Liu Ruiyi. (2018) Generation Logic, Type Construction and Meaning Decoding of Network Video Spectacle. Modern Communication (Journal of Communication University of China),04:86-91.
18. Li Ailong. (2019) Intelligent capitalization and Capital Intelligence. Academic Circles,10:67-74.

Research and analysis of the development of news film and television based on the integrated media environment

Yi nong Tian[1,*], Guanwen Diao[1]

[1] College of Imaging, Dongguo University

Abstract. The expansion of the digital integration mode of contemporary digital cinema from creation to projection and the change of cinematic language both herald the breakthrough of technology once again and the opening of a new era of digital cinematic art. From the aesthetic point of view, digital cinema has overturned and dismantled some of the theoretical essentials and important concepts of traditional film aesthetics about film ontology, and the changes and progress of digital cinema's generation principle, audiovisual environment and media properties have profoundly influenced film art creation and audiences' aesthetic approach. Digital reproduction and regeneration can realistically simulate reality and transcend reality to construct the story world in film, forming a higher level of aesthetic composition and aesthetic concept. The innovation of film media and the transformation of digital film machines enable audiences to develop an embodied experience at a super close distance or without distance, thus falling into the "immersion of the here and now" and "playable story". The digital anthropomorphic film phenomenon experience and the immersive viewing mode of multimedia sound and image have perfectly realized the freedom and transcendence of digital film aesthetic experience and become an important force to promote the development and evolution of film aesthetics. It is the original purpose of this dissertation to explore the relationship between digital film technology development and film aesthetic discourse under the new digital technology conditions, to sort out the aesthetic implications and aesthetic characteristics of digital films, to explain the film concepts and aesthetic issues in the digital era, and to try to complete the systematic construction of contemporary digital technology film aesthetics.

1 Introduction

With the rapid development of digital film technology and computer graphics generation technology, film production has entered the digital era, and the whole process from production to projection of images has undergone a radical change [1]. Digital cinema can obtain high-definition and high-fidelity audio and video files by using digital cameras to complete live-action filming, or by carrying out digital modeling and digital adjustment through virtual cameras, image graphics processing software, graphics video processing software, etc., to realize the process of associating live images with models, synthesizing and integrating dynamic images, and seamlessly editing images to obtain virtual images [2]. The digital film can add digital scenery and color processing at will, and the digital image creation can be modified for an unlimited number of times. The composition, perspective, color expression, space, texture, environmental climate and other picture generation and artistic effects can be designed according to the needs of the plot, and the pursuit of fake or exciting and shocking audio-visual spectacle. As the technical means of film, digital technology continues to develop the interactive function of film images with audiences in the process of generation and dissemination, and with 3D, VR, AR, high-resolution, high frame rate and other high-tech means fully penetrate into the various processes of the film industry. Digital images can also be produced after traditional film shooting by digital intermediate film from film to magnetic, and then supported by magnetic-to-film technology to accommodate different projection equipment and screening needs [3].

2 TECHNOLOGICAL EVOLUTION AND MEDIA INNOVATION OF DIGITAL CINEMA

The technical nature of film and the artistic nature of film are complementary and dialectically unified. The technical means of film are the prerequisite and foundation for the forward movement and development of film art, and the innovation and progress of film art in turn promote the innovation and enrichment of technology [4]. The technological evolution and media innovation of digital cinema have profoundly influenced the development of film art and the aesthetics of contemporary audiences. The subversion of the principle of digital image generation has brought a new way of artistic expression to digital cinema, giving rise to many new explorations of film technology practices, and the

* Corresponding author: yit23480@gmail.com

 https://doi.org/10.1051/shsconf/202316702014

continued exploration of the properties of the film medium has created new ways of digital aesthetics [5].

2.1 Image generation: from traditional lenses to digital anaglyphs

In terms of the developmental flow of digital technology involvement in film, some digital technology has long been used in the traditional film production field, such as program control applied to film developing machines, digital dubbing machines, cameras, etc. The local digital use in film hardware and production processes was a useful attempt by the film industry to improve production efficiency, but of course at first digital technology did not have a profound impact on film production and artistic creation until the creation and development of CG technology provided frame-by-frame generated computer animation and precise algorithms to reshape images for film creation and production [6].

2.2 Media innovation: the development of "digital integration" in the film industry

In a complete sense, digital cinema refers to "films in which traditional optical, chemical, or physical processing techniques are partially or completely replaced by digital technology in the filming, post-processing, and distribution of the film" [7]. In other words, digital cinema is a digital system, which includes shooting, transmission, encoding, distribution and screening, etc. These processes follow the sequence from digital shooting and production to distribution and broadcasting, but they are interrelated and influence each other, which is very different from the relatively independent but non-interfering production process in the traditional film era, and it makes the whole film industry closely integrated as a whole [8].

3 TECHNOLOGICAL CHANGE TRIGGERS BREAKTHROUGHS AND RECONFIGURATIONS IN FILM AESTHETICS

3.1 Subversion and Abandonment: The Breakthrough of Digital Cinema on Traditional Film Aesthetics

The editing method of digital film has further changed the "passive reception" proposed by montage into "active participation", forming a new media form of more complex montage. This has led to a non-linear narrative path for film creation, allowing film ideologies to fall into the fragmentation of modernity and the more fragmented interpretation of post-modernity, digital mimicry is completely different from the "restoration of material reality" required by documentary aesthetics to pursue the reproduction of reality, digital mimicry emits the falsehood of artificial intervention at all times, but has the form of "hyper-reality". "The ontology of the photographic image as a means of artistic montage and photorealism has since lost its ontological meaning and fallen into a situation of collective dissonance [9]. The powerful technical nature of digital cinema brings about the consideration of aesthetic issues such as virtual existence, cinematic reality, and the reproduction of reality in cinema, which will inevitably lead to the revision of our understanding of the ontology of cinema and the further construction of the aesthetics of digital cinema [10]. In summary, digital cinema has dismantled important concepts of traditional film aesthetics, and the generative principle, audiovisual environment and media properties of digital cinema have affected the creation of film art and the aesthetic approach of the subject. On the one hand, the ontological issues of classical film aesthetics have changed in the digital era, and the subversive nature of image production has made Bazan's "ontology of the photographic image" disappear and Eisenstein's "aesthetics of montage" become less effective. Although the meaning of the camera does not seem to have changed much in the viewer's perception due to the mimetic logic of digital mimicry, it has completely changed the original meaning of the image from the perspective of creative image transformation and mimicry generation, and digital film has formed a new ontological form due to its digital reproduction and regeneration. On the other hand, modern film aesthetics is concerned with the relationship between the film image and the audience and its changes, digital film is no longer free from the interference of the image itself as the traditional relationship between the image and people in the era of mechanical reproduction, and the ideographic function of digital mimesis makes the distinction between "referent" and "referent" of Metz blurred and indistinguishable. "The digital film technology can simulate reality, express reality, and transcend reality, forming a higher level of aesthetic composition and aesthetic concept, and the audience's aesthetic consciousness and subjective motivation have been enhanced as never before. Both the artistic and aesthetic aspects of cinema, realistic mimesis and aesthetic transcendence have become important forces driving the development and evolution of film aesthetics.

3.2 Integration and Progression: The Landing Point of Aesthetic Reconstruction of Digital Technology Cinema

The change of film media formed by digital technology has caused the problems of image ontology and mimetic reality, and the reproduction and transcendence of reality by digital mimicry and image transformation of virtual reality will inevitably bring about new aesthetic forms and the consequent changes in audience aesthetic consciousness and aesthetic psychology, which cannot be reasonably explained by traditional film theories, thus the artistic and aesthetic problems of films in the digital era are bound to be completely new constructions, film aesthetics had to respond to the demands of this technological change and reconstruct film aesthetics in the era of new digital technologies [11].

As far as the objects of film aesthetics are concerned, the interventions of Bazan and Eisenstein from the technical perspective, Balazs from the film culture perspective, Hugo Minsterberg, Rudolf Einheim, Jean Mitri, Metz and others from the psychological perspective all make us ponder that film aesthetics has always In fact, film theory needs to rely on the audiovisual system under technical support in order to form film aesthetic expressions and then abstract them into general concepts and laws. The changes in film media formed by digital technology have given rise to the problems of image ontology and mimetic reality, and the reproduction and transcendence of reality by digital mimesis and image-transformed virtual reality are bound to bring about new aesthetic forms and the consequent changes in audience aesthetic consciousness and aesthetic psychology, which cannot be reasonably explained by traditional film theory, thus the artistic and aesthetic problems of films in the digital era are bound to be completely new constructions. Therefore, film aesthetics has to respond to the requirements of this technological change and reconstruct the film aesthetics of the new digital technology era.

This reconstructive approach is based on.

First of all, the changes in film language and artistic communication call us to return to the basic aesthetic scope of film to find out how digital film shows new characteristics that are different from traditional film aesthetics, and then we can make an abstract overview of the new film aesthetic expression and aesthetic pursuit. First, digital cinema needs to be based on the essence of images, to rethink the rewriting of traditional images' real characteristics by digital mimicry, the significant aesthetic changes in the relationship between time and space, narrative concept, light and shadow modeling, sound and picture relationship, and to develop a comprehensive aesthetic interpretation of digital mimicry's "super-real" concept of reality. Secondly, the fusion of reproduction and representation formed by the digital transformation of the long shot and montage also presents a balance between realism and formalism, which needs to be summarized in the process of close reading of the text and comparison with traditional film language to explore the new aesthetic form of digital mimicry in the reconstruction of reality; thirdly, classical film aesthetics studies the relationship between film and theater, literature, music, painting and other art disciplines. Digital cinema gradually regards film as a special form of digital media and devotes itself to studying the relationship with interactive media, especially interactive TV, online games and digital animation. Whether this deviates from the scope of film as art or the symbiotic needs of art and media in the digital age, we need to think deeply about the relationship between digital technology and film art, identify the artistic nature of digital film in technical reflection, and explore The relationship between digital technology and film art requires us to think deeply about the relationship between digital technology and film art, to identify the artistic nature of digital film in technological reflection, and to explore the aesthetic value of film amidst digital transformation. Secondly, in the comprehensive shift of contemporary video art to image view, from the perspective of classical film aesthetics, the change of bearing medium and image generation of digital film marks the subversion of traditional film ontology, which not only dismantles the two traditional film ontological concepts based on montage and photorealism, but also enables us to see that after the culture of consciousness, "the renewal of art gives up the search for deep meaning and turns to superficial pleasure satisfaction. After the culture of transcendental consciousness, we can see that "the renewal of art has abandoned the search for deep meaning and turned to superficial pleasure satisfaction. The "visual shock" and "entertainment carnival" seem to confirm the conclusion of "the end of art", especially the digital film that focuses on showmanship deviates from the runway of "grand theory" and makes the aesthetics of digital cinema more beautiful. The "grand theory" runway has allowed sensuality to revel while aesthetics tends to be blunted. The "grand theory" includes many studies on film culture and social criticism, such as ideological theory, psychoanalysis, Frankfurt School, feminism, postmodern culture, etc. The reason why they occupy the main position of film theory is "the inevitable result of the logical development of film theory research itself It is the attempt of film theory to seek new academic resources and theoretical directions through 'externalization'".2 In contemporary times, grand theory has developed into the promotion of landscape theory and visual culture, and the performance of digital film aesthetics with its emphasis on the function of displaying the spectacle fits well with the critique of spectacle shock in these "culturalisms". In his critique of "grand theory," Baldwell suggests that "a powerful challenge to grand theory is an intermediate range of inquiry, one that facilitates a shift from empirical specificity to more general argumentation and an exploration of implication. This fragmentary focus on problem-solving reflection and research is very different from the nebulous discourse of grand theories and from the piling up of information" (3). This "middle range" in fact requires us to return to a contemporary film aesthetics that focuses on aesthetic experience, rather than fleeing from it in the visual shock of spectacle ecstasy, and it is worth noting that Baumgarten's position in creating the discipline of aesthetics was to consider aesthetics as a study of "the perfection of perceptual awareness. It tends to discover beauty through the accumulation of sensual perception and the imagination and reflection of "rational-like" thinking, rather than exploring the deeper meaning of beauty in the mode of thinking of rational analysis. In our view, the retreat from grand theories does not mean abandoning these cultural self-proliferations and their contribution to film studies as a critical method, but rather thinking about these aesthetic impacts based on the new issues arising from contemporary film practice in the context of the technological means and historical breakthroughs of digital cinema as a disruptive change today, such as a profound understanding of the epistemological impact and aesthetic impact on film theories such as authorship, film machines, and ideology. In order to better understand the aesthetics of digital cinema, it is necessary to understand the epistemological impact and aesthetic transmutation of film theories such as authorship, film machine and ideology. Again, in terms of aesthetic concepts, the new media properties of digital era films have brought about an increase in audience interaction and an awakening of subject consciousness, which is particularly obvious. The freedom of digital mimetic writing has fundamentally changed the intrinsic relationship between cinema and the aesthetic subject. What we need to consider is not a secular aesthetics that focuses on perception, but how aesthetic perception enables contemporary digital film aesthetics to achieve aesthetic transcendence, because perception "is not, first of all, an event in the world that one can explain in terms of causality, but as a re-creation and a re-constitution of the world at every moment "4 . Thus, in the contemporary atmosphere of "aestheticization of daily life", how to break the rational chains of film aesthetics, return the aesthetic sensibility to its proper importance, treat the aesthetic process as a vivid and novel digital experience rather than a philosophical meditation, and more fully reflect the combination of consciousness and body to their own aesthetic activities. This also requires the establishment of a new aesthetic concept of digital cinema, leading contemporary audiences to become "aesthetic people" who gain aesthetic freedom in the digital age.

4 CONCLUSION

Looking back at the history of film development, the technical nature of film has played an important role in the development of cinematic art since its inception. The film has undergone the innovation process from silent to sound to digital surround sound, from black and white to color to digital color mixing, from film to digital intermediate to full digital integration, and each technological progress has brought endless vitality and vigor to the artistic communication and aesthetic expression of the film, while each qualitative change and leap in film art has been deeply marked by the technology, which is symbiotic and intertwined with the technology. Compared with previous technological advances, digital technology has a disruptive and transformative meaning, bringing film art into a new aesthetic ecology. Such high-tech means as digital recording and sound-mimicking technology, color mixing management, CG imaging technology, 3D/IMAX projection technology, VR technology, sensors and multimedia technology have brought film art creation into the mimicry stage of the simulation era, transforming the aesthetic elements of the basic aesthetic categories, forming the aesthetic characteristics of film in the digital era, giving digital film aesthetics a unique vision and connotation, and bringing freedom and transcendence to the aesthetic experience of the audience.

REFERENCES

1. Wei Jian, Ding Hao, Huang Zina, Han Jiale, Cao Yujia, Wang Yan. A preliminary study on the development strategy of news gathering and production for meteorological film and television in Guangxi[J]. Meteorological Research and Applications,2018,39(04):95-98.
2. Jiang Boyan. Guided by Xi Jinping's thought of socialism with Chinese characteristics in the new era to promote the high-quality development of journalism, publishing, broadcasting, film and television work [J]. Inner Mongolia propaganda ideology and culture work,2018(09):10-13.
3. Ni Guanghong. Focusing on the transformation of new and old dynamics to promote the high-quality development of the press, publishing, broadcasting, film and television industry [J]. Modern audiovisual,2018(02):1.
4. Yan Shiwei. The current situation and cooperation development of China-Africa cooperation in journalism, communication and film industry[J]. Communication and Copyright,2018(02):160-162.DOI:10.16852/j.cnki.45-1390/g2.2018.02.061.
5. Zhang Chunlang, Wang Liya, Chi Ming. Thinking about the path of improving the public opinion leadership of traditional broadcasting in the new era--in the context of the "Thirteenth Five-Year Development Plan" of press, publication, broadcasting, film and television[J]. Sound and Screen World,2018(01):5-8.
6. Tu, Ming-Hua, Wang, Xiao-Jie. The idea of the development of radio and television in the new era - four key words to interpret the "Thirteenth Five-Year Plan" of press, publication, broadcasting, film and television [J]. Sound and screen world,2018(01):8-11.
7. Wei Jian, Research on the current situation of meteorological film and television news reporting in Guangxi and development strategies. Guangxi Zhuang Autonomous Region, Guangxi Zhuang Autonomous Region Meteorological Service Center, 2017-12-26.
8. Yang Peng. A case study of media integration development of traditional media--Fujian Radio and Television Group news channel as an example[J]. Southeast Communication,2017(12):8-10.DOI:10.13556/j.cnki.dncb.cn35-1274/j.2017.12.004.
9. The "Thirteenth Five-Year Plan" for the development of press, publishing, broadcasting, film and television [J]. Young journalists,2017(31):2.
10. LU Juan, ZHANG Junchang, GUO Tancheng. Insisting on cultural orientation to build core competitiveness--analysis of key words in the "Thirteenth Five-Year" Development Plan of Press, Publication, Radio, Film and Television[J]. China Publishing, 2017(21):6-11.
11. Li Siqu. Carrying cultural self-confidence to create a new era of industrial growth pole--a reading of the "Thirteenth Five-Year Plan" for the development of press, publishing, broadcasting, film and television [J]. China Publishing, 2017(21):12-19.

Optimization study of media creation ideas based on the new media environment

Hui wen Zhao[1,*], Yi nong Tian [1]

[1] College of Imaging, Dongguo University

Abstract. Nowadays, new media are being used in more and more fields, most notably in the communication industry, with endless online media and ever-evolving cell phone terminals appearing in our lives all the time and becoming an essential part of our lives. Under the influence of this background, the creation and dissemination of art have gradually become dependent on new media and have begun a profound transformation. On the one hand, the new media has promoted the circulation of art, making the originally niche art categories to the mass market; on the other hand, the expanding market and audience also play an important role in influencing art creation. Therefore, the creative thinking of art creators nowadays will also change accordingly with the change of the background of the times.

1 Introduction

New media is a form of communication that uses digital technology to provide information and services to users through computer networks, wireless communication networks, satellites and other channels, as well as terminals such as computers, cell phones and digital TV sets. From a spatial perspective, "new media" refers to the media that can be globalized across geographical boundaries, supported by digital compression and wireless network technology, with its large capacity, real-time and interactivity, as opposed to "traditional media". In the twenty-first century, political stability, economic prosperity, and the continuous development of information technology have led to the advent of the information age [1]. New media as a product of contemporary society is penetrating every aspect of life, and the new media era has quietly arrived. New media drive the development of art, and the creation and dissemination of art gradually depend on new media to begin profound transmutation. The creation of dance has also been influenced by new media, and the thinking of dance creators has become more diverse under the influence of new media [2]. "In modern dance creation, we can see the new thinking of choreographers, feel the diversity of their creations and the richness of their performance forms". After the creation and development of new media, dance creation has been elevated to a new level. Therefore, this paper will take the dance choreographer under the new media as the research object, take the dance works as the entry point, analyze what changes are happening in the dance creation under the new media, and how the creative thinking of the dance choreographer as the subject of creation is changing in the new media, so as to explore the profound influence of the new media on the creation of dance art [3].

2 RICHNESS OF SUBJECT MATTER SELECTION

The constant emergence of new media has brought different creative thinking, of which dance choreography has also been affected [4]. Moreover, the emergence of new media makes dance choreographers give a different impetus to the shaping of dance images, the organization of dance language, choreography design, the organization of dance teaching, and so on. "And the choreographer needs to keep up with the times and be innovative in his thinking when creating dances in order to be more attractive and be able to better demonstrate the artistic charm of dance." The emergence of new media can make the traditional choreography more innovative and can make full use of new media technologies and resources to update dance creation. This paper focuses on the new changes in the creative thinking of dance choreographers in the context of new media [5].

The selection of materials is the starting point for the choreographer in the creation of the dance, and in the selection of materials for the dance, the creative thinking of the choreographer must be expanded as much as possible, so that the dance performance can be presented in the most excellent form, highlighting the artistic charm of the dance [6]. The original dance choreographer in the selection of materials because of the limitations of science and technology, beyond the reality of the selection of materials is often less, but the emergence of new media technology makes the choreographer's selection of materials with diversity, the use of new media technology and resources can be closer to reality in the selection of materials for dance. The special nature of choreography allows the choreographer to make the choreography directly realistic in its selection [7]. Dance choreographer's

* Corresponding author: 595009233@qq.com

thinking is not only dependent on reality, so the selection of materials can be based on reality, but also the imagination of the creation of the subject matter, in relying on the thinking of literature to create dance works, but also need to have commonality and at the same time fully explore their own independent personality, so that the creation of dance can be innovative. In the selection of new media dance, the same is based on reality, and then combined with new media technology, some images in reality, or some images in imagination, using technological means to express, the original some can not be expressed in the theater dance content, "borrowed new media to express, to give the audience a kind of real space and new media interlaced feeling, can better lead the audience into the dance, so that the audience through the new media into the dance, feel the artistic charm of dance and infectious power" [8].

As a dance choreographer must have innovative thinking, independent individuality of opinion, for the content of the dance you want to create for the selection of materials, you must carry out in-depth analysis, in order to borrow the new media to better performance of dance. New media is a bridge for choreographers to create something new, and borrowing new media can make dance performances more relevant [9]. When many choreographers create dance, the use of new media still occupies a small part and does not form a big concept. However, new media provide choreographers with the opportunity to reconnect with the body, theater, and dance in their dance selections, bringing profound artistic value and creative meaning to choreographers [10].

3 THE DIVERSITY OF CREATIVE MEANS

The emergence of new media has enabled dance creation to be combined with new media and to be more diversified in terms of creative means. "The original choreography relied on sculptures, murals, and other forms to fix the dance form, enabling the use of prose and poetry for the spatial expansion of the dance context when creating, enabling the creation and performance of dance art to be more diverse and pluralistic" [11].

When choreographers think about creation, they are not only limited to their previous thoughts, but also become more active and richer in their creative means by combining new media. As we mentioned before, various new media dance, turning dance into "digital", technology becomes part of the choreographer's creation, creating a new and strange chemical reaction. In particular, the dance company work "Pixel" is considered a dance masterpiece. In their self-introduction they have mentioned "placing the body at the heart of the challenges of technology and artistry, adapting the latest technological tools to create permanent poetry, and allowing a visual language based on playful pleasure to give nourishment to the imagination." It is ample proof that the new generation of new media dance explorers are clear that the technical tools must be reworked and used for dance, so their choreography becomes more enjoyable.

4 THE ATTENTION OF THE WORK FEEDBACK

In the use of new media, the feedback of dance works becomes more important. The audience can directly reflect their impressions of the dance by watching and enjoying the dance through new media, and can have the most real and intuitive feelings about the dance works, so the choreographer can better understand the impact of the works on people and people's evaluation of the dance works. The emergence and use of new media technology has made dance creation more perfect and innovative, and it is also a new artistic exploration when it comes to the perfection of the individual dancer's life value. New media does not ignore the intrinsic value and power of dance, enabling dance works to be closely integrated with the audience and enabling more people who love dance creation to participate in dance. For example, the dance TV program "Dance Dance" has enabled some civilian dance creators to realize their dance dreams through the TV platform. Therefore, when dance creators perform dance works, "pay more attention to the feedback of the dance works, so that the choreographer can be more practical in the level of thinking, timely understanding of the feedback of the creation of dance, and timely changes in the idea of dance creation".

The creation of dance works ultimately returns to the audience, so each choreographer is borrowing new media to create and then present his or her own work, paying more attention to the analysis of how the final presentation of the dance work will impact. While the new media became an enabler, it was also able to quickly allow them to identify certain problems in the choreographer's approach to dance creation. Dance choreographers have to skillfully apply new media for timely updates when creating their own works in order to create more excellent works. For example, the excellent dance creator Yang Liping has spent her life interpreting the life of a peacock, but at each stage, the peacock she has interpreted has a different artistic feeling. From the very first "Peacock Princess" to "Spirit of the Bird" and then "Love of the Bird", Yang Liping has a deeper knowledge of peacocks. The new media better show Yang Liping's inner thoughts, creating a beautiful dance scene and bringing the audience a different artistic experience. In 2012 CCTV Spring Festival Gala Yang Liping and Wang Di's double peacock dance, under the reflection of new media technology, let people see the beauty of peacock. All this perfect performance is inseparable from the hard work of Yang Liping, the dance creator, who understands the needs of the audience to be able to create such a wonderful dance.

Fig. 1. Dance "Bird's Love

5. A PRELIMINARY STUDY ON THE CREATIVE THINKING OF DANCE CHOREOGRAPHERS UNDER THE THRESHOLD OF NEW MEDIA

The emergence of new media has brought different creative thinking to people, among which dance choreography has also been affected. And the emergence of new media makes dance choreographers bring different impetus to the shaping of dance images, the organization of dance language, choreography design, the organization of dance teaching and so on. "And dance choreographers need to keep up with the times and be innovative in their thinking when creating dances in order to be more attractive and to be able to better display the artistic charm of dance" ②. The emergence of new media can make traditional dance choreography more innovative, and can make full use of new media technology and resources to update the dance creation. This paper focuses on the new changes in the creative thinking of dance choreographers under the threshold of new media.

The first section of the richness of the selection of the subject matter

The selection of materials is the starting point of the choreographer in the creation of dance. When selecting materials for dance, the creative thinking of the choreographer must be expanded as much as possible, so that the dance performance can be presented in the most excellent form, highlighting the artistic charm of dance. The original choreographer in the selection of materials because of the limitations of science and technology, beyond the reality of the selection of materials is often less, while the emergence of new media technology makes the selection of materials for dance choreography become diversified, the use of new media technology and resources can be closer to reality in the selection of materials for dance. Because of the special nature of choreography, the choreographer can directly make the dance creation realistic when selecting materials. The choreographer's thinking is not only dependent on reality, so the selection of materials can be based on reality, but also the imagination of the creation of the subject matter, and in relying on the thinking of literature to create dance works, but also need to have commonality and at the same time fully explore their own independent personality, so that the creation of dance can be innovative. In the selection of new media dance materials, the same is based on reality, and then combined with new media technology, the reality of some images, or imagination of some images, the use of technological means to express, the original some can not be expressed in the theater dance content, "borrowed from the new media to express, to give the audience a sense of reality and new media intertwined space It can better lead the audience into the dance, let the audience enter the dance through new media, and feel the artistic charm and infectious power of the dance" ①.

When the choreographer creates the dance, the materials can be real events, literary works, historical events, myths and legends, etc. As one of the artistic sources for the choreographer, these can restore the situation as much as possible with the help of new media technology, or the imagination can be fully illustrated, driven by the thinking, so that the choreographer can innovate his own thinking concept and present the dance to be more The choreographer is driven by the mind to innovate his own ideas and present the dance in a more realistic way. When the audience enjoys the dance, they can fully integrate into the scene that the choreographer wants to express and feel the artistic charm and value of the dance. For example, Qi Zhifeng's dance work "Holographic Shadow Dance - Growing Up" is based on a person's dance learning experience, which is a very realistic character growth process. The choreographer uses modern dance and contemporary dance elements to tell the life journey of Qi Zhifeng, a national standard dancer, from his teenage years, youth to his twilight years. Qi Zhifeng's teenage experience is played by young dancers Zhao Kuang and Zhou Jiajing, while Qi Zhifeng disguises himself as an initiation dance teacher. The melodious violin is like a song, telling the story of the dancer's childhood spent in the practice room.

2. Diversity of creative means

When choreographers think about creation, they are not only limited to their previous thoughts, but also become more active and richer in creative means by combining new media. As we have mentioned before, various new media dances turn dance into "digital" and technology becomes part of the choreographer's creation, creating a new and strange chemical reaction. In particular, the company's work "Pixel" is considered to be a dance masterpiece. As mentioned in their self-introduction, "placing the body in the challenging core of technology and artistry, adapting the latest technological tools to create permanent poetry, and allowing a visual language based on playful pleasure to give nourishment to the imagination." It is ample proof that the new generation of new media dance explorers are clear that the technological tools must be reworked and used for dance, so their choreography becomes more playful.

6. NEW MEDIA INTRODUCTION

According to the changes of new media users and audience groups, the evolution of new media can be divided into elite media stage, mass media stage and personal media stage.

1. Elite media stage

At the beginning of the new media, only a few groups of people had access to the new media and used it to disseminate information, most of them were professionals in the media field with high cultural quality and social class status, so this was the elite media stage.

The term "elite" first appeared in France in the 17th century, referring to a select few or outstanding people who surpassed most people in intelligence, character, ability, property and other aspects, and played a vital role in social development.

The people who used new media in the early stage belonged to a minority group in the media audience group. They had an avant-garde media communication consciousness and also had more advanced and richer media resources, and were the first group of people who benefited from the new media.

2. Mass media stage

The development of new media entered the mass media stage when it was developed and popularized on a large scale. Until now, new media, mainly mobile media such as cell phones, have been enjoyed by the majority of audiences, and the use of new media to convey knowledge and information has become a norm of media communication.

The development from elite media to mass media is inseparable from the decrease of communication cost brought by the progress of media technology. New media has become a kind of mass media with cheaper communication cost, more convenient communication method and richer information communication content, and its communication content and form have even changed people's way of life and their understanding of media essence to some extent.

3. Personal media stage

Along with the continuous development and popularity of new media technology, individuals who did not occupy media resources and platforms in the past and who have media expertise began to gradually express their opinions and views through the Internet and show them to the audience through the platform, which is a sign of the arrival of the personal media stage.

7 CONCLUSION

The emergence of new media has given new creative ideas and methods in the creation of dance choreography and has made it faster for the dissemination of dance. This paper takes the definition, causes and applications of the new media era as the entry point to explore the changes in the creative thinking of dance choreographers under the influence of new media, to study the impact of new media on dance creation, and to analyze the advantages and disadvantages brought by new media to dance creation. This paper explores the new ideas of dance creation brought by new media in the context of various dance programs, such as CCTV dance competition, TV dance competition "Dance Competition", "China Good Dance" and "World of Dance", etc. It is hoped that by analyzing the evolution of the creative thinking of dance choreographers in the era of new media, we can broaden the creative thinking and direction of dance choreographers, enhance the value of dance creation, enable dance choreographers to avoid harm under the influence of new media, help dance choreographers to create better, and provide a certain theoretical reference basis for the creative development of Chinese dance. The search for the art of dance is endless, and choreographers must have the creative thinking to keep up with the times in order to bring forth new life. As an apprentice on the path of dance creation, the author must recognize the problems faced by dance creation nowadays and put the new media, an emerging technology, to better use. In the future, we will continue to explore and innovate dance choreography in our work and study. I believe that with the aid and influence of new media, dance choreographers will have a stronger sense of innovation and create better and better dance works.

REFERENCES

1. Hu Xiaohang. Exploring the strategies of art book editors to adapt to the new environment of media integration[J]. Tomorrow's Style,2022(24):179-182.
2. Hu Jing. Analysis of drama film and television creation in the context of new media[J]. Culture Industry,2022(19):52-54.
3. Fan Jingti. Research on the redesign of algal well patterns under the threshold of music visualization[D]. Nanjing Art Institute, 2022. DOI:10.27250/d.cnki.gnjyc.2022.000262.
4. Sun Ting. Characteristics of interactive film and creation ideas[J]. Western Radio and Television,2022,43(07):116-118.
5. Liao Jing. Exploring the influence of new media on the creation of drama, film and literature [J]. The House of Drama,2022(06):21-22.
6. Li Jinghan. The creation ideas of micro-documentaries in the new media environment--Breakfast China as an example[J]. Audiovisual,2020(12):62-63. DOI:10.19395/j.cnki.1674-246x.2020.12.031.
7. Li Yiwei. The influence of new media on the creation of drama, film and literature [J]. Popular Literature and Art,2020(14):176-177.
8. Zhao Junjie. Research on the development of urban propaganda film creation from the perspective of new media[D]. Dalian University of Technology, 2020. DOI:10.26992/d.cnki.gdlqc.2020.000509.
9. Chen Lizhu. The creation of drama film and television in the context of new media[J]. Drama House,2020(12):70-71.
10. Li Dan. Research on the application of MG animation in folk craft micro-documentaries [D]. Guangxi Normal University,2019.
11. Jin Yinhui. Research on dynamic VI design creation in the new media context [D]. Tianjin University of Technology,2019.

Heterogeneous isomorphism -Spatial Gene Transcreation Design for Traditional Villages in Hunan

Jin Lin[1,3] Chuan He[1*] Shuang Shi[2,3] Jiongfeng Song[1]

[1]School of Architecture, Changsha University of science & Technology,Changsha, Hunan 410114,China
[2]School of Civil Engineering, University of South China, Hengyang, Hunan 421001, China
[3]School of Transportation Engineering, Hunan Institute of Traffic Engineering, Hengyang, Hunan 410000,China

Abstract. This study analyzes the spatial genes of traditional villages by taking Zhang Guying village in northern Hunan Province as an example.Starting from the theory of heterogeneous isomorphism, we discover the methods and paths of identifying, presenting and translating the spatial genes of traditional villages in Hunan region.Through the gene identification extraction method, it analyzes the spatial and cultural genes of Zhang Guying village and obtains the genetic elements and composition of the traditional villages.Based on the design model of gene translation, the "Duanwu family" is systematically analyzed, and it can build a traditional village regeneration and development system.In view of this, his research relies on the application of "heterogeneous isomorphism" theory, which can be obtained from the gene translation method, isomorphic design model.It can promote the regenerative continuation of traditional villages and low-carbon ecological development.

1. INTRODUCTION

Traditional villages are profound historical, humanistic, artistic and social values that have been preserved over a long period of time, and still have their spatial form, livelihood and production intact.Since 2012, more than 10,000 villages have received attention and policy support in the six batches of traditional Chinese villages that have been published to date.

Originally from the Greek, "gene" is the unit of DNA sequence that controls heredity in biology.Based on the genetic properties of genes, sociologists have explored the ways in which culture is passed on in conjunction with the laws of genetic change.In 1950, the American scholars Alfred L. Kroeber and Clyde Klukhohn proposed the theoretical idea of "cultural genes", which is subdivided into cultural units that are similar to genes and can be passed on in an iterative manner.Based on this, this paper examines the identification of traditional village genes and their regeneration and translation from the "heterogeneous isomorphism" theory of Gestalt psychology.

2.HETEROGENEOUS ISOMORPHISM CONCEPT

As one of the core theories of Gestalt psychology, "heterogeneous isomorphism" is a theory that studies the perception of people and their external environment.The perception includes not only the "representation" reflected in the appearance of the form, but also the deeper "meaning" of the thing [1].The new combination of the elements of a thing produces a whole that has the essential characteristics of the original parts and gives a new meaning to the thing, called 'heterogeneous isomorphism'.The interaction between"individual-behavior-environment"can produce "force", and the place force has convergence and completeness, and this process can be regarded as "isomorphism".The heterogeneous isomorphism theory uses the path of "identification-extraction-processing" to build a linkage between human psychology, emotional memory and cultural environment, enabling people to perceive the deeper "meaning" in the genetic characteristics of the village.

3. OVERVIEW OF THE STUDY AREA

3.1 Research villages

Zhang Guying Village is located in the eastern part of Yueyang City, Hunan Province, named after its founder "Zhang Guying" , and has a history of development of about six hundred years ago.A large number of buildings from the Ming and Qing dynasties remain inside the village, and the total construction area can reach 50,000 square meters.Among them, the three buildings of "Dangda"gate, "Wangjia"and "Shangxin" houses , which are cleverly laid out, are called the "Folk Palace" by the outside world.In addition, Zhang Guying Village has a rich history and culture and has been selected as one of the first "Famous Chinese Historical and Cultural

* Corresponding author:Chuan He. Email:13100315098@168.com

Villages" and "List of Traditional Chinese Villages" .Thus, this paper selects Zhang Guying village as the research object to study the spatial genes of traditional villages in Hunan.

3.2 The "heterogeneous" phenomenon of traditional villages

With the tourism development of Zhang Guying Village, the spatial form within the village has experienced expansion and spreading. The original traditional village space has undergone a characteristic evolution in terms of spatial form and logical relationships.The "heterogeneous" phenomenon is mainly reflected in the following four aspects:(1) the architectural style, which lacks the identification of genetic elements. The existing buildings are difficult to renew and their functions are seriously differentiated, and it is difficult to control new buildings.(2) In terms of habitat, the internal environment of village dwellings is poor and lacks adaptive upgrading.(3) In terms of spatial form, the surrounding commercial development is disorderly, and the village space lacks protection and development paths, resulting in the destruction of the original form.(4) In terms of natural ecology, environmental problems are highlighted and there is a lack of adaptive ecological development.it explores the re-living path of Hunan traditional villages from the theory of heterogeneity and homogeneity, which can promote the recombination of traditional village genes and build a development path of "cultural continuity, ecological friendliness, spatial suitability and economic sustainability" , Figure 1.

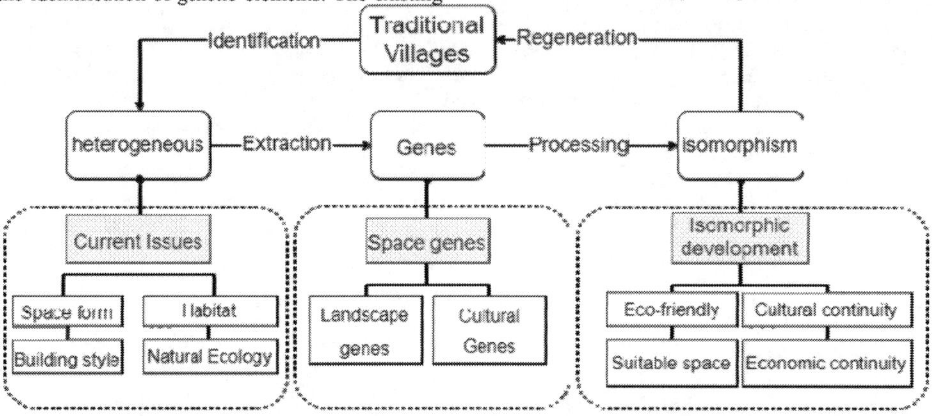

Fig.1 "Heterogeneous homogeneity" process framework

3.3 The "isomorphic" connotation of traditional villages

The traditional village connotation is reflected in the continuation of simple family customs, unique customs and behaviour, and its main carrier is the village pattern, architectural style and spatial form [2].The innovative whole is highlighted by the heterogeneous homogeneity focusing on the combination of elements, which can be explored in depth for their traditional village spatial genetic elements.In the writings of the Finnish architect Juhani Pallasmaa, it is stated that "In an impressive architectural experience, space, time and matter together constitute the sensory dimension, which integrates penetrating consciousness into the physical place [3]."It is clear that the human sensory dimension will play a key role in the identification and continuation of traditional village genes.

4. GENETIC ANALYSIS

4.1Systematic classification of genes

Genes have multiple manifestations of spatial carriers and cultural connotations, and are an important means of systematically identifying elements of traditional villages [4].In view of this, research was carried out to obtain elemental unit genes for the natural ecology, street texture, historical architecture and folk culture of Zhang Guying village.A systematic classification of traditional village genes, which can be divided by the village as a whole into two subsystems: spatial texture and material spirit.From the point of view of the genetic regeneration of traditional villages, "heterogeneous homogeneity" can be seen as a process of "multiple and complex group evolution" and "individual combination of the system as a whole".As a result, through the systematic classification of the genetic elements of traditional villages, it can be used to optimize village conservation and regeneration strategies ,Table 1.

Table 1 Analysis of gene elements translation in villages

Type	Factors	Mode	Translating Strategies
Space genes	**Region:** Topography Climate Hydrology Vegetation	A B	Enhancing the overall landscape appearance
	Villages: Entrance Street Public Road	B C	Create a good public space in the village

Type	Factors	Mode	Translating Strategies
	Building: Space Structure Decoration Function	A B C	Innovative traditional "courtyard" model.
Cultural Genes	**Value:** Family Customs Totems Festivals	A B	Conservation of traditional culture.
	Construction: Technique Layout Scale Content	B C	Innovative territorial construction systems
	Religion: Grade Confucianism Friendship Covenant	A	Preserve the overall cultural environment of the village.

Note: A = direct recognition of inheritance, B = extraction and processing of translations, C = innovative regenerative translations

4.2 Elemental analysis of spatial genes

The "spatial gene" is the information base for storing and recording the spatial characteristics of traditional villages, which includes different types of genetic elements such as regional morphology, village texture, building clusters and building monoliths [5].As a result, spatial genes can be identified in four parts, which are analyzed in detail below ,Figure 2.(1) In terms of geographical form, the villages are mainly "surrounded by mountains and water, with farming in the front and houses in the back".(2) In terms of the village texture, the main streets and lanes are on the outside, while internal corridors connect the houses, forming a clear spatial axis of "main-secondary -branch" .(3)In terms of architectural clusters, the interior of the village is dominated by Ming and Qing dynasty buildings, while most of the other dwellings are built around the "dragon-shaped"mountain.(4)In architectural monoliths, spatial genes reflect a complex logic through the resolution of cognitive structures and diagrammatic language[6].In the"Dangda" gate building group, the primary and secondary spaces are logically related, and the courtyard and patio are the core, which can be regarded as the "single gene" , Figure 3.In addition, the two sides of the main axis are connected by corridors, and the Chinese character "Feng" spreads to both sides in order to form the community genes.

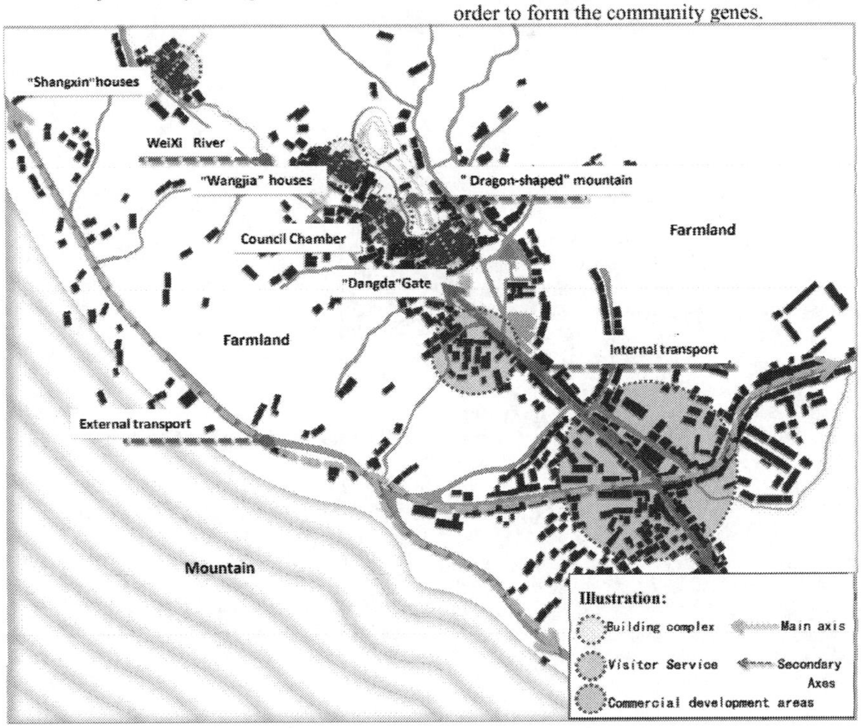

Fig.2 Overall analysis of Zhang Guying Village

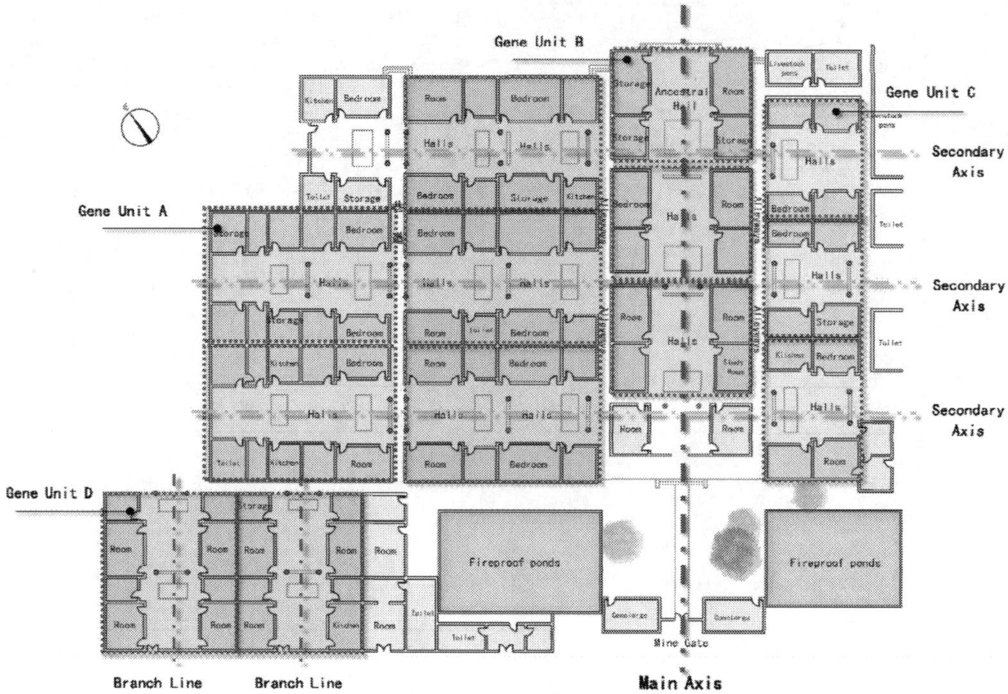

Fig.3 Analysis of the"Dangda" gate building group

4.3 The multiple dimensions of cultural genes

The semiotician Lotman, who studied both "space" and "time", found that most memories arise from culture while they can acquire a collective memory presence [7].In this way, the genetic dimension of village culture is explored in accordance with the theory of " heterogeneous isomorphism ".(1) The dimension of value beliefs, which consists of traditional concepts and graphic language selected and passed down through history.(2)The dimension of construction law, which is reflected in the spatial creation of regional dwellings, public spaces and streets within the village.(3) The dimension of the patriarchal system, which is a manifestation of the concentration of power and the hierarchy of ritual and law, and the strict hierarchical order became an internal factor in the spatial development of Zhang Guying village [8].Therefore, a focus on the "cultural genetic" dimension will enable the continuation of rural culture while highlighting the logic of regeneration of traditional village spaces.

5. TRANSCREATION DESIGN

5.1 Project Overview

The "Duanwu Family" rural resettlement housing project is located in the northern part of Miluo City, Hunan Province, with a construction area of approximately 40,000 square meters.The overall area involves six surrounding villages, with nearly four hundred rural families demolished and relocated there, with a good community atmosphere and neighborly relations.The new form and style of the village settlement with its depth of connotation has been created as a new impetus for the development of the countryside.In view of this, the case study of the "Duanwu family" can highlight the advantages and value of genetic transcription of traditional villages.

5.2 Genetic trait identification

According to the Gestalt psychologist Kurt Koffka, simple combinations do not represent the totality of things, but rather the combined whole, with its connectedness, is the perfectly unified structure [9].Here, the village of Zhang Guying has unique genetic qualities and abundant cultural resources that have created a hereditary spatial order and cultural identity.In terms of the identification of genetic elements, by subdividing and extracting the genetic traits of traditional villages, two major genetic bodies were obtained and core elements were extracted , Table 2.Its gene subjects can be divided into Characteristic and Introverted genes.

Table 2 Identification of Genetic Traits in Village

Elements	Contents	Identifying Features
Characteristic genes: Space Building Ecology	The interior of the original village is surrounded by water and roads, while the interior is connected by corridors	Street scale public space natural relationship
	The building complex has a regular plan structure and orderly spatial arrangement, while the axis relationship is clear and distinct.	Space pattern Construction Techniques Ethnic Decoration
	A village with a pleasant environment and a village pattern by the mountains and water	Environmental and Settlement Form, Combination of House and Courtyard
Introverted genes: Faith Etiquette	The villagers have been based on Confucianism for generations, and a family tradition of filial piety and integrity prevails.	Filial Piety Family Style Frugality Cultural Identity
	Architectural layout of the hierarchical order, courtyard relationship, which reflects the order of respect, filial respect for ancestors and other rituals.	Space rituals Hierarchical Relations Construction Systems

5.3 Gene element isomorphism

Most of the villages in northern Hunan have similar contradictions in development, with problems such as "hollowing out" and poor living environment. The "Duanwu Family" serves the villagers relocated from the surrounding villages, and is designed to restore the local "flavor" and revitalize the spatial vitality of the villages.Based on the identification of genetic traits, the common elements are reorganized, reshaped, and the attributes are continued, in order to highlight the "meaningful" parts of traditional villages.(1) In terms of public life, we create iconic public places in the form of towers to satisfy the collective life of individual villagers and to revive the original memory of the village.(2) In terms of habitat, the two-story or partial three-story approach is used to satisfy the basic needs of regional residences with good spatial comfort.(3) In terms of the natural environment, the spatial form of "by the mountain with the water" is adopted, and it is built near the "Golden Hook" reservoir and facing the "upside-down Golden Hook" hill, which satisfies the basic elements of ecological genetic isomorphism.

5.4 Native gene translocation

The genes of the vernacular have the capacity for heredity, evolution, mutation and selective elimination, so the cultural and spatial elements preserved in traditional villages today are of high value and significance.Therefore, the design of spatial gene translation should be the process of "identification-extraction- presentation",and regeneration has become the core carrier of spatial construction nowadays.The "Duanwu Family" project uses Zhang Guying Village as the source of gene extraction, and its gene translation and presentation can be divided into the following three aspects.(1) In the continuation of vernacular culture, the creation of space inside the village satisfies the way villagers behave and gives the village a spatial carrier for cultural activities and cultural transmission [10].(2) In terms of spatial form, the spatial plan is developed by the series of "patio" and the progression of the Chinese character "Feng", which presents the genetic characteristics of the traditional village of "top merging and bottom dividing, multiple rows moving together",Figure4.(3) In terms of cultural rituals, six building forms have been designed through a layered presentation of "spatial classification -cultural continuity-formal evolution".Through spatial progression, combination and comparison, its multiple categories of monoliths are arranged in a staggered and orderly manner, creating a sense of enclosure in the space of the settlement,Figure5.The final design of the genetic translation of the "reinvention"will promote the regeneration of the countryside and the transmission and continuation of local culture.

Fig.4 Analysis of the spatial form of the building

 https://doi.org/10.1051/shsconf/202316702016

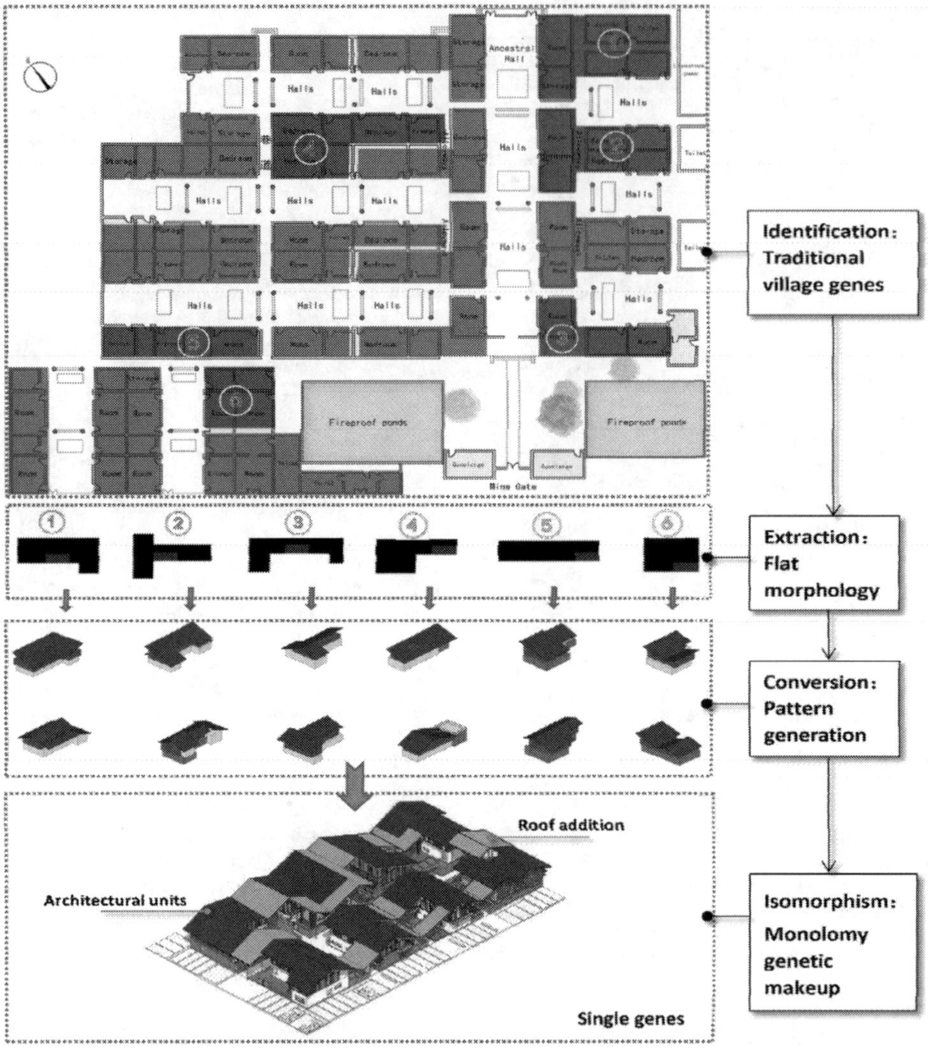

Fig.5 Design analysis of the "Duanwu Family"

6. CONCLUSION

In the face of the current problems of natural ecology, habitat, cultural discontinuity and "hollowing out" of traditional villages, it is urgent to explore adaptive regeneration mechanisms.Therefore, this paper uses the heterogeneous isomorphism theory to analyze and research the gene translation of the "Duanwu Family" rural construction project using Zhang Guying village as a case study of gene identification and excavation, and its conclusions are summarized as follows.

(1)Based on heterogeneous isomorphism theory, this research analyzes the spatial and cultural genes of Zhang Guying village to obtain the genetic elements and dimensions of traditional villages in Hunan.From the "individual - behaviour -environment" perception model,the"Recognition - extraction - category" gene translation process was discovered, which provides the basis and conclusion for the regeneration of traditional villages and gene translation.

(2)The analysis of the spatial and cultural genetics of Zhang Guying village has led to a geographically adapted approach to transcreation design.One is to obtain the genetic logic of traditional villages based on the identification of characterizing and introverted genes, combined with genetic characteristics and spatial regeneration.The second is the focus on the 'heterogeneous' elements of traditional villages, combining local natural ecology, local culture, habitat and public activities. It can build up a design path of "extraction and identification - translation and homogeneity" to meet the ecological, cultural and livability needs of different people.

ACKNOWLEDGMENTS

Foundation item: This paper is one of the stage results of the key scientific research project of Hunan Provincial Department of Education, 《Research on the Evaluation of the Effectiveness and Renewal Technology of Village Construction in Hunan Region in the Context of Ecological Livability》(21A0189) 、Natural Science Foundation of Hunan Province, 《Research on modular optimization design of village public space in Western Hunan from the perspective of form vitality coupling》(2022JJ30605).

REFERENCES

1. Bo Y,Jin L,Chuan H.Research on urban ecological park design based on perceptual dimension[J]. Journal of Changsha University of Technology(Natural Science Edition),2022,19(03):129-136.DOI:10.19951/j.cnki.1672-9331.2022.03.013.
2. Chuanqiao Z,et al.,Protection and development of traditional villages under the concept of "urban double-cultivation"--The traditional village of Democracy Village, Jiuxiang Town, Hanyuan County as an example[C]//.In:Zhengzhou, China. 2019 Proceedings of Urban Development an Planning.2019:1046-1052.
3. Juhani Pallasmaa.The Eyes of the Skin Architecture and the Senses[M].John Wiley &Sons Ltd.2005,:18-25.
4. Weihan Z, Dingqing Z. Landscape genetic analysis of traditional settlements along the ShuRoad(Southern section of Shaanxi)[J].New Architecture,2018(06):38-41.
5. Jing X, Qingqing Z,Tianyi X,et al.Multi-scale identification and dual-system interpretation of traditional settlement landscape genes: A study of Qiang settlements in western Sichuan[J].Small Town Construction,2021,39(02):57-65+84.
6. BingHua Z, YuHang K,et al.The analysis of traditional village liju diagram from the perspective of pictorial language: the example of Yuliang village in Mingxi County, Fujian Province[J].New Architecture,2022(06):145-151.
7. Lotman, Y. M,Ann Shukman.Universe of the Mind:A Semiotic Theory of Culture[M].Indiana University Press.1990.
8. Shiyu D.From morphological fractal to orderly fractal: the example of Zhang Guying Village in Yueyang City, Hunan Province[J]. Huazhong Architecture,2012,30(09):162-164.DOI:10.13942/j.cnki.hzjz.2012.09.034.1
9. Kurt Kauffka. Principles of Gestalt psychology [M]. Translated by Li Wei, Beijing. Beijing University Press, 2010.
10. Jin L, xiong X,et al."Spatial Conservation and Development of Dong Ethnic Settlements under Environent Behavior Studies." Journal of Sociology and Ethnology4.7(2022).DOI:10.23977/JSOCE.2022.04071

Food as a medium: The framework and cross-culture implications of short videos of overseas Chinese sharing Chinese food in foreign countries

Keke Zhang*

School of Journalism and Communication, Zhengzhou University

Abstract. Focusing on foreigners' experience of Chinese food, a group of overseas Chinese released a series of short videos on social media about sharing Chinese food with their family and friends abroad. By analyzing the framework of such videos at high, medium and low levels, we find that the food narratives of overseas Chinese reflect both their main consciousness of spreading Chinese food culture and follow the basic logic of platform traffic. Therefore, elevating this cross-cultural practice to foreign communication requires them to shift from spontaneity to consciousness in cultural communication. On the other hand, we can also see the coexistence and intermingling of Chinese and foreign cultures, which reveals that we should turn to a communication view of civilizational diversity and inclusiveness, and transcend civilizational superiority with civilizational coexistence.

1. INTRODUCTION

Short food videos present the culture that originally exists in color, taste and sound in a visual way, attracting a large number of creators with their unique attitude towards life, aesthetic value and cultural connotation. Among them, overseas Chinese with dual cultural identities, by virtue of their geographical proximity, focus on foreigners' experience of Chinese food publishing a series of short videos about sharing Chinese food abroad on social media. Here, the food symbols integrated with special cultural identities have a common charm that transcends time and space and countries, and act as a medium of emotional communication, which is part of the "food" of daily life from the surface structure, but a manifestation of cultural values from the deep structure. The food sharing behavior of overseas Chinese is a kind of cross-cultural practice, reflecting the clues of Chinese food culture spreading and exchanging in heterogeneous cultures. At a time when cultural exchange and mutual appreciation between Chinese and foreign cultures are becoming more and more important, studying the construction framework of their videos helps to explore the logic behind their creation and deconstruct their practice of sharing Chinese food overseas. It can also provide new inspirations for the exchange and mutual appreciation between Chinese and foreign cultures While exploring and optimizing the communication path of Chinese cuisine culture to the outside world.

2. STUDY DESIGN

2.1 Sample selection

Through comparison, it is found that overseas Chinese mainly post their food sharing videos on domestic social media, with Bilibili being the main place. At the same time, some overseas Chinese also synchronize a small number of videos on international social media such as YouTube. In other words, the content of food sharing videos posted by overseas Chinese is consistent on both domestic and international platforms. Based on this, this study selects ten overseas Chinese accounts that operate on both Bilibili and YouTube. And based on Bilibili, after eliminating the videos that are not related to food, each of them selected the 10 videos with the highest number of views, totaling 100 videos, to analyze their overall framework and explore their internal logic.

2.2 Analytical framework

In 2000, Zang Guoren put forward the theory of low-level, medium-level and high-level frames when discussing media frames. The high-level structure refers to the abstract meaning or main idea of the event, which is expressed in the title, introduction and introductory sentence. The mid-level structure describes the main event, including the history, result, attribution and commentary of the event. The low-level structure is the language or symbols, including grammatical structure, rhetorical use and so on. Specifically, this study analyzes

*Corresponding author. Email: 1186660157@qq.com

overseas Chinese food videos at three structural levels: high, medium and low, based on Zang Guoren's three-level framework theory and combined with the textual characteristics of food videos. The high level structure focuses on the theme setting and emotional tone of the videos; the middle level structure deconstructs the main characters, content selection and narrative strategies; the low level structure analyzes the footage, language symbols and rhetorical use of the videos.

3. RESEARCH FINDINGS

3.1 High-level framework

3.1.1 An interculturality that transcends oneself

From the high-level framework, the theme of the videos is mostly the daily dinner of friends and relatives, and 99% of the videos create a beautiful atmosphere of family harmony and neighborhood harmony. From the viewpoint of dining habits, the videos show a kind of "East meets West" dining scene. Unlike the Chinese "sit around, eat together" dining habits, the West pursues a "meal sharing system", that is, one person, one plate, feel free to add food. The video shows a combined "sit around, share and eat" dining scene, where a communal spoon and chopsticks are used to sit around and eat Chinese food together. In terms of cutlery, the use of Western-style knives and forks and Chinese chopsticks is the norm, and it does not disrupt the harmony of the scene. On the whole, this is a respect for the cultural differences and an acceptance of their specificities, reflecting a kind of intercultural mindset, that is, looking at things from a connected point of view, going beyond the usual "subject-object" model to pursue a "subject-subject" coexistence.[1] Interculturality is the result of Habermas's application of intercultural thinking to culture, which encompasses mutual understanding, respect, and coexistence between cultures.[2] The video does not deliberately highlight the differences between Chinese and foreign food habits, but puts the particularities of Chinese food in the context of dialogue and takes the common sense of human beings' pursuit of a warm and beautiful life as the basis of coexistence, thus constructing a story framework of sharing Chinese food in a harmonious and different way.

3.1.2 Beyond the spread of food culture

The sharing of food inevitably involves the dissemination of food culture. In the video, overseas Chinese play the dual role of food sharer and Chinese culture popularizer. Here, food as a medium of communication spreads food culture and brings about the spread of other diverse cultures at the same time. Yuqi in Finland, has repeatedly shown foreigners learning Chinese kung fu while waiting for their meals; Bingbing in Northern Europe considers cheongsam as a must-have outfit for dinner parties; and Eva in France enjoys Chinese guzheng music after dinner. The food here has gone beyond the limits of color, sound and taste, and extended the spread of Chinese martial arts, Chinese costumes, Chinese music and other cultures linked by the act of "gathering". It is for this reason that we can not only gain meaning from the food itself, but also from the dress and behavior of the people who enjoy it, which gives us a glimpse of their attitude towards heterogeneous cultures. Food is not only an important tool to connect with individual feelings, but also carries the feelings of family and nation, thus becoming a medium for the dissemination of diverse cultures beyond food culture, thus building a narrative framework that is born out of food but beyond it.

3.2 Mid-level framework

3.2.1 The Other's perspective in defamiliarization narration

"Defamiliarization" originated from Russian formalism. It refers to stripping away the obvious things in a work to create a sense of freshness and curiosity for the audience, so as to enhance the level and degree of feeling for the work.[3] Food, as a common symbol in life, tends to fall into the rut of bland narratives. However, short food videos of overseas Chinese take a unique perspective of "how foreigners evaluate Chinese food", making people feel fresh among familiar things. Statistics show that 91% of the videos have multiple scenes, going beyond individual food tasting sessions to show the process of making the food and the stories of the people behind it. In one video, Yu Qi in Finland shows her family working together to dig the soil, bury the chicken, fire it, put it out of the pit and try the meat, where the Finnish family plays the dual role of experiencing Chinese food and making it, while the audience experiences the "birth of the chicken" through the hands of others. It is through this defamiliarization presentation method that the short food videos of overseas Chinese update the audience's understanding of food, get rid of the restriction of habituation, and adopt a unique and creative way to let people get new discoveries in the things that are "familiar".

3.2.2 Dissipation of stereotypes in attitude change

In terms of the evaluation of food, 1% of the videos indicate that they do not like it, 5% consider the food to be average and 94% show an overall liking. In addition to the explicit attitudes of liking, the videos don't deliberately avoid general comments about individual foods. Rice noodles, white fungus, and tripe are often hard to be accepted for foreigners. From the overall presentation of the content, the evaluation of the food is relatively objective. In addition, 14% of the videos show a change of attitude, from disbelief and rejection before tasting to shock and love after tasting. For example, they changed their dislike of Chinese hot breakfast after trying the bun and soya milk. This strategy of resorting

 https://doi.org/10.1051/shsconf/202316702017

to contrast shows the implication of reducing the "other stereotype" of culture. Stereotypes are relatively simplistic attitudes that ignore differences in detail and can be seen as expectations, beliefs or overgeneralizations about the characteristics of other cultural groups.[4] There is no doubt that the foreigners in the video have certain stereotypes about Chinese food due to their own culture, but it is after the tasting of the food and the explanation of the culture behind it that these stereotypes are dissipated. Thus, this strategy of resorting to contrast in the video shows the effectiveness of food culture communication while highlighting the deliciousness of Chinese food.

3.3 Low-level framework

3.3.1 Cultural identity resorting to language

Language originates from a specific culture and is a symbol of a particular cultural identity. The use of linguistic symbols in the video can be understood in two ways. Firstly, the use of Chinese dialects by overseas Chinese themselves indicates their own cultural identity while evoking a sense of cultural identity shared with the audience. For example, Mimi from Sichuan often refers to her German husband as "gua wa zi". Here culture is an unconscious habit, an act, and an expression of cultural identity. Secondly, the use of Chinese by foreigners shows the cultural acceptance from other countries. The foreigners in the video often use poor Chinese to pronounce the names of dishes and praise the deliciousness of the food. Here the language functions as a form of emotional communication, reflecting an effort to approach a heterogeneous culture. Overall the use of Chinese language symbols in the video shows the overseas Chinese's affirmation of their home country's cultural way of life and the foreigners' acceptance of Chinese culture.

3.3.2 Symbolic attraction to the audience

From the use of audiovisual symbols, 83% of the videos use special effects, and 77% of the videos use both simultaneous sound and background music. The live sense of visual symbols enhances the audience's sense of participation, while the vividness of auditory symbols plays a role in setting the atmosphere and rendering emotions. In terms of creative content, 93% of the videos do not contain commercials, while 7% of the videos have commercials. In terms of title grammar, 34% of the titles used exclamatory sentences, 18% of the titles were in the form of questions, and 16% of the titles used the compound form of exclamations plus questions and thus completed the setting of suspense. For example, "When the British in-laws met Shanxi cuisine, they were too scared to eat the rice cakes!" "The Iranian sister-in-law is here for dinner again! This time, I gave her a whole Chinese chicken claw, see if she dares to come back next time?" In addition, the use of words such as "can't stop", "conquered", and "shocked" is a strategy to exaggerate to enhance the attractiveness. In terms of the choice of dishes, hot pot, which is representative of Chinese cuisine, appears most frequently, while "challenging" foods such as snail noodles, preserved eggs and chicken feet satisfy the audience's curiosity. The abundance of information nowadays also means the relative lack of attention, and the competition for virtual space and media time makes the content on the platform follow the thinking of traffic, and obviously the production of overseas Chinese food short videos also implies this logic.

4. THE DOUBLE MEANING OF THE FRAMEWORK CONSTRUCTION

The framework involves selection and prominence, and the food narratives of overseas Chinese reflect both their subjective consciousness of spreading Chinese culture and follow the basic logic of flow. On the one hand, the visualization of daily life reflects the awareness of overseas Chinese to share their good life, the act of sharing Chinese food reflects the cultural subject consciousness of overseas Chinese to break through the individual, and the harmonious and different images of Chinese and foreign food culture exchange reflect the group's identification with the cultural identity of their home country and the awareness of cross-cultural dialogue. On the other hand, the use of audiovisual symbols, the setting of title grammar, the selection of dishes and the insertion of advertisements show the commercial attributes and fan effects of overseas Chinese food short videos embedded in digital platforms. Thus we can see that overseas Chinese who have implanted national emotions have consciously practiced cultural communication, and with dual cultural backgrounds they see the possibility of Chinese and foreign cultural exchange and integration, but still follow the mindset of platform traffic in the process of visualizing and presenting their food sharing behavior. Therefore, how to balance online traffic and real-life practice, so that overseas Chinese can reap commercial benefits and motivate their cultural communication is the most important thing to consider when amplifying the role of this group in foreign communication.

5. THE EFFECTIVENESS OF CULTURAL COMMUNICATION

Jürgen Habermas once said in "The Past as the Future" that "different cultural types should transcend the limits of their respective traditions and fundamental values of their forms of life and respect each other as equal partners in dialogue.[5] The short food videos of overseas Chinese show the exchange and collision between two heterogeneous cultures, and also provide a possibility to examine our own culture from the perspective of the other. From the comments on the coexistence of rejection and love of Chinese food, we can see that only by trying to understand the differences and common points can we break away from ourselves. The

Chinese-Western dining scene reminds us that cultural differences do not always impede dialogue, and that unity and diversity are more like two sides of an equally important coin. This reflects that the ultimate goal of cultural communication is not to be accepted completely but to gain understanding and respect on the premise of recognizing differences. This inspires us to establish a new concept of communication based on the diversity, equality and inclusiveness of civilizations, so that we can "transcend civilizational barriers with civilizational exchanges, transcend civilizational conflicts with mutual appreciation, and transcend civilizational superiority with civilizational coexistence" and thus construct a new practice of international communication of Chinese culture in the context of civilizational exchanges and mutual appreciation.[6]

6. CONCLUSION

Food symbols are not only regional, but also have spiritual commonality, which is both the expression of internal culture and the sharing of universal meaning. Taking food as a medium can not only effectively spread the charm of food culture, but also bring an extension effect for the spread of higher spiritual culture. We can see the commonality and integration of cultures in the short videos of food released by overseas Chinese, which reveals that we should turn to a communication concept of civilizational diversity and inclusiveness, and use cultural exchange to facilitate cultural communication. However, this group does not have the ability to structure narratives yet, and the communication practice faces the challenge of flow thinking. Therefore, to elevate the cross-cultural narrative of overseas Chinese to foreign communication requires not only the official attention to this group but also their own transformation from the spontaneity to the consciousness of cultural communication.

REFERENCES

1. Gao Weihua,Shi Tian. The construction of cross-media story worlds under interstitial thinking[J]. Modern Communication (Journal of Communication University of China),2021,43(08):89-93.
2. Luo Xuan,Long Xiao Yi. A review of research on interculturality theory[J]. Modern Business Industry,2020,41(02):85-86.DOI:10.19311/j.cnki.1672-3198.2020.02.044.
3. [German] B. Brecht. Brecht on Drama [M]. Translated by Ding Yangzhong, Beijing: China Drama Press, 1990: 25.
4. Sun Y. Chun. Introduction to cross-cultural communication [M]. Beijing: Peking University Press, 2008:10.
5. [German] Jürgen Habermas. The past as future [M]. Translated by Zhang Guofeng, Zhejiang: Zhejiang University Press, 2001: 215.
6. Zhang Hengjun. Building a new pattern of international communication of Chinese culture in the exchange and mutual appreciation of civilizations[J]. Foreign Communication,2022(09):12-15.

The Construction of Cross-Cultural Communication Strategies Amid the Interaction of Narrative Subject ——the identification of the other and the telling of the self in A Long Cherished Dream

Fenglin Xing[1], Yi Li[2,a*]

[1,2]Nanjing Normal University, Nanjing, Jiangsu, China

Abstract: *A Long Cherished Dream*, a documentary centered on the poverty alleviation campaign in China, differs from other documentaries on the same subject in that it was shot by Malcolm Clarke, a British director. Therefore, this documentary is featured by the engagement of *the other* into the Chinese context. Condensed in its cross-cultural communication strategy were the identification of *the other* and storytelling of *the self*, constructing a third-dimension Chinese story and Chinese culture amid the interaction between different subjects. This paper, taking *A Long Cherished Dream* as an example, analyzes the narrative strategies of the text from the perspectives of different subjects in hopes that certain references and reflections could be provided for the cross-cultural narrative exchange of documentaries.

1. INTRODUCTION

1.1. Background

The progress of media technology has provided new opportunities for the exchange and sharing of human civilizations. However, it is undeniable that Western countries still enjoy a dominant place in today's global cultural communication context, and at the same time, China's efforts for cross-cultural communication have been regarded as rigid due to the ideological gap between China and the West.

As film art brings people all-round sensory satisfaction of both images and audio, the language of film, with no clear boundaries of ethnic and culture, distinguishes itself from ordinary language. Therefore, it is easier to stimulate people's emotions and trigger their in-depth thinking via image communication. Documentary, as one form of film art, is more speculative and philosophical than other forms of images in addition to its function of recording the real world.

Amid the progress of documentaries on China, their narrative subjects have seen a diachronic change from subjective narration to narration by the other and then the mixed narration of the two. In recent years, the emergence of co-production documentaries has gradually broken the previous division between the subject and the object. The director teams with renowned foreign directors as the mainstay focus on ordinary Chinese and let them tell Chinese stories that belong to the self. Such construction as breaks the dual narration, combining sensibility and rationality, and emotional and rational thinking, has objectively produced a positive interaction with Chinese culture and thus can be seen as a new attempt for the cross-cultural communication of Chinese culture.

The co-production of documentaries on Chinese themes has attracted the attention of many a scholar. Starting from individual cases, they have analyzed these documentaries from text narration to communication context, attempting to summarize the strategies and key points for the success of cross-cultural communication of such documentaries. For example, it is due to the creative breakthroughs in narration, themes, and characters that *Chinese Restaurant*, the first domestic food documentary centered on overseas *Chinese restaurants*, has realized the connection between national culture and world culture. [1]In addition to the content, the construction of the transmission path is equally important. To further enhance the cross-cultural communication capacity of Chinese documentaries, it is necessary to take the international audience as the base and place equal emphasis on "go out to sea by borrowing ships" and "go out to sea by building ships". Also, certain adjustments must be made to the discourse game strategy and a framework featured by shared culture and cognition needs to be developed. By doing these, we can further enhance the cross-cultural communication ability of Chinese documentaries.[2]

1.2. Significance

As a typical co-production documentary, *A Long Cherished Dream*, filmed by Oscar-winner, British

* Corresponding author: a792878600@qq.com

director Malcolm Clarke, focuses on the stories of ordinary people in China in the new era. From remote imagination to close observation, from interpretation to dialogue, this paper will take this documentary as an example to analyze the characteristics of documentary communication strategies constructed amid the interaction of subjects so that certain references and reflections might be provided for more cross-cultural image practice in the future.

2. BEYOND IMAGE: FROM THE GAZE OF THE OTHER TO THE PRESENCE OF THE OTHER, MUTUAL REFERENCE PROVIDING MULTI-DIMENSION THINKING

The narrators in cross-cultural images can be subdivided into the main character within the image text and the story builder outside the image. Since the discussions here are held from the perspective of cultural communication of China, the main character within the image text could be regarded as *the self*, and the story builder outside the image, *the other*.

Generally, the story builders mentioned in this paper refer to foreigners telling Chinese stories. The forms of images are ever changing with the development of information technology. From the perspective of diachronic changes, the group characteristics of foreigners telling Chinese stories show a trend of diversification and decentralization. From professional film directors to foreign Internet celebrities and KOL (Key opinion leader) in the era of short video and we-media, the continuous enrichment of the identity of the other has exerted a certain impact on the telling of Chinese stories. Also, the difference of the identity of *the other* contributes to the difference of the telling of Chinese stories. Edward Twitchell Hall Jr divides the environment in which human languages exist and are used, namely the context, into high context and low context to study the differences between cultures. According to Hall, the culture of the US and Europe belongs to the low-context culture: a lot of information needs to be clearly encoded so that it could be transmitted. Therefore, information exists in language rather than context. Chinese culture, however, belongs to high-context culture where the transmission of information exists in the context or is internalized by individual doers. That means language plays much a smaller role in terms of high-context culture. [3] According to this synchronic perspective, the identity of *the other* telling Chinese stories can be roughly divided into Westerners from low-context culture, Easterners from high-context culture, and overseas Chinese. The notion " identity " itself comes from the symbolic representation system and hence contains certain significance. Identity has always been an important topic in cultural studies. Different or multiple identity identification provides a presupposition and premise for the cross-cultural communication of images.

Malcolm Clarke, the story builder of *A Long Cherished Dream*, is first of all a British from a low-context background while the stories took place in China, a country of high-context environment. The ensuing difference and collision give the image the gene of interaction and fusion. Secondly, Malcolm Clarke is a world-renowned director whose works have won many international awards, including an Oscar, in his 40-year career. Therefore, his identity as the other is blessed with "halo effect", providing enough topic heat and credibility for *A Long Cherished Dream*.

Presence means that the body is in the place where things happen and go on. *Presence* firstly observes the subject, namely the other analyzed above. It also observes *the situation*. In sociology, social situations make up the environment in which people live and act. In the study of communication, Meyrowitz believes that the media information network has produced new social situations. The combination of sites, props, dialogue, and music in the image forms the situation in time and space, which provides the background for the characters' behaviors. [4]

Based on this, the *presence of the other* in *A Long Cherished Dream* has two meanings. First, Malcolm Clarke, a Westerner holding filming activities in the Chinese social scene, can have the most direct understanding of Chinese society as he goes deep into the realities of life and to the scene. Both poverty alleviation campaign and progress in the new era are social issues within Chinese ideology. That means only when one gets to know real social stories can he better eliminate the strangeness and stereotypes brought by cultural separation. Aside from that, as the director of this film appears in the image and faces the audience in the way of physical presence, telling his perception in the shooting process and understanding of these stories.

A Long Cherished Dream has four episodes, each of which focuses on a Chinese man who has worked his way out of poverty. At the end of each episode, Malcolm Clarke will sit in front of the camera for a few minutes, during which his opinion of the main character is given. Yet the monologue of the director is interspersed with relevant images of the characters in the story. Such comparisons present us with "Chinese impressions" generated amid the mutual care of the subject and object. The main characters are usually not good at direct expression as the Chinese of high-context culture show more actions, body language, and meaningful silence in front of the camera. But the director's monologue is a direct expression of praise and reflection. This collision of high and low contexts not only constructs a multidimensional space beyond the subject and object, but also arouses the audience's thinking in different contexts. The director's "presence of *the other*" satisfies the imaginative presence of people in the same context as him, evoking sympathy.

3.WITHIN THE IMAGE: DELEGATE THE RIGHT TO SPEAK AND FOCUS ON DAILY LIFE

In Chinese documentaries, the connotation of "foreigners telling Chinese stories" can be seen as composed of "foreigners in the state of nature telling Chinese stories" and " foreigners from the perspective of construction telling Chinese stories " . [5] These two narrative perspectives can be understood as the game of ideological discourse power between China and the Western world. The former takes the Western world as the center and the Chinese stories it tells are merely based on its own interests. Yet the latter tries to incorporate *the other* into *the self*, thus spreading the macro image of a nation. Both narrative modes, however, build images with grand discourse. Even in the Chinese stories told from the perspective of construction, *the self* mentioned in them is not *the self* of subjectivity. In the case of *A Long Cherished Dream*, what we see is more of a narrative pattern centered on narration of daily life, which refers to a kind of imaginative expression based on individual life experience, a small narration of daily trifles, and a real self-elaboration of living individuals. For documentaries, the narration of daily life in the image world is a kind of constructed text. [6]It is different from the previous narrative pattern in that people in the image no longer exist as symbols or backgrounds of the overall representation structure. Neither do they serve the defined meaning in the structure. Instead, they construct their own daily existence with their life of triviality and human touch.

But different from the daily life in the objective world, the daily life in the image world is organized with certain narrative logic, which is derived from the in-depth description technique of anthropology. The four stories in *A Long Cherished Dream* are all titled after the names of their main characters: Kai Yong, Lin Bao, Huai Fu and Zi Xu. Although every Chinese character has its corresponding explanation in the dictionary, the director Malcolm Clarke let them present their own meaning via images. From the "poverty alleviation cadre" Kai Yong to Lin Bao and Huai Fu, who strive to move from the countryside to the city, and to the rural entrepreneur Zi Xu who has been fighting for his career for many years, many ordinary people like them have jointly shaped the local image of China and, at the same time, carried out their own life practice.

In addition, the telling of *the self* is also reflected in the lines -- the personal emotion is expressed in dialogue and monologue rather than commentary. Living in a high-context culture, Chinese people always find emotional expression implicit. Due to the arduous task of poverty alleviation, village official Kai Yong often fails to take care of his own family, burdening his wife a lot. In the documentary, there are few scenes of his wife and all her support for her husband's work is condensed into a single sentence: "I wish I were the poverty-stricken family assigned to him." When it comes to the core of the documentary, Malcolm Clarke, even facing possible conflicts between high-context narrative subjects and low-context audience subjects, still uses blank narration with no intervention. This design precisely leaves space for the imagination of the audience and enables *the self* who are narrating to truly express himself.

4. AFTER THE IMAGE: REALIZING THE SYMPATHETIC COMMUNICATION AMID CLOSE RELATIONSHIP AND RURAL CHINA

Co-production of the documentary itself, aiming at the audience at home and abroad at the same time, carries the responsibility of cross-cultural communication. In this context, an obvious problem faced by creators is how to minimize the cultural image discount caused by secondary encoding and decoding. Michael Wood, the director of BBC who has completed several co-produced documentaries, bypasses the problem of translation by adopting all-English dialogues coupled with body language. In doing this, he tries to realize the discourse extension from China to the rest of the world with this " cultural proximity " . [7] Distinguished from Wood's cultural proximity strategy, Malcolm Clarke pays more attention to the expression of original Chinese image and relies on emotional proximity to realize the extension of communication discourse.

A Long Cherished Dream, though themed on the poverty alleviation efforts in China, has in its narration an emotional tone not affected by time and space. First, the four episodes of this documentary are invariably presented in the simplest way of communication -- intimate interpersonal communication. Modern society is filled with onerous interpersonal communication, but in the documentary, Malcolm Clarke only shows the most intimate interpersonal communication of the main characters, which may be between people: father and son, husband and wife, and different generations; or between people and space: Lin Bao and the truck she spends her days with, Huai Fu and the performance venue he relies on. Even if the media extends people's senses, the communication between "bodies", scarce in today's fast-paced society, remains the most primitive and intimate of all. Once the emotional resonance is captured, relationships of intimacy could unleash tremendous power that goes beyond time and space to connect the other side of the globe.

Secondly, the presentation of " hometown " and "homesickness" in the documentary. Whether it is the poverty alleviation campaign or the establishment of a well-off society, the pace of urban modernization emphasizes "moving forward". However, from the old people who refuse to be relocated in the first episode Kai Yong, to Lin Bao's and Huai Fu's returning home after their success, to the last episode of Zi Xu, where the successful entrepreneur attends his farmland, they all show a feeling of nostalgia for returning to their native land. As a developing country, China retains many rural features, which is a kind of tradition from the past, a response to the question "where am I from?", and in a sense, a representation of a fading pastoral utopia. As a man living in a developed country, Malcolm Clarke sees more urban civilization homogenized by globalization. However, it seems that we can also see the "pastoral time" described by the British in the history books. Therefore, this kind of inquiry and recollection can be shared by all mankind. What it reflects is that people have doubts and thinking about the industrial civilization which develops at an unusually fast speed. It is precisely because the film contains more profound human emotional issues in addition to the Chinese context that the topics belonging to China have more possibilities to be expanded.

5. ENLIGHTENMENT: CHINESE DOCUMENTARIES SHOULD CONSTRUCT A THIRD-DIMENSION IMAGE DIALOGUE STRUCTURE

Bakhtin proposes the theory of intrinsic dialogue and divides the intrinsic dialogue of discourse as the object of interpretation into three tiers, of which the third-tier dialogue refers to the dialogue between the speaker (author) and the super addressee (also known as the third party and the comprehencher) in the long term. [8] Bakhtin believes that this third-party position is a completely special one as it is an absolutely just responsive understanding. The anticipation should be in the mysterious distance, or in the distant historical time. And the author of the expression can consciously predict the existence of the highest "superaddressee" to varied degrees. [9]In the narrative logic of *A Long Cherished Dream,* there is a "responsive understanding of absolute justice", which exists in a "text calling structure" that is based on the integration of "the perspective of the other" and "the perspective of the other" yet transcends the two. If Chinese documentaries are to achieve a new breakthrough in cross-cultural communication, they should draw more narrative strategies from the dialogue theory so as to establish a brand-new image of China.[10]

First, the construction of a space where Chinese themes and Western themes can communicate. Subject is the prerequisite of image construction, and the determination of subject is the concretization of the purpose of dialogue. Documentaries with cross-cultural communication as the core task should have both the cultural characteristics of *"the self"* and the cultural synesthesia of *"the other"* in the selection of subject, strike a balance between the normalization and defamiliarization, and provide a premise space of the third dimension for dialogue.

Second, the construction of a space where high context and low context can communicate. In constructing image narrative language, a narrative chain with multiple discoursal features should be formed as far as possible. By doing this, different identities and perspectives can be presented simultaneously, so that Chinese appeals from the perspective of *"the self"* and foreign appeals from the perspective of *"the other"* can be responded and play their roles simultaneously, thus forming a third position of "intersubjectivity".

Third, the construction of a space where implications can communicate. The ultimate purpose of image presentation is to convey a certain point of view or a certain spirit or image. However, such communication, once led by "hegemony", will backfire more often than not. Chinese spirit belongs to the national spirit, but it also represents a kind of human spirit, which can be shared and enriched by human civilization. When refining the implication of images, we should learn to leave blank space for more comments, so as to form a real dialogism and inclusiveness of great country culture.

6. CONCLUSION

In a word, against the background of globalization, the narration of Chinese stories is undergoing the transformation from "the telling by the self" to "the identification of the other". However, in order to give full play to the cross-cultural value of Chinese stories, image narration should achieve a telling between the host and the guest jointly built by the two. The co-production documentaries represented by *A Long Cherished Dream* provide a new way of telling stories. We should learn from the strategies and apply them in the narrative construction of more documentaries, facilitating the Chinese stories' going global.

References

1. Zhang Yifei,Cui Hongtu.Cross-cultural Narration and Communication Strategies of the Documentary Chinese Restaurant[J].Contemporary TV,2020,(12):81-83.
2. ChenBilong,TangNing,Ju Huilin.On the Cross-cultural Communication Strategies of China Documentary[J].China Television,2022,(04):70-76.
3. Hall T.Beyond C.Garden City[M].New York:Doubleday,1976:10.
4. Meyrowitz,J.,No Sense of Place:The Impact of Electronic Media on Social Behavior,Beijing,Tsinghua University Press,2002:16.(in Chinese)
5. Meyrowitz,J.,No Sense of Place:The Impact of Electronic Media on Social Behavior,Beijing,Tsinghua University Press,2002:16.(in Chinese)
6. Zhang Xiaoqin,Liang Junjian,From Daily Life to Audio-visual Narrative Text: Practice and Exploration of Documentary Image Narration[J].Modern Communication(Journal of Communication University of China),2018,40(02):112-116.
7. Wang Hengqian,Shao Zhenqi.Cross-cultural Narration of Michael Wood's China Documentary[J].TV Research,2022,(06):92-94.
8. Zhao Yanhong. Theoretical Exploration and Realistic Focus of Dialogue Communication [J]. Southeast Communication, 2018,(07):49-52.
9. Ling Jianhou. Linguistic Thought in Bakhtin's Discourse Theory [J]. Russian Teaching in China, 2001.
10. Jiefei Zhao. Interpreting the Cultural Symbols in

 https://doi.org/10.1051/shsconf/202316702018

Chinese Documentaries: Taking a Bite of China and Beautiful China as Examples. Frontiers in Educational Research (2020) Vol. 3 Issue 11: 41-46.

A Review and Prospects of Chinese Character Research in the Last Thirty Years

Yunbo Niu*

Jinling Insititute of Technology, Nanjing, China

Abstract. At this stage, the study of Chinese characters is undergoing a transformation of modern applications and technology, and research in this field is both deeply rooted in the traditional Chinese study of script and flourishing with new vitality. This paper analyses the five major components of Chinese character research, namely, the study of archaic characters, the study of the phonology, morphology and meaning of Chinese characters, the study of the application of Chinese characters in the computer field, the study of the cognitive processing of Chinese characters, and the study of Chinese character teaching, and the three methods of Chinese character research, namely, exegetical examination, corpus database and empirical research methods. This paper summarizes the literature on Chinese character research, analyses the trends in Chinese character research, and summarizes the review and outlook of Chinese character research in the past thirty years.

1 Introduction

The Chinese characters were created at the turn of the Xia and Shang dynasties, and are the only writing system of their own origin to have developed into a phonetic script, which has shown great vitality over the centuries. The study of Chinese characters began in the pre-Qin era, and the Shuowen Jiezi is the most comprehensive collection of traditional script research. In the twentieth century, under the impact and influence of Western academic culture, the infiltration of knowledge from disciplines such as education, information processing and psychology, and the application of research methods, the field of study of script has undergone an applied and technological transformation, and this traditional discipline has taken on a new life.

Based on the statistics and combing of literature over the past thirty years, this article discusses in detail the content of Chinese character studies, some basic theories and perspectives in Chinese character studies, the main methods of Chinese character studies, the interdisciplinary characteristics of Chinese character studies, and looks forward to the development trend of Chinese character studies, with a view to contributing to the disciplinary construction of Chinese character studies.

2 Literature and Statistics of Chinese Character Studies

The search was conducted on the China Knowledge Network using the subject term "Chinese characters", with the subject area restricted to Chinese language and writing and the journals selected as Peking University Core and Chinese Science Citation Index (CSSCI). A total of 12,616 papers were searched by hand. Figure 1 shows the number of papers published on the topic of "Chinese characters" by year, and Figure 2 shows the number of papers on specific categories of Chinese character research.

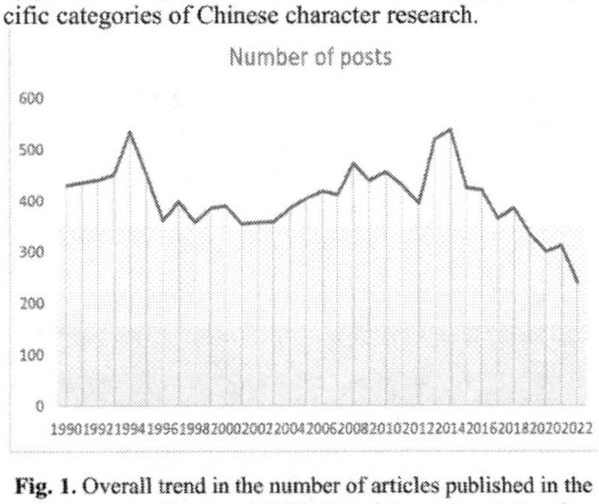

Fig. 1. Overall trend in the number of articles published in the textual studies literature, 1992-2022（the self-made form）

As can be seen from Figure 1, the literature can be roughly divided into four stages: the first stage was 1990-1994, when the number of publications tended to increase, reaching a small climax with 533 publications in 1994; the second stage was 1995-1997, when the number of publications decreased sharply compared to 1994, and was even lower than the number of publications before 1994. The third stage is 1998-2012, although there are small twists and turns, the number of articles published is more or less the same, with a slow growth trend, with more articles published in 2008 and 2010; the fourth stage is from 2013 to the present, starting from 2013, the number of core jour-

* Corresponding author: 2373733780@qq.com

nal papers increased to a similar number of articles published in 1994, and the number of articles published is significantly higher than the previous stage, but only lasts until 2014 The number of articles published in the following years decreases significantly, indicating a decrease in the enthusiasm of the academic community for the study of Chinese characters. The trend in the number of articles has been flat overall over the thirty years, but 2022 is the lowest number of articles in the last thirty years, which also shows a downward trend in the enthusiasm for research on Chinese characters.

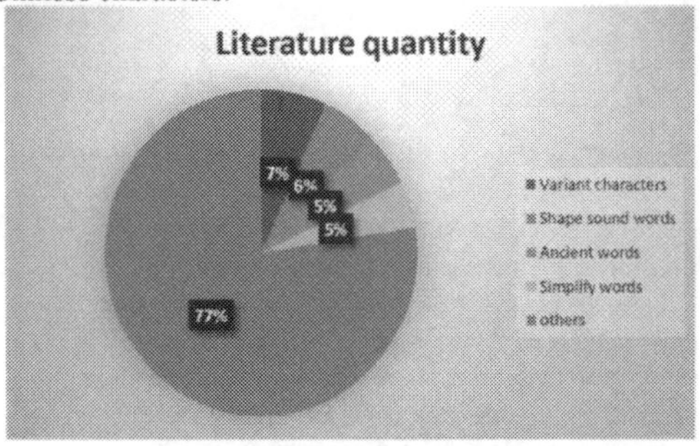

Fig. 2. Thematic distribution of textual studies literature, 1992-2022（the self-made form）

As can be seen from Figure 2, the distribution of research content in Chinese character literature is dominated by variant characters, morphophonetic characters, ancient characters and simplified characters. Scholars have made the three elements of Chinese characters, morphology, phonology and meaning, the focus of discussion, and have paid close attention to issues such as the teaching of Chinese characters, Chinese character writing, Huiyi characters and the Six Books. The art of calligraphy and reform of Chinese characters are also major areas of research, while new directions have emerged in the study of Chinese characters, such as Chinese characters and computer linguistics, recognition and processing of Chinese characters, and other psycholinguistic directions.

According to the content of literature research, it can be roughly divided into the following five areas: (1) research on ancient characters, (2) research on the morphology, phonology and meaning of Chinese characters, (3) research on teaching Chinese characters to foreigners, (4) research on the application of Chinese characters in computers, and (5) research on the cognitive processing of Chinese characters. According to the division of Chinese character research methods, they can be roughly divided into exegetical examination, database research and empirical research methods.

3 Analysis of the content of Chinese character studies

3.1 The study of ancient writing

The study of oracle bones is an important part of the study of ancient writing, and there are many studies of oracle bones, mostly focusing on the relationship between oracle bones and the evolution of Chinese character form and structure, and the preservation of oracle bones, focusing mostly on the discovered oracle bones, and less on the continued search for the discovery of oracle bones. The study of oracle bone inscriptions is basically combined with the history of the Shang dynasty, with analysis of the Shang dynasty's simple books and documents, and with an examination of the Shang dynasty's shrine culture. There are also studies that construct the functions of the components of the oracle bone script and explore the cultural transmission of the oracle bone script in the context of digital media (Wang Qian 2022) [1].

Scholars in the last three decades have studied ancient scripts in greater depth. Zhang Qingsong (2017) [2] conducts a comprehensive and in-depth examination of a large number of variant characters in the book of characters arranged according to the morphological divisions of Chinese characters, the Zhengzitong, written by Zhang Zili, and makes a relatively objective summary and evaluation, which has high academic significance and theoretical value. Gao Xingquan (2015) [3] analyses textual phenomena in the evolution and standardisation of characters, such as orthography and variants, ancient and modern, passages and falsehoods, and traditional and simplified calico.

Xia Nengquan (2015) [4] analyses the causes of differences in variants in ancient documents, and Wang Lijun (2015) [5] proposes basic principles for optimising contemporary Chinese character systems based on variants of traditional and simplified characters.

In the 1980s, scholars developed a new understanding of the 'abundance of pseudonyms in ancient books'. Studies in both exegesis and ancient literature have noted the existence of a large number of pseudoborrowed characters in the ancient books of the pre-Qin dynasty and have come to a new understanding, linking pseudoborrowing to the rationality of Chinese characters.

Chen Yachuan (1982)[6], in his study of the Six Books, simplified characters and the teaching of Chinese characters, suggested that the use of the Six Books to diagram the structure of Chinese characters could not be said to be a genuine departure from modern Chinese character forms, nor could it reflect an overview of the structure of Chinese characters, a view shared by researchers in the early twentieth century, when the term 'modern Chinese characters' emerged and scholars began to The term 'modern Chinese characters' emerged, and scholars began to link the study of 'modern Chinese characters' to the traditional 'six books' theory, further deepening and refining the concepts of the earliest monograph on Chinese characters, Shuowen Jiezi.

3.2 Study of the morphology, phonology and meaning of Chinese characters

Chinese characters are two-dimensional phonetic-semantic scripts with a certain degree of rationality in their configuration, with a wide range of symbols and complex structures. Li Yunfu (2014) argues that Chinese characters are difficult to learn because of their characteristics, which are manifested in three aspects: form, function and rationale, and compares these three attributes of Chinese characters with different languages from the same perspective, providing new ideas for the study of Chinese characters [7]. The common "sound, form and meaning" are the basic elements of Chinese characters, but they are always closely related in the study of Chinese characters, with the form being the external visual aspect, the rationale being the explanation of the cause of the form, and the function being the value of the form's existence. It is difficult to distinguish between them.

Form.

Number of strokes, parts and orthography

In addition to traditional script morphology, which investigates the strokes and parts of Chinese characters, many scholars have also conducted research on the morphological elements of Chinese characters with the help of corpus and psychological experiments. You Haojie (2003) [8] empirically investigates the findings related to stroke count, part count and topology with the help of a corpus of mediated languages. By examining the effect of stroke count on the visual word processing of second language learners of Chinese at different language levels, it was revealed that second language speakers of Chinese differ from native speakers in their processing, with speakers of languages with shallow orthography tending towards analytical processing and native speakers with deep orthography tending to adopt holistic processing (Zhou Ying and Jiang Nan 2023) [9]. A paper-and-pencil test illustrates that the budding development of orthographic awareness among international students and the development of part-position awareness precede part-awareness (Hao, Meiling 2007) [10]. A study of the distinguishing features of Chinese character parts, part position information, the structural symmetry features of Chinese characters on the processing mechanism of Chinese character recognition, and the influence of part features and structural types on international students' Chinese character writing (Hao Meiling ， Fan Huiqin2008) [11].

Variant characters and a large number of pseudonyms still exist in the Chinese character system, and Chinese character orthography has been emphasized in recent years. The main elements of Chinese character orthography include the restriction of traditional characters, the abolition of variant characters, and the use of standard simplified characters, so that the simplification of Chinese characters has become an important trend in the evolution of Chinese characters in the past fifty years. Chen Shuangxin and Dong Yue (2016) put forward several focal issues concerning the current research and application of modern Chinese characters, namely the issue of traditional and simplified characters and the issue of variant characters [12]. Scholars have also done a great deal of research on the collation of variant character lists, such as Shao Wenli's survey and analysis of the first batch of variant character collation lists and pointed out the main problems and causes [13][14], and this trend is reflected in the 1956 Chinese Character Simplification Scheme, the 1977 Mandarin Chinese Character List, the 1986 Modern Chinese General Character List, and the 2013 Master List of Simplified Characters. With the gradual popularisation of Mandarin, the trend towards simplification of Chinese characters is also increasing.

Due to the gradual simplification, digitisation and diversification of Chinese characters, the study of Chinese characters has developed from the traditional study of glyphs to a multifaceted trend, and the intersection of multidisciplinary theories and methods has become increasingly evident.

Rationale.

Lin Yan (2006), based on a dialectical analysis of the views of Saussure, Jacobson, Greenberg, Xu Guozhang, Shen JiaXuan and many other linguists, affirms that there is both arbitrariness and rationality in the linguistic symbol system, and that linguistic symbols are a unity of arbitrariness and rationality [15]. The Chinese characters are a combination of phonetic, morphological and semantic characters, which have a scientific rationality.

Study of the "building blocks" of the rationale unit.

The components of Chinese characters are pictograms, signs, ideograms or phonograms, and those that have lost their function due to changes in character form or language are called marks. Chinese characters are a flat, square-shaped symbol system that uses ideographic components (including pictographic, denotative and marking components) as well as phonetic and notational components to record Chinese morphemes and syllables. The main characteristics of Chinese characters are their square shape, their multifunctional and complex rationale, and the uncertainty of their recording function. According to Li Yunfu (2014), the rationale unit is often called a component, which refers to a form that has a certain function in constructing a character to express a speech position [7].

Rationale study.

The study of the rationale of Chinese characters has evolved, and the general view is that it is a process of 'weakening of rationale - reconstruction of rationale - return of rationale'. The simplification of Chinese characters also weakens the rationale. However, most of the knowledge of the rationale of Chinese characters comes from ancient texts, such as the Six Books Theory in the Shuowen Jiezi, and although there are a few studies that have kept pace with the times and reflect the diversity of Chinese character development, they are relatively traditional and outdated, and need to be investigated by scholars.

Functionality.

Chinese characters are a combination of sound, form and meaning; the form is subordinate to the character itself, while the sound and meaning are the elements that express the language. Only by understanding the relationship between the form, sound and meaning of a Chinese character can the function of the character be fully utilised. If the relationship between the sound and the meaning of the character is not understood, the character's function cannot be fully exploited. Therefore, in the study of Chinese characters in terms of their function, scholars have focused on the sense of the meaning and sound of the morpheme.

The morphophonetic advantage of Chinese characters was gradually established in the course of their evolutionary development. Early on, the function of Chinese characters was mainly to record phonetic sounds, but later on they embarked on the path of morphophoneticization, which has continued since the formation of the Spring and

Autumn period and the Warring States period. During this process, a large number of morpho-sounding characters were created by adding morphemes (ideograms). This suggests that Chinese characters cannot fully meet the needs of their corresponding linguistic units through the recording of phonology alone, which is determined by the characteristics of the Chinese language. The function of Chinese characters is characterised by the uncertainty of their recording function and the difficulty of fully corresponding words. In terms of morphology, the process of morphophonetisation of the Chinese character symbol system is clearly a process of flourishing. This is the reason why Chinese characters parted ways with other ancient writing systems and took a unique path of development, and why Pinyin cannot replace Chinese characters.

At present, scholars still do not pay enough attention to the symbolic system of Chinese characters, and their attitude towards pronunciation cognition appears to be "emphasizing the meaning symbols but not the phonetic symbols". The reasons for this are: the large number of phonetic symbols, the weak ability to form characters, the fact that the frequency of the same phonetic symbol is much lower than that of the meaning symbols, and the fact that the common use of phonetic symbols limits the adoption of the rule-based approach to the perception of Chinese character pronunciation (Wan Yexin 2007) [16]. Many scholars have reached diametrically opposed conclusions about the perception of the formation of the morpho-phonetic advantage of Chinese characters. One is that Chinese characters have increased their epigraphic component, and the other is that the tendency to mnemonicize Chinese characters has been suppressed. It has been argued that the latter is more consistent with historical facts.

A functional understanding of the morphology, phonology and meaning of Chinese characters not only provides a better understanding of the morphological aspects of Chinese characters, but also, and more importantly, a better understanding of how the relationship between language and writing is perceived and dealt with. This understanding is still relevant to the teaching of Chinese characters today.

3.3 Research on teaching Chinese characters to foreigners

Teaching philosophy: Chinese character orientation and language teaching.

Research on teaching Chinese characters to foreigners has flourished as a result of China's rising international status and comprehensive strength. The teaching of Chinese characters to foreigners in China started late, and it is only since the Fifth International Symposium on Chinese Language Teaching held in 1996 that Chinese character teaching has received great attention. However, in the whole process of teaching Chinese as a foreign language, the teaching of Chinese characters is still in a weak link.

There has always been a controversy between the "word-based" and "character-based" concepts in Chinese language teaching theory. Influenced by the acquisition of English as a second language, the 'word-based' teaching philosophy dominated Chinese as a foreign language in China until the 1990s, when the 'word-based' teaching philosophy originated, with Bai Lesang and Zhang Pengpeng writing 'The Enlightenment of Chinese Language and Script'. It was only in the 1990s that the idea of "word-based" teaching emerged and became more and more vocal. In response to the controversial development of Chinese character teaching, the first question is whether, in the overall design of Chinese character teaching, the tasks of writing and reading can be separated and reasonably matched to the tasks of language teaching; the second question is whether the process of reading can be elevated to a cognitive approach to Chinese characters, and whether the clarification of word-word correspondence can be developed into a comprehensive understanding of the relationship between words, morphemes and words (Wan Yexin 2007)[16]. In 2008, Shi Zhengyu proposed a teaching philosophy based on the basic framework of "word-phoneme and Chinese character"[17], which is different from the new teaching philosophy of "word-based" and "word-based", and represents a breakthrough in theoretical research on Chinese character teaching. However, most of the prevailing methods of teaching Chinese characters still follow the "word-based" concept, with the phonetic and semantic structure of Chinese characters being the main focus, such as the "language" view and the "language teaching" view, which take the specificity of Chinese characters as the origin. The "language" view and the "language teaching" view are based on the specificity of Chinese characters.

In the early 1950s, the 'language first, then text' approach was discussed and experimented with in relation to the teaching of language and Chinese characters, but due to the poor results, it was changed to the 'language in parallel' approach, which is still used today (Wan Yexin 2015) [18]. However, in the past decade or so, the 'language integration' approach has been under constant review. There is a consensus that the disadvantages of this approach - that the teaching of Chinese characters is subordinate to the teaching of words, that the teaching of Chinese characters is essentially based on the teaching of writing and the correction of misspellings, and that it is not conducive to the learning of Chinese characters by foreign students - have largely been identified.

At the same time, there is a significant minority who believe that the disadvantage lies first in the inability to follow the system of the Chinese characters themselves, first in their individual form and then in their combined form; from the simple to the complex, from the easy to the difficult, in a gradual manner. Others, however, have noted that the simplicity of teaching characters is bound to limit the choice and arrangement of words and sentences in textbooks, and have therefore reintroduced the idea of 'language separation'. The difference is that Li Yunfu (2014) further emphasises the importance of teaching and practising the functions of Chinese characters as the main task [7], but this is not the case in the current situation of teaching Chinese characters, and there is no corresponding teaching content.

The innovation of teaching models and teaching methods for teaching Chinese as a second language must, in the final analysis, be based on the characteristics of Chinese language and Chinese characters, and must be integrated with the characteristics of Chinese language and Chinese character applications, rather than being rigidly applied to ready-made language teaching methods designed for Indo-European languages. The study of Chinese characters in terms of their phonetics, morphology and meaning is integrated into the study of teaching Chinese characters to foreigners and becomes the cornerstone of the study of teaching Chinese characters to foreigners. There are several innovative directions of research in the teaching of Chinese characters to foreigners, with Chinese characters as the base, teaching philosophy, teaching content, teaching methods and teaching tools all having a focus.

Teaching content: Chinese character materials and literacy scales for the international student community.

The globalisation of the Chinese Language Proficiency Standard levels for international Chinese language education proposed in 2021 is important for the development of Chinese character studies. Chinese character research should also have a global perspective, thus promoting the development of literacy scales for the international student population and the innovation of Chinese character teaching materials. For example, in the field of Chinese character acquisition, the Rasch model and item response theory were used to demonstrate that the reliability and validity of the literacy scales developed for international students were generally good, and Wu Qiuping, Hong Wei and Deng Shulan (2017) [19] constructed test indicators through regression models, and the form of Chinese character recognition measurement provided a new perspective for placement tests. The study has significant practical implications for teaching Chinese as a foreign language. In 2019, the Handbook of Chinese Character Recognition and Reading [20], and in 2021, the Handbook of Chinese Character Writing [21] will be published, which will be of reference value for the assessment of Chinese character teaching, acquisition research and the development of measurement tools.

The design of Chinese character teaching materials has also provided innovative directions in recent years. Zhao Jinming (2008) argues that in order to break through the bottleneck of Chinese character teaching, the first step should be to clarify misconceptions about Chinese characters and to establish a scientific view of Chinese characters [22]. Chinese characters themselves are a complete scientific system. Zhao Jinming (2009) [23] goes on to deepen the concept proposed that the relationship between teaching Chinese and teaching Chinese characters is a key issue that teaching materials must deal with. How to lift the fear of Chinese characters and break through the bottleneck of teaching Chinese characters is a problem that the textbook should focus on and address. Wan Yexin (2015) [18] takes the same attitude, changing the pursuit of "literacy quantity" to learning from the past, which complements the aforementioned literacy scale for international students, echoing the enhancement of the cognitive function of Chinese characters and the transformation of the cognitive pathway of Chinese characters; the student-centred approach and the teaching materials as a supplementary platform are meant to break the bottleneck of Chinese character teaching. This shows that scholars already have a philosophy of valuing Chinese character teaching and Chinese character teaching materials.

However, the current implementation of Chinese character teaching materials remains unreasonable, and Xu Rui （2016） argues that due to the lack of initial attention to the teaching of Chinese characters, Chinese character teaching materials are more lacking compared to other aspects, with some focusing on pictorial teaching methods, others on Chinese character radical teaching methods, and few comprehensive and systematic teaching materials, and the difficulty level of Chinese characters is not differentiated, making it difficult to teach Chinese characters[24].

Wan Yexin (2015) suggests that using exercises and games as the main body of teaching materials can maximise word recurrence [18]. Efforts are made to realise the principles of practicality, fun and science in Chinese character teaching materials. After proposing a direction of change for Chinese character teaching materials, the subsequent practical teaching has not followed suit. As mentioned above, the synchronisation of languages has made the teaching of Chinese characters dependent on the teaching of Chinese, and whether Chinese characters should be taught as a stand-alone subject has been the focus of debate among scholars, with one advocating that Chinese characters should be taught as a stand-alone subject (Zhao, 1999; Liu, Yanmei, 2002; Ma, Yanhua, 2002; You, Haojie, 2003[8]; Chen, Jing, 2010) and the other opposing that Chinese characters should be taught as a stand-alone subject. The other is against the establishment of separate classes for teaching Chinese characters (Shi Dingguo and Wan Yexin, 1998) [25].

The systematic and comprehensive nature of Chinese character materials and the direction of their focus are closely related to whether or not Chinese characters are taught as a stand-alone subject. At present, the content of the Chinese character textbook dominates the curriculum and the development of a comprehensive and systematic textbook will help to make the Chinese character curriculum independent.

Teaching methods.

The more popular methods of teaching Chinese characters, such as stroke teaching, component teaching, morpho-sounding character teaching and character family theory teaching, are closely related to the development of Chinese character studies and are complementary to each other. The focus of academic attention is on the discussion of 'traditional teaching methods' and 'new teaching methods'. Some representative student-centred, cognitive approaches to teaching Chinese characters include the two-brain approach to literacy, the peculiar association approach to literacy, and the tethered classification approach to literacy. These new teaching methods also provide international Chinese language teachers with new ideas for innovative and effective teaching of Chinese characters.

Liu Yanmei's (2009) [26] experimental study of Chinese character strategy training proved that it is necessary to develop Chinese character learning strategies to help students learn Chinese characters and to reduce the burden of writing them from memory. It is important to focus on the teaching of 'phonetic notes' in the teaching of Chinese characters to avoid the phenomenon of 'emphasising form over sound'.

3.4 Research on the application of Chinese characters in computers

Study of Chinese Character Formation.

The study of Chinese character conformation is a fundamental research in computerized Chinese character information processing. In recent years, research on data compression techniques with high compression ratios has been carried out in China, of which superposition and contouring techniques are two important ones. This is also an innovative point of combining Chinese character conformation research with computer technology.

Study of Chinese character morphological analysis.

Chinese character research is mainly related to the tendency to digitise Chinese characters, such as Chinese character morphological analysis techniques, which have received enthusiastic attention in Japan and China. Japanese research has used this technique to analyse Chinese characters in addition to many Japanese textual applications, such as the analysis of mapping relationships based on Chinese character morphological features and shell aspects to predict the formal structure of sentences of a particular grammatical form. Research is also being carried out abroad on the structural description of kanji forms, which abstractly decomposes kanji forms into interconnectable defined parts or elements, and decomposes kanji figures into separate parts through text parsing and image recognition to achieve the goals of structural parsing, text recognition and graphical analysis, which are important for the computer analysis of kanji forms.

3.5 A study of the cognitive processing of Chinese characters

The academic community has focused on the cognitive mechanisms in the process of Chinese character acquisition from a psycholinguistic perspective. Dong Xiaoyu (2003) proposed a theoretical concept of a 'multi-level Gestalt bi-directional processing model' for Chinese character recognition, suggesting that Chinese character cognition involves a 'whole-part-whole' process [27]. Peng Danling (1997) points out that parts are also a unit of Chinese character recognition. Compared to the characterisation of strokes, the analysis of parts occurs at a higher level [28]. Zhang Zhenjun, Ding Guosheng and Chen Baoguo (2011) used the age effect on Chinese character acquisition to complete a teaching study on the phonological integrity hypothesis [29]. Bai, Lili and Chen, Baoguo (2011) used a word category judgment task to examine the effect of age of acquisition on word category information processing [30]. Although there are many related studies, such as an experimental study of foreign students' recognition of morphemic characters (Chen Hui and Wang Kuijing 2001) [31], an experimental study of foreign students' awareness of morphemic phonological cues (Jiang Xin 2001) [32], and the role of phonological information next to sounds in international students' Chinese character learning (Hao Meiling and Shu Hua 2005) [33], there is still little progress.

These studies provide a paradigm for empirical research on the cognition of Chinese character acquisition, which is of great significance as a good foundation for future research on Chinese character acquisition in the field of psychology.

4 Overview of Chinese character research methods

The field of textual research has continued to expand, the theoretical system of textual studies has been constructed, the research content has become extensive, and the research methods have become more scientific and rigorous. Throughout the past three decades, the main methods of Chinese character research have involved: exegetical examination, corpus data, and empirical methods.

4.1 An exegetical examination

Wang Yankun（2006）[34] divided the methods of precept research into three categories, the first of which --- the method of exploring the meaning of words by word form, including three subcategories: seeking meaning according to shape, seeking meaning by sound, and dialect proof; The second major category - methods of seeking the meaning of words by virtue of context, including three sub-categories of inferring meaning from the text, clarifying meaning from the compound language, and seeking meaning in the text; the third major category - methods of seeking the meaning of words by virtue of relevant documentary materials, including three sub-categories of seeking meaning from the original text, knowing meaning from different texts, and seeing meaning from the set of examples. Modern scholars have become more systematic and structured in their approach to exegesis, strengthening theoretical summaries and guidance. Exegetical methods have also gradually become more modern, with the use of computers and the Internet as key features. The study of Chinese characters using exegetical methods is mostly the study of ancient texts, which are closely linked to Chinese cultural traditions.

4.2 Corpus data

The number of corpora available for Chinese character research is constantly improving and increasing. The data corpus system is becoming more and more scientific and systematic, and the large-scale electronic text database through scientific sampling and processing has greatly improved the efficiency of scholars' Chinese character research, and is a solid foundation for the development of Chinese character digitization. At present, most foreign research on Chinese character shape analysis is based on database design, which enables fuzzy retrieval of Chinese character shapes. The database system stores information on Chinese character shape characteristics, forming computer-recognisable shapes that can be used to support multi-character recognition and multi-character text analysis applications, and can also develop applications relating to the recognition of Chinese characters from pinyin to Chinese character shapes.

4.3 Empirical Methods

In the last thirty years, empirical methods for Chinese character research have been largely based on questionnaires and reflective summaries of literature on practical teaching research, but there have also been innovative breakthroughs. For example, the combination of eye-movement research and scientific experiments in the field of psychology has greatly improved the efficiency of Chinese character research by setting up instruments and procedures for computational science and using eye-movement devices to test the patterns of Chinese character acquisition.

Eye-movement research involves using instruments to measure the eye movement patterns of users as they observe and use products, and analysing them. In the eye-movement study of second language learners reading phonetic texts at different levels of Chinese language proficiency [35], it was concluded that although the learners' reading speed increased as their Chinese language proficiency improved, there was still a large gap with native speakers. The main focus was on learning Chinese characters, and even beginners did not pay too much attention to reading phonetics at the expense of writing Chinese characters. Even the frequency of attention to pinyin decreases as Chinese proficiency improves. This may be related to the fact that Chinese characters are ideographic, thus not only addressing the fact that Chinese character teachers do not know whether their students neglect to master Chinese character forms in the acquisition process, but also enhancing the depth of research into the ideographic features of Chinese characters.

The use of eye-movement technology is a hot trend in the study of bilingual word processing in Chinese second language learners. As Chinese language proficiency increases, second language learners rely less on analytical processing and more on holistic processing, resulting in more efficient word recognition, but high-level second language learners have not yet made the transition from analytical to holistic processing. Previous studies have only examined the 'word length effect' and the 'stroke count effect', which selected mostly native speakers, with inconsistent results, and fewer studies of Chinese second language involving the stroke count effect. Within the framework of holistic and analytical processing, there are two studies that specifically focus on the effect of stroke count in Chinese bilinguals with reference to native Chinese speakers, which examined the effect of double-word stroke count in bilinguals, and the other examined the effect of single-word stroke count in this group, but neither of these studies. However, neither of these studies examined the effect of language level on the stroke count effect.

5 Shortcomings and Prospects of Chinese Character Research

In summary, there have been many categories of Chinese character research in the last thirty years, and interdisciplinary research has been innovative. However, the late start of Chinese character research and the lack of attention it has received have led to the fact that there are currently many ideas or perspectives on Chinese character research that are not clearly implemented at a later stage, and that the results of the proposed ideas are less likely to be validated. Moreover, due to the differences in root concepts and academic backgrounds, the same area of Chinese character research may yield very different results. Solutions to the same problems are similar or repeated, and there is a great deal of repetition of ideas within fifty years, concentrating on Chinese character morphology, component orthography, and language teaching philosophy. Although there have been breakthroughs in scientific experiments and in the field of cognitive psychology in the cognition of Chinese character acquisition, they are not obvious and need to be concretely implemented in practical Chinese character research.

Of course, in the light of the current state of Chinese character research, the author also has the following perspectives on the future of Chinese character research. Some Chinese universities and research institutions are currently carrying out transformative research on Chinese characters. For example, the Laboratory of Chinese Character Research and Modern Applications at Beijing Normal University has built a multidisciplinary platform for integrated research on Chinese characters. The first is a multidisciplinary approach to innovative Chinese character research. For example, it has been developing a platform for organising and building resources for ancient and modern Chinese characters; researching automatic character acquisition technology and developing a system for annotating the attributes of Chinese character structures; and developing a platform for humanistic computing and data mining based on Chinese character big data. The second is the construction of a Chinese character big data resource base: a repository of Chinese character morphological resources and evolutionary relationships over the ages; a repository of Chinese character holographic attributes; a repository of academic resources for ancient and modern Chinese character research; a language corpus and knowledge map based on Chinese character morphology, phonology and meaning. There are also products and services for modern applications of Chinese characters: the development of a big data service platform for Chinese character constructions; the development of a big data service platform for canonical texts; the development of a knowledge base of Chinese sentence patterns and an automatic graphical sentence analysis system; and the construction of a Chinese character education platform.

6 Concluding remarks

In the last thirty years, Chinese character research has tended to be multi-perspective and diversified, with research content going deeper, research theories and research methods taking on an interdisciplinary character, and innovations and breakthroughs increasing, but in the specific teaching of Chinese characters, the pedagogical effects are not significant, and the visual development of tools for Chinese characters is weak. More and more young scholars are devoting themselves to the study of Chinese characters, and with the simplification, diversification and digitalisation of Chinese characters, the study

 https://doi.org/10.1051/shsconf/202316702019

of Chinese characters combined with computer applications and the field of psychology has taken a new direction. We must not only actively explore the theoretical directions of Chinese character research, but also put our feet on the ground to implement concrete measures in the teaching of Chinese characters to show the light of hope for Chinese character research.

References

1. Wang Qian. Digital empowerment to cross-media narrative: The dilemma of oracle culture communication and new ideas[J]. News Lovers,2022, No.539(11):66-68.
2. Zhang Qingsong. A Study of Variant Characters [J]. Language Construction,2017(27):82.
3. Gao Xingquan. Script evolution, script phenomena in standardization - orthographic, heterogeneous, ancient and modern, common and false, traditional and simple calligraphy [J]. Chinese Calligraphy,2015, No.023(22):104-107.
4. Xia Nengquan. An analysis of the causes of the differences between the variants of Wang San and Guang Yun[J]. Seeking, 2015, No.273(05):140-144.
5. Wang Lijun. Basic principles of contemporary Chinese character system optimization[J]. Application of Spoken and Written Language,2015, No.93(01):10-18.
6. Chen Yachuan. The six books, simplified characters and the teaching of Chinese characters [J]. Language Teaching and Research, 1982(01):85-104.
7. Li Yunfu. Characteristics of Chinese characters and teaching Chinese characters to foreigners[J]. World Chinese Language Teaching,2014,28(03):356-367.2014.03.006.
8. You Haojie. The effects of number of strokes, number of parts and type of topology on the mastery of Chinese characters by learners from non-Chinese cultural circles[J]. World Chinese Language Teaching, 2003, No. (02):72-81.
9. Zhou Ying, Jiang Nan. A study of Chinese second language learners' double-word processing [J]. World Chinese Language Teaching,2023,37(01):111-126.
10. Hao Meiling. The sprouting and development of international students' awareness of Chinese orthography[J]. World Chinese Language Teaching, 2007, No. (01):29-39.
11. Hao Meiling, Fan Huiqin. The influence of component characteristics and structure type on Chinese character writing of international students[J]. Languages Teaching and Research,2008, No.133(05):24-31.
12. Chen Shuangxin, Dong Yue. Several current focal issues concerning the research and application of modern Chinese characters[J]. Language Teaching and Research,2016, No.181(05):106-112.
13. Shao Wenli. Analysis of the measurement of the first 410 groups of characters in the first batch of the "First List of Variant Characters"[J]. Language and writing applications, 2004(04):18-25.
14. Shao Wenli. The main problems of the first batch of variants and their causes [J]. Language and writing applications, 2003(01):47-55.
15. Lin Yan. Arbitrariness and rationality are both essential properties of linguistic symbols[J]. Language and Translation, 2006(01):8-13.
16. Wan Yexin. From the study of Chinese characters to the teaching of Chinese characters - several problems in the process of understanding the Chinese character symbol system [J]. World Chinese Language Teaching, 2007, No. (01):5-15.
17. Shi Zhengyu. A preliminary study of word-phoneme and Chinese character teaching [J]. World Chinese Language Teaching, 2008, No. (02):109-118.
18. Wan Yexin. How to break the "bottleneck" of Chinese character teaching: A study of Chinese character teaching materials with the example of Chinese Characters - Cognition[J]. World Chinese Language Teaching,2015,29(01):130-142.2015.01.011.
19. Wu Qiu-Ping, Hong Wei, Deng Shu-Lan. The application of Chinese character recognition in the entrance placement test for Chinese bilinguals - an empirical study on constructing simple Chinese proficiency discrimination indicators[J]. World Chinese Language Teaching,2017,31(03):395-411.
20. Fu Yanbai. Chinese Character Recognition and Research Manual,2019(01):12.)
21. Handbook of Chinese Character Writing Assistance published [J]. World Chinese Language Teaching,2021,35(03):421.
22. Zhao Jinming. Teaching Chinese as a Second Language: Concepts and Models [J]. World Chinese Language Teaching, 2008, No. (01):93-107.
23. Zhao Jinming. Teaching Environment and Chinese Teaching Materials[J]. World Chinese Language Teaching,2009,23(02):210-223.
24. Xu Rui. Modern Languages (Academic Comprehensive Edition),2016(07):156-1579
25. Shi Dingguo, Wan Yexin. A survey report on the teaching of Chinese characters to foreigners[J]. Language Teaching and Research, 1998(01):36-48.
26. Liu Yanmei. An experimental study on the necessity, teachability and effectiveness of Chinese character strategy training[J]. World Chinese Language Teaching,2009,23(02):280-288.
27. Chen Chuanfeng, Dong Xiaoyu. Structural symmetry characteristics of Chinese characters and their recognition and processing mechanism[J]. Languages Teaching and Research,2003(04):58-63.
28. Peng Danling, Wang Chunmao. The basic unit of Chinese character processing: Evidence from the stroke number effect and the part number effect[J]. Journal of Psychology,1997(01):9-17.
29. Zhang Zhenjun, Ding Guosheng, Chen Baoguo. The age effect of Chinese character acquisition: A test of

the phonological integrity hypothesis[J]. Psychological Development and Education,2011,27(06):577-583.

30. Bai Lili, Chen Baoguo. The effect of age on word-like information processing in Chinese character acquisition[J]. Psychological Science, 2011,34(02):343-347
31. Chen Hui, Wang Kuijing. An experimental study on foreign students' recognition of morphemes[J]. World Chinese Language Teaching, 2001, No. (02):75-80.
32. Jiang Xin. An experimental study on foreign students' awareness of epigraphic clues in morphophonetic characters[J]. World Chinese Language Teaching, 2001, No. (02):68-74.
33. Hao Meiling, Shu Hua. The role of voiceside speech information in Chinese character learning among international students[J]. Languages Teaching and Research,2005(04):46-51.
34. Wang Yankun. The development of exegetical methods [J]. Language Studies,2006(04):7-10.
35. Xu Jingjing, Jiang Xin. An eye-movement study on the reading of annotated texts by second language learners with different Chinese language levels[J]. World Chinese Language Teaching,2021,35(04):548-561.

"Large fat sighings" of a "large bulky figure": identity-building and formation of new order in *Persuasion*

Yukang Zhang[1], *Mengyang* Wei[1,*]

[1]Department of Foreign Language Studies, NingboTech University, China

Abstract. This essay provides a multifaceted analysis of Jane Austen's novel *Persuasion* through the lens of narrative consciousness, rhythm and sexual anxiety. Drawing on Lacanian theory, it examines the self-identification processes of the heroine Anne Elliot and Mrs Musgrove against, and with relation to patriarchal norms. By analyzing the dynamics presented in Mrs Musgrove's "large fat sighings", the essay explores the effect of incongruity between the human body and spirit; and it further examines Mrs Musgrove's Sphinx-like figure, who inspires desire and inhibitions in the Oedipus-like Captain Wentworth with a significant impact on the formation of community and order. The essay investigates long-neglected subjects presented in Austen's novels and enriches the complex portrait of Mrs. Musgrove as both Mother and Sphinx.

1 Introduction

1.1 Research background

1.1.1 Introduction of Persuasion

Persuasion is the last complete novel by Jane Austen, published posthumously in 1817 and titled by Henry Austen.

For more than two centuries, *Persuasion* has been an enduring favourite among common readers and held in high esteem by scholars, with its twisted and moving love story where a young couple overcome social impediments when pursuing the marriage of true minds.

1.1.2 Former moralist interpretations

Since Austen's novels are consistently set around the confined subjects of courtship and marriage of the genteel class, her works are considered to concern the sense of restraint, domestic obligation and the English national identity (Johnson, 1997: 217)[1]. Therefore, moral rectitude is often embedded inside the readers' imaginations of Jane Austen (Fu, 2018: 70)[2].

As the apprentice of Samuel Johnson, Austen is indeed a moralist, and in fact, praises and preaches the virtue of reason and morality in her novels. This dominant view has significantly influenced the range and perspectives of critical studies on Austen's novels in academia (Geng, 2011: 58-61)[3]; *Persuasion* is no exception; the preaching spinster is often considered pristine and her works sexless. The majority of the studies on *Persuasion* are focused on the dialectical connection between reason and passion, or rather sense and sensibility, presented in the complex and intricate entanglement between Anne and Wentworth (Geng, 2011: 58-61)[4].

1.1.3 Recent Lacanian readings

Despite the long neglect of the application of Lacanian theories in the field of Austenian studies, recent decades have witnessed a fervent upsurge of studies on Austen's novels in the context of Lacanian theory, as Austen's novels "connects her powerfully to some central Lacanian concerns" (Garofalo, 2019: 61)[5]. The self-identification process of the heroine Anne Elliot has been analysed within the Lacanian framework of "the instant of glance," where she attempts to locate herself in the imaginary register, comparing herself against the perfect ideal models constructed by society (Michalski, 2021: 77)[6]. The same pattern could also be applied to the Mrs Musgrove's identification as a mother and woman in the patriarchal society.

1.2 Research purpose

1.2.1 Influence of narrative consciousness

As mentioned above, David Sigler (2019)[7] explored, through a close reading of the dynamics presented in the joke, the effect of the relationship between narrative consciousness on building identities and shaping the community. This essay aims to further study the effects of the narrative consciousness in its triangular relationship with Mrs Musgrove and the readers, which can also be interpreted into the triangular structure of the joke, consisting of the joke teller, the listener, and the

* Corresponding author: nkweimengyang@163.com

butt of the joke. Comprehension of this joke demands interactive involvement from the readers, on whom the danger of being put in the position of the butt of the joke is forced. And this susceptibility can be further interpreted as an argument for the legitimacy of the ridicule, that one who ought to be less rigid and more moderate in his demands of the common people, if in their case, they fail to live up to the "sublime ideals" (Eagleton, 2019: 24)[8].

1.2.2 The unharmonious rhythm of Mrs Musgrove

This research takes the "large fat sighings" of Mrs Musgrove as the focal point. The comedy against the "large bulky figure" of Mrs Musgrove that express mental sorrows, is truly a marvellous demonstration of the incongruity between the human body and spirit. This conflict between the physical and spiritual will be further explored in this essay with a focus on its influence on the formation of community and order.

1.2.3 The effect of sexual anxiety

Within the theory of "the anxiety of influence," Mrs Musgrove can be interpreted as the Sphinx that represents "natural" and "sexual anxiety" (Bloom, 1997: 39)[9]. By studying this scene from the viewpoint of sexual anxiety, more profound insight and acute glimpses of Jane Austen's real speculation about the enduring topic of identity and identification, can be offered to scholars and researchers in academia. Furthermore, this research would lead to a more accurate portrayal of Jane Austen, instead of the mere stereotype of a middle-class spinster in the pre-Victorian era.

This study is also meant to present a more balanced examination of the long-overlooked subjects and elements like female lust and desire in Jane Austen's novels, which can be significantly valuable for interpreting subtle implications and deeper themes, which Austen so delicately set under the cover (Gilbert & Gubar, 2000: 146)[10]. Projected onto the characters that Austen creates, it is possibly Austen's own anxiety as a writer in the face of all her predecessors (Zhang, 2019: 158)[11]. In a word, this research can provide an illuminating and complex insight into the existing research in the Austen academia.

2 Narrative consciousness

2.1 The meta-narrator

2.1.1 The criticism of the joke

Persuasion is a novel about "loss, desire, sexual difference, and social competition," whose dynamic tension is vividly demonstrated in the scene where Mrs Musgrove is being ridiculed (Michalski, 2021: 77)[12].

"They were actually on the same sofa, for Mrs Musgrove had most readily made room for him;—they were divided only by Mrs Musgrove. It was no insignificant barrier indeed. Mrs Musgrove was of a comfortable substantial size, infinitely more fitted by nature to express good cheer and good humour, than tenderness and sentiment; and while the agitations of Anne's slender form, and pensive face, may be considered as very completely screened, Captain Wentworth should be allowed some credit for the self-command with which he attended to her large fat sighings over the destiny of a son, whom alive nobody had cared for.

Personal size and mental sorrow have certainly no necessary proportions. A large bulky figure has as good a right to be in deep affliction, as the most graceful set of limbs in the world. But, fair or not fair, there are unbecoming conjunctions, which reason will patronize in vain,—which taste cannot tolerate,—which ridicule will seize" (Austen, 2012: 67)[13].

The "large fat sighings" and "comfortable, substantial size" of Mrs Musgrove, as ridiculed by the narrator, which may strike modern readers as fat-shaming, is an unusually harsh joke in Austen's works. The joke about changes in physical appearance has drawn fierce criticism and is considered to be a "savage caricature" (Mudrick, 1952: 212)[14], or "infamous proclamation" (Heydt-Stevenson, 2005: 183)[15], among others.

2.1.2 The insertion of readers forced by the joke

However vicious such remarks may be, according to George Orwell, "the aim of a joke is not to degrade the human being, but to remind him that he is already degraded" (Orwell, 1968: 285-286)[16]. Joke as it is, Mrs Musgrove here as a fictional character within the story narrated, obviously cannot be reminded by the narrator; the reminding must be directed at the readers, who are also susceptible to the same danger of being ridiculed.

To comprehend this joke, the readers must have seen themselves as a norm in relation to the figure of Mrs Musgrove, which results in their identification not only with the joke teller but also with the butt of the joke (Sigler, 2019: 47)[17]. Hence the joke may not directly apply to Mrs Musgrove, but still, through the readers, indirectly still succeeds in "reminding" the butt of the joke (to be), not the fact of already being degraded, but the possibility of being degraded, for the readers are subject to the same social norms that ridicule Mrs Musgrove. Hence the butt of the joke in this case now includes both the fictional characters and the reader.

2.2 The super-narrator

2.2.1 Justification of the caricature

As to "the narrating consciousness," after ridiculing Mrs Musgrove, it acknowledges "a right to be in deep affliction" but also justifies the ridicule it has earlier made, which is considered a double structure of the meta-narrator and the super-narrator. In the question of the function of the super-narrator, when it simply mocks the meta-narrator and "justifies its cruelty," Sigler believes such a setting makes the narrator a butt of the joke, and a community is created through the dynamic operation of the joke, under the regulation of the superego (Sigler, 2019: 45-49)[18]. This is a process, as described by the Lacanian philosopher, where "the ego ideal itself turns out to be the partial object" (Zupančič, 2008: 32)[19].

However, "the narrating consciousness" can be considered to consist of, instead of the meta-narrator and the super-narrator, simply one more complex narrator. The super-narrator in Sigler's analysis strikes an impression as a more "omniscient" narrator than the meta-narrator, with a more extensive view.

2.2.2 The idea of "pastoral"

Writing of William Empson, Terry Eagleton argues that the pastoral in Empson's view could be understood as "a large-minded plebeian wisdom," which asks oneself to live up to the "'high' human values" and accept the fact that others may not (Eagleton, 2019: 24)[20]. From such a perspective, the narrator practices sublime wisdom by first vindicating the "right to be in deep affliction" of a body of whatever "personal size" and later accepting the fact that reason, the "high" value, cannot be forced upon others. On Sádor Ferenczi, Eagleton remarks that "yet we are amused because we acknowledge the force of the inhibition in the very act of violating it" (Eagleton, 2019: 13)[21]. Therefore, by suggesting the possibility of the vindicated right being violated instead of criticizing this violation as a taboo to be inhibited, the narrator denies, or at least, deduces the power of such inhibition and the pleasure of breaking it, shaping the community of dynamic power balance.

Nonetheless, such a strategy also has its very own limitations, for the "irreverence," in Eagleton's term, signified by the justification of the ridicule that Mrs Musgrove suffers, actually, reinforces the "social norms by demonstrating how remarkably resilient they are, how good-humouredly capable of surviving any amount of mockery" (Eagleton, 2019: 14)[22]. Therefore, though the performative process of jokes could "creates community through its very operations" (Sigler, 2019: 47)[23], it is still susceptible to the social norms of a larger community.

3 Rhythm

3.1 General incongruity between body and spirit

In the field of humour studies, among the theories that repeat most frequently, there is incongruity theory (Buijzen & Valkenburg, 2004: 147)[24]. Simon Critchley writes, "What is funny, finally, is the fact of having a body" (Critchley, 2002: 62)[25], which Terry Eagleton immediately amends, "more precisely, one might claim, the incongruity involved in neither quite having a body nor quite being one" (Eagleton, 2019: 21)[26]. The "large fat sighings" uttered by the "large bulky figure" of Mrs Musgrove, best demonstrate such a comic spirit, for the humour of the joke derives from the incongruity as a result of the thingy "large bulky figure" sighing out human sorrows, which is exactly what Wyndham Lewis describes as "physical bodies, behaving as persons" (1928: 247)[27]. In the latter part of this essay, the incongruity between the physical and spiritual will be further explored, through an analysis of the "large fat sighings" as a rhythmic pattern that connects the body and mentality.

3.2 The unharmonious bodily rhythm

3.2.1 The sorrow of spirit in flesh

The comical effects of the "large fat sighings" in building such social communities can be further explored, as a rhythmic pattern.

The definition of the verb "sigh" varies according to dictionaries, as defined as "to take and then let out a long deep breath that can be heard, to show that you are disappointed, sad, tired, etc" by Oxford English Dictionary[a], or "let out a deep breath, as a way of expressing feelings such as disappointment, tiredness, or pleasure" by Collins English Dictionary[b]. The act of sighing, different from a more regular breathing pattern, emphasises on the process of letting out an audible breath. It is obvious that when the verb "sigh" is applied, the stress is arranged upon the process of breathing out in the rhythmic pattern of breathing.

According to Henri Bergson, the readers connect with the characters through the established "regularity of the rhythm" and the readers would expect the "periodic returns of the measure," feeling the "graceful" pleasure of "holding the future in the present" (Bergson, 2001: 12)[28]. The unbecoming conjunction of the figure and sorrow seems to divide the poor fat woman into two distinct parts. As the mindful sorrow of Mrs Musgrove seems to be "beside" her very own physical self, she becomes the "oddly plural" (Bevis, 2013: 21)[29]. These hardly rhythmical nor graceful "large fat sighings" out of "a large bulky figure," therefore, reinforce the ambivalence between the physical and the spiritual, and as mentioned in the beginning, "the incongruity involved in neither quite having a body nor quite being one" and hence broken is the illusion of mastery.

3.2.2 The Bakhtinian fool

Ridiculed as Mrs Musgrove is, for she is a butt of the joke though among many others, she is a Bakhtinian fool, "extremely mocked by the objective narrator" and "used as a mask for the author to criticize society" (Nodeh, 2013: 2)[30]. Her "large bulky figure," screening the admirable, pleasant-to-eyes Anne Elliot for Captain

---

[a] OED (2023) Available from: https://www.oxfordlearnersdictionaries.com/definition/english/sigh_1?q=sigh [Accessed 27 March 2023].
[b] Collins English Dictionary (2023) Available from: https://www.collinsdictionary.com/dictionary/english/sigh [Accessed 27 March 2023].

Wentworth, the rightful male, "makes strange the world of social conventions" (Bakhtin, 1981: 404)[31]. The "unbecoming conjunctions" of Mrs Musgrove's "personal size" and "mental sorrow" which as commented by the narrator, "reason will patronize in vain." "Reason" here signifies a Cartesian cogito. When reason patronizes effectively, the "huge mathematical machine" keeps on extending and the world becomes still and lifeless (Bloom, 1997: 39)[32]. And therefore, becoming a butt of the joke, Mrs Musgrove nullifies reason.

Schopenhauer writes that "humour depends upon a subjective, yet serious and sublime mood, which is in voluntarily in conflict with a common external world very different from itself" (Schopenhauer, 1887: 282)[33]. And when, as in this scene, the dominating value of reason is denied and through the humour that derives from, the "diverting" scene where, in Schopenhauer's term, "this strict, untiring, troublesome governess, the reason, for once convicted of insufficiency," the fool overturns her master (Schopenhauer, 1887: 280)[34].

4 Sexual anxiety

4.1 Mrs Musgrove as mother

Unlike Mrs Bennet and Miss Bates, the other Bakhtinian fools in Austen's works, Mrs Musgrove is special as a mother who can be said to have never cared about her son until his death, which is odd and peculiar.

The answer to the neglect Dick Musgrove suffers alive may be found in the narrator's description of Mrs Musgrove as "more fitted by nature to express good cheer and good humour, than tenderness and sentiment." "Good cheer" and "good humour" are the amiable qualities that engage one in pleasure-seeking while "tenderness and sentiment" are the ones for motherhood. Hence the neglect that Dick Musgrove suffers from can be understood as the result of Mrs Musgrove's denial of motherhood by nature.

"For the human subject to emerge it must not simply be conscious of its own distinctiveness but must be recognized as a human subject by another," which Hegel described as the dialectic of "Lordship and Bondage" (Homer, 2005: 23)[35]. Therefore, despite Mrs Musgrove's rejection against, she will always be identified as a mother as long as she is recognized so. However, according to Elias Canetti, "a mother is one who gives her own body to be eaten" (Canetti, 1962: 404)[36]. Blackwell puts forward the idea that in the film adaptations, Anne Elliot's anorexia signifies her resistance to being a mother (Blackwell, 2004: 160)[37]. Though such anorexia is demonstrated more obviously in the film, there is a base in the text to justify such an assumption, for in this exact scene Mrs Musgrove's "large bulky figure" indeed blocks Anne's "slender form."

Then in contrast to the slenderness that resists childbirth, the bulky figure of Mrs Musgrove is part of the flesh regained through the death of her son and it is not until the death brings back the flesh that the son has "eaten," did Mrs Musgrove realize what a child should mean to a mother. To Mrs Musgrove, the process of being identified as a mother is "a struggle of desire and recognition," as the death of her son is also the death of herself (Homer, 2005: 24)[38]. As her son dies, the alienating image of Mrs Musgrove perceived as a mother, dies along. Therefore, the "profitable" Dick Musgrove, "whom alive nobody had cared for," has become the more amiable "poor Richard."

4.2 Mrs Musgrove as Sphinx

4.2.1 Mrs Musgrove's self-identification

Now to Captain Wentworth in the scene, Mrs Musgrove, screening Anne Elliot, has become the Sphinx that is, in Harold Bloom's term, "natural" and "sexual anxiety" (Bloom, 1997: 39)[39]. "The tripartite structure" of Captain Wentworth, Mrs Musgrove (in the middle), and Anne Elliot, creates "an intricate and eroticized interplay of desires and inhibitions" (Sigler, 2019: 47)[40]. Anne starts to be reconciled with Wentworth, according to Blackwell, for "unburdening her body from the maternal role" and relaxing her nerves with the examples of "the childless Crofts" (Blackwell, 2004: 160)[41]. And blocking Anne, as any Sphinx would do, Mrs Musgrove is also a muse to Captain Wentworth, with a "bulky figure" which bearing less motherhood now bears more flesh, the dislodging of the Sphinx must inspire him.

According to the text, Captain Wentworth "attending" such "large fat sighings" of a woman less fitted "by nature" to express "tenderness and sentiments." Such "large fat sighings" can be considered sexual, for it is a most vivid sensual expression, to which "a large bulky figure has as good a right... as the most graceful set of limbs in the world", that are, as described, in need of "self-command" to "attend to." Erotic and sexual as they are, the sighings, which are the presence of a man, seem to locate in the gap between autoeroticism and heterosexual coitus and can best identify with masturbation, which nevertheless "seems to challenge the historicizing impulse" (Sedgwick, 1991: 820)[42]. Mrs Musgrove here is one perfect example of, in Sedgwick's term, "The Muse of Masturbation" (Sedgwick, 1991: 818)[43]. The masturbation of the mother, who has lost a son, is set free from the narrative of reproduction, and therefore seems to affine with "ahistorical or history-rupturing rhetorics," which is a challenge to the history as which the heterosexual history "masquerades so readily" (Sedgwick, 1991: 820)[44].

On Lacan's thought experiment about prison and identification, Derek Hook remarks, "I can have no full understanding of what I 'objectively' am" (Hook, 2013: 128)[45]. And so cannot Mrs Musgrove, whose notion of "self" relies upon the alienated image others perceive; her identity as a mother is truly at the mercy of others.

4.2.2 Mrs Musgrove as a muse with her riddle

When the history of reproductive heterosexuality is challenged by the masturbation of the woman no longer fertile, which is "no insignificant barrier indeed," Captain Wentworth now, like Oedipus, faces the blockage of Sphinx. However, unlike Oedipus, "who did not know that the Sphinx was his Muse," Captain Wentworth, attending to her sighings, realizes that the answer to the riddle-like sighings, "non-motherhood by nature," is also what the Muse offers, as it is put forward that a man's desire is activated only in a triangular relation to the desire of other people (Bloom, 1997: 39; Michalski, 2021: 78)[46].

But this is not the only thing she offers, that prevents him from a tragic fate, like the one of Oedipus. A large part of Oedipus' tragedy can be attributed to his Cartesian cogito, through which he fails to know the world (Yan, 2021: 120)[47]. Captain Wentworth, however, avoids such a fate when he carefully listens to the sighings, "which reason will patronize in vain." As the consequence of the omniscient power of reason being challenged, the authority of the Cartesian cogito is also challenged and the social norms, which should regulate society, can no longer be trusted. The denial of the proper functioning of reason challenges the world where everything is supposed to be orderly and though oppressively, meaningful. When the meaning of grand narrative is deprived by the ill-functioning reason, the world seems to, to a certain extent however minor, turn into a Bakhtinian carnival that "scrambles distinctions and revels in the collapse of meaning" (Eagleton, 2019: 10)[48].

Therefore, in a way, the fool has succeeded in adding to the world the touch of carnival, and as limited as the influence of one so ridiculed may be, she provides "a momentary respite" from suffocating blandness (Eagleton, 2019: 10)[49].

5 Conclusion

It has become a truth universally acknowledged that Austen's novels are closely associated with Lacanian studies and in desperate want of discusses the application of Lacanian theory in the analysis of Jane Austen's novel, *Persuasion*. The double triangular structure, one that consists of Mrs Musgrove, Anne Elliot, and Captain Wentworth, which is in another of Mrs Musgrove, the reader, and the narrator, fits perfectly into another triangular structure of the joke-teller, listeners, and the butt of the joke, which provides a rich context where multiple perspectives on the interpretations of Austen's power. Austen delicately arranges the triple triangular structures, so that readers could directly interact with characters through identification. The application of Lacanian theories has proved itself to push the Austenian studies beyond the mere moralist value of restraint and domestic obligation.

The idea of incongruity and the ambivalence between the physical and spiritual embedded in the "large fat sighings" of Mrs. Musgrove, disrupt the regularity of the rhythm and reinforce the oddity of having both body and spirit. The author also discusses Mrs. Musgrove's role as a mother and the sexual anxiety she creates as a Sphinx-like figure, blocking Anne Elliot's slenderness and inspiring Captain Wentworth. Finally, the author explores Mrs. Musgrove's identity as a muse of masturbation and how her notion of self relies upon the alienated image others perceive. The incongruity of the large fat sighings has succeeded in shaping and bestowing readers, characters, and the narrating consciousness, new evolving identities, onto which the past and the future project.

ACKNOWLEDGMENTS

This study is supported by the Talent Introduction Project of Ningbo Tech University, project code: 20230322Z0049.

References

1. Johnson, Claudia L. Austen Cults and Cultures[A]. In Edward Copeland and Juliet McMaster. The Cambridge Companion to Jane Austen[C]. Cambridge: Cambridge University Press, 1997.
2. Fu, Xiaohui. The Victorians' Uses of Austen: A Review on Jane Austen and the Victorian Heroine [J]. New Perspectives on World Literature, 2018, No.282(06): 64-71.
3. 4. Geng, Liping. On the Dialectical Views of Love in Persuasion: A Case Study of "Emotional Literature and National Morality"[J]. Journal of PLA Academy of Arts, 2011(02): 58-61.
5. Garofalo, Daniela. Abandoned by Providence: Loss in Jane Austen's Persuasion [A]. In Daniela Garofalo (ed.). Lacan and Romanticism[C]. Albany: State University of New York Press, 2019.
6. 12. 46. Michalski, Isabelle, and David Sigler. Logical Time in Austen's Persuasion: Desire and the Unproductive Anxious Interval[A]. In Michael Kramp (ed.). Jane Austen and Critical Theory[C]. New York: Routledge, 2021.
7. 17-18. 23. 40. Sigler, David. Jane Austen: Comedy Against Happiness[A]. In Erin Goss (ed.). Jane Austen and Comedy[C]. Lewisburg: Rutgers University Press, 2019.
8. 20-22. 26. 48-49. Eagleton, Terry. Humour[M]. New Haven and London: Yale University Press, 2019.
9. 32. 39. 46. Bloom, Harold. The Anxiety of Influence: A Theory of Poetry[M]. New York: Oxford University Press, 1997.
10. Gilbert, Sandra, and Susan Gubar. The Madwoman in the Attic: the Woman Writer and the Nineteenth-Century Literary Imagination [M]. New Haven and London: Yale University Press, 2000.
11. Zhang, Sumei. The Printed Publication and Austen's Dual Cultural Identity Construction[J]. Foreign Literatures, 2019, No.155(03): 59-69+157-158.

13. Austen, Jane. *Persuasion*[M]. London: Penguin Group, 2012.
14. Mudrick, Marvin. Jane Austen: Irony as Defense and Discovery[M]. Princeton: Princeton University Press, 1952.
15. Heydt-Stevenson, Jill. Austen's Unbecoming Conjunctions[M]. New York: Palgrave Macmillan, 2005.
16. Orwell, George. As I Please 1943-1945: The Collected Essays, Journalism and Letters of George Orwell, Volume 3[M]. New York: Harcourt Brace Jovanovich, Inc., 1968.
19. Zupančič, Alenka. The Odd One In: On Comedy[M]. Cambridge, MA: MIT Press, 2008.
24. Buijzen M., Valkenburg Patti M.. Developing a Typology of Humor in Audiovisual Media[J]. Media Psychology, 2004, 6(2): 147-167.
25. Critchley, Simon. On Humour[M]. London: Routledge, 2002.
27. Lewis, Wyndham. The Wild Body[M]. New York: Harcourt, Brace and Company, 1928.
28. Bergson, Henri. Time and Free Will: An Essay on the Immediate Data of Consciousness[M]. New York: Dover Publications, Inc, 2001.
29. Bevis, Matthew. Comedy: A Very Short Introduction[M]. Oxford: Oxford University Press, 2013.
30. Nodeh, Soghra. Dialogic Narrative Discourse in Austen's Emma: A Bakhtinian Review[J]. k@ta, 2013, 15(1): 1-8.
31. Bakhtin, Mikhail Mikhaĭlovich. The Dialogic Imagination: Four essays[M]. Austin: University of Texas Press, 1981.
33. 34. Schopenhauer, Arthur. The World as Will and Idea Volume II[M]. Boston: Ticknor and Company, 1887.
35. 38. Homer, Sean. Jacques Lacan[M]. London and New York: Routledge, 2004.
36. Canetti, Elias. Crowds and Power[M]. New York: Viking Press, 1962.
37. 41. Blackwell, Bonnie. A Slender Girl and a Thin Plot: Starvation and Spectacle in Persuasion[J]. Literature Interpretation Theory, 2004, 15(2): 153-180.
42. 43. 44. Sedgwick, Eve Kosofsky. Jane Austen and the Masturbating Girl[J]. Critical Inquiry, 1991, 17(4): 818-837.
45. Hook, Derek. Towards a Lacanian Group Psychology: The Prisoner's Dilemma and the Trans-subjective[J]. Journal for the Theory of Social Behaviour, 2013, 25(2): 115-132.
47. Yan, Di. The Quarrel between Poetry and Philosophy in Oedipus Tyrannus[J]. Foreign Literature Review, 2021, No.139(03): 112-130.

An Analysis of the Protection and Inheritance of Intangible Cultural Heritage from the Perspective of Cultural and Creative Industries: A Case Study on the "Huojing" in Bashu

Xili Gan*

Chengdu Polytechnic, Chengdu, Sichuan, China

Abstract: Taking the Intangible Cultural Heritage of the "Huojing" in Bashu as an example, this article discusses the protection and inheritance of intangible cultural heritage under the perspective of cultural and creative industries. Intangible cultural heritage is a product of the agricultural civilization era, which is currently incompatible in the era of high-speed development of the Internet and digital economy, leading to difficulties in its protection and inheritance. Cultural and creative industries are emerging industries in modern times, which integrate with other industries through unique innovative forms and operational methods. It is particularly important to fully tap cultural resources and transform them into a source of creative industry to enhance competitiveness. Intangible cultural heritage provides cultural materials and creative sources for the cultural and creative industries. In turn, the cultural and creative industries bring unprecedented development opportunities to intangible cultural heritage, promoting economic benefits, providing innovative mechanisms, and a platform for integration into modern society.

1.Introduction

The earliest known written record of the "Huojing" dates back to the Han Dynasty in China. The ancient Shu people used this traditional handicraft and intangible cultural heritage skill to dig wells for salt and cook with natural gas, as well as to smelt metals. This was a great cultural achievement of the Shu people in terms of cultural transmission and innovation. In today's technologically advanced society, this emblematic symbol of BaShu culture deserves our attention for further research, protection, and inheritance. This article will use the cultural heritage of the "Huojing" in Linchuan as a background, and the earliest recorded use of natural gas for production and daily activities in the BaShu "Huojing" culture as a case study. By combining intangible cultural heritage with cultural creativity, the article will analyze how to enhance the expansion of intangible cultural heritage, explore how to broaden the path of inheritance, and extract how to integrate traditional forms with modern resonance, integration and advancement. This article will also explore the path to further active protection of intangible cultural heritage through market orientation, re-creation, and re-invention.

2.The "Huojing" in Bashu is an important cultural heritage.

2.1 The historical value of the "Huojing" in Bashu.

The recorded history of the "Huojing" in Bashu refers to the burning phenomenon of natural gas wells in ancient texts. The term "Huojing" actually refers to a general term for a cluster of wells, which includes not only simple gas wells, but also wells that extract both water and gas or oil and gas. The ancient Chinese technique of deep well excavation was applied, mainly for the extraction of brine water or salt wells. Later, the development progressed to include the extraction of natural gas, known as "Huojing." The history of "Huojing" can be traced back to at least the Han Dynasty. In the Western Han Dynasty, the "Huojing" in Linqiong were used to produce salt, which is the earliest record of natural gas development and utilization in the world. According to historical documentation on the development of science and technology, the earliest recorded use of natural gas occurred in England in 1668, but "Huojing" preceded it by more than 1600 years.

Mr. Ren Naiqiang, the master of the collection of Huayang Guozhi, said in the "Annotations to the Huayang Guozhi School": "(Linqiong) Huojing field, in the second century BC there was a Huojing. Those who built salt wells in Pichi dug out natural gas. ...... This

*E-mail: ganxili419@qq.com

shows that 2,000 years ago, natural gas was already available in China's Bashu region, and the historical documents containing this scientific miracle are very precious. [5] "Second century BC" and "2,000 years ago" refer to the time period of the Qin and Han dynasties, and Sichuan in the Qin Dynasty probably already had a "Huojing". According to Zhang Hua's "Natural History" of the Western Jin Dynasty: "There is a Huojing in Linqiong, five feet in width, two or three feet deep, and the well is in the south of the county, and in the past, people used bamboo and wood to throw fire. Zhuge Prime Minister looked at each other, and the fire turned stronger. Hold the pot and cover the well, boil salt to get salt. The fire was extinguished when it entered the house, and it will not be burned again. " [4] With the deep gouging of salt wells, natural gas buried in the ground with abundant content is naturally developed and utilized. Jin Ren Zuo Si's "Shudufu" has the sentence "The Huojing sinks in the dark spring, and the high smoke flies in the sky", which is the ancients' deliberate description of the Sichuan Huojing, which shows the prosperity of the Huojing in those years.

Yang Xiong, a native of Chengdu, was a leading figure in the national academic community during the late Western Han Dynasty. During the Yongshi period of Emperor Chengdi of Han (16-15 BCE), he wrote the "Shudu Fu," which includes the phrase "Tongliang Jintang, Huojing Longqiu." The term "Huojing" in this context refers to the Huojing town in Qionglai today. According to Liu Liang, a Tang dynasty scholar who annotated the "Shudu Fu," "There is a Huojing in Shudu, located southwest of Linyou County. Huojing also refers to salt wells. To light the Huojing, people first use household fire. After a moment, there is a rumbling sound like thunder, and it bursts into the sky with a radiance of ten miles. They then use a bamboo tube to collect the light without any charcoal." This passage suggests that during the Western Han period, people in Chengdu had already begun using natural resources, such as the natural gas from Huojing, to boil the brine from salt wells. This was a remarkable feat of ancient technology by the Shu people. Furthermore, numerous salt well paintings have been unearthed from Eastern Han tombs in Chengdu, such as the Yangzishan in 1954 and Huapaifang in Qionglai in 2014. These paintings vividly depict the widespread and busy scene of boiling salt using Huojing during that period. It is true that the "Huojing" of Huojing town was already in existence during the Western Han period, and it is not a mere falsehood that the town boasts the "world's first well." From this origin, in the Sichuan region, numerous well groups, gas wells, oil gas wells, and salt wells have subsequently developed.

2.2 Intangible Cultural Heritage in the "Huojing" Culture of Bashu

The Sichuan Basin was an inland lake millions of years ago, and large amounts of saltwater and natural gas are still stored underground. Sichuan was the earliest region in China to excavate salt wells. The "Huojing" cultural heritage in the Sichuan-Chongqing region refers to the exploration, excavation and utilization of wells, and includes the application of multiple intangible cultural heritage technologies. Dr. Joseph Needham of the UK believes that China was not only the first country in the world to develop natural gas, but also that the drilling and excavation techniques used in oil exploration today are definitely inventions of the Chinese, with much evidence to support its use as early as the first century BC to the first century AD in Linchuan, Sichuan. According to "The Chronicles of Huayang • Chronicles of Shu," Li Bing, who served as the Governor of Shu in the period of Emperor Xiaowen of the Qin Dynasty (around 250 BC), "recognized the water veins of Qi, bored salt wells in Guangdu, and various reservoirs and ponds. Shu then flourished with abundant health." "In the third year of Emperor Xiaoxuan's reign, the Wenshan commandery was abolished and a northern military commander was appointed. Twenty salt wells were then dug in Linchuan and Pujiang, and salt-iron officials were added." "The Chronicles of Huayang" records Linchuan County: two hundred miles southwest of the county. Originally inhabited by the Qiang people, they were relocated to Shang County by Emperor Qin Shihuang. There is the Bu Pu River, which joins the Huojingg River. There is the Huojingg, whose light shines up to the sky at night. The people were fond of the light, and threw their home fires into it. The flames roared for a while, and the fire burned for dozens of miles. The gas was contained in bamboo tubes and pulled for a whole day without going out. The well has two types of water. When the well water is boiled with the fire, one bucket of water can produce five dou of salt. When boiled with household fire, there is not much salt produced. There is an ancient Shishan mountain and a stone mine, with stones the size of garlic. When the stones are burned, they become flowing cast iron, which is very strong. An iron official was appointed as a result. There is an ancestral temple for iron, and during the reign of Emperor Wen of the Han Dynasty, copper and iron mountains were given to Deng Tong, a minister. Deng borrowed money from the wealthy Zuo Wangsun and received a thousand horses a year. Therefore, Wangsun's fortune accumulated to billions, while Deng's money was spread throughout the country. Wangsun's daughter, Wenjun, could play the qin. At that time, there was a minister named Sima Changqing, who visited Wangsun's house with the county magistrate Wang Ji. Wenjun then eloped with Changqing. During the Han Dynasty, the county residents Chen Li served in Ba commandery, Zangke, and Tianshui as the governor with unique governance. The Chen and Zheng families are also known for their large families.

At the end of the war, a number of wells, including the "Guangdu Salt Well" in Western Shu, were successfully drilled, and their application served as a precursor to the modern large-scale development and utilization of oil and natural gas, occupying an important position in the history of new energy utilization in the world. Looking back at Su Shi's record of the production process of the Zhuotong well in the "Shu Salt

Commentary", "Hewn with round-edged chisels like a bowl, and several tens of Zhang deep; with giant bamboo to remove the knots, male and female ends connect to form a well, and with a brine barrier to separate the fresh water, the saltwater spring rises up." It has been verified that the drilling technology of the Zhuotong well was more than 800 years earlier than that of the West. The drilling technology of the Zhuotong well is also recognized as the "father of world oil drilling" and is hailed as the "fifth great invention of ancient China." The invention of this drilling technology not only greatly improved the level of salt production, but more importantly, during the process of mining brine, humans discovered the deep-buried oil and gas resources, which opened a brand new page for the world energy revolution. The Zhuotong wells in various salt areas of Sichuan include salt wells, oil wells, and water-oil wells. The production technology of the Zigong well salt deep drilling and extraction (traditional production technology of the Zigong well salt) is one of the intangible cultural heritage projects in China. It includes drilling design procedures, pre-drilling preparations, drilling, repairing, salvaging, gas-brine identification, and displaying premonitory signs of gas-brine resources during drilling, ultimately forming a deep salt well. Therefore, technologies such as Li Bing's large mouth well salt drilling technology, well salt production technology, Zhu Wangsun's metal smelting technology during Emperor Hanwen's reign, and the well salt technology of the Zhuotong well invented in the Northern Song Dynasty, which involve drilling tools, drilling equipment, clear hole tools, salvage tools, and sweep hole tools... the development and use technologies of these hundreds of types of equipment are all classified as intangible cultural heritage projects in the "Huojing" culture of Bashu.

2.3 Examining the Current State of the Intangible Cultural Heritage Protection and Inheritance through the Perspective of the Ancient Town of Huojingg.

Based on the above data and analysis, it can be concluded that the Huojingg Cultural Heritage and the ancient town of Huojingg where it is located is recognized as the earliest place in the world to discover and use natural gas. This is a town with both cultural history and technology. The ancient town of Huojingg was founded during the Northern and Southern Dynasties period of the Northern Zhou dynasty (557-581 AD), with a county history dating back to 689 AD. It is a civilization ancient town with a history of nearly 1500 years. Today, several local households in the ancient town continue to use the traditional method of transporting natural gas through bamboo pipes to boil water, cook food, and make rice. In the nearby cornfields of residents' homes, there are still large stone jars measuring two meters square covered with a diameter of more than one meter, large wooden barrels with a diameter and height of more than one meter, and large iron pots with a diameter of nearly two meters, which are used for separating saltwater and natural gas. The technology process of using fire to boil salt from wells in Huojingg Town 2000 years ago is reenacted. Many of these unknown and neglected intangible cultural heritages that are on the verge of being lost are precisely found in the folk. They are so low-key that their existence is almost imperceptible, but their historical and cultural connotations far exceed many famous ancient towns.

Currently, the spread and protection of intangible cultural heritage (ICH) in different regions of China exhibit uneven development, and this issue is also prominent in the Bashu region where the "Huojing" ICH culture is situated. Some of the factors contributing to this imbalance include the uneven development of tourism in different areas, unequal approval rates of ICH projects, and uneven dissemination pathways for ICH platforms. Among these factors are examples such as the "Zigong brine deep drilling and pumping technology," which has been listed in China's National List of Intangible Cultural Heritage, and the establishment of Zigong Salt Industry Historical Museum, as well as attractions such as the Zhoutongjing Site Park in Da Ying County, which has been rated as a 3A national tourist attraction. However, there are still underdeveloped sites such as the deserted Lincang "Huojing" ancient town, the Baiyun Salt Well Site, and the Chengdu Ancient Salt Well Site. Taking Chengdu’s ancient town as an example, incomplete statistics have shown that people's enthusiasm for the town is as high as it is for Sichuan hot pot. Chengdu's ancient town attracts those who appreciate the nostalgic atmosphere and tranquility of life. Apart from the Huanglongxi, Luodai, Anren Shuguo area, there are over twenty small and large ancient towns in Chengdu. The following table shows the real-time ranking of visitor traffic in Chengdu’s ancient town attractions, (data taken from the internet):

Table 1: Ranking Table of Real-time Visitor Traffic in Chengdu Ancient Town Scenic Spots that are Captured in Real-time.

Ranking	Name	Footfall index	Traffic congestion index
2	Kuanzhai Alley Ancient Street Quarter	8.02	2.189
4	Jiezi Ancient Town	7	5.859
8	Pingle Ancient Town	6.2	9.393
10	Jinli Street	4.52	3.2
12	Luodai Ancient Town Scenic Area	4.08	1.186
13	Sandaoyan Scenic Area Ancient Town	3.97	2.519

18	Wenshu Monastery Ancient District	2.72	4.053
21	Huanglongxi Ancient Town	18.67	6.451
35	Dayi Liu's Manor Ancient Town	0.44	1.412
40	Luodai Ancient Town	0.02	1.211

Based on real-time data, it is evident that the public has a high degree of affection towards ancient town scenic spots. However, some hidden gems like Huojingg Ancient Town did not appear in the ranking. Should these undiscovered ancient towns be neglected? Of course not. However, with the increasing number of ancient towns, the issue of homogenization arises. The inherent image of ancient towns in China, the superficial indulgence in food, drink and entertainment, and the thousands of years of historical charm are still unknown to many. To protect and inherit intangible cultural heritage in the future and to seek differentiation in the development of ancient town tourist sites, new development models need to be pursued, and individualized survival principles found. How can we use the new innovative and inheritable ways of cultural and creative industries to carry forward the less popular intangible cultural heritage and make the cultural phenomenon of "Huojingg" in Bashu "a hot spot"?"

3.The mutual value exploration between cultural creativity and intangible cultural heritage.

3.1 The Significance of Cultural and Creative Industry Development of Intangible Cultural Heritage

Intangible cultural heritage is a brilliant achievement that has been accumulated and created through the long history of various peoples and ethnic groups in their life and production practices, and has been passed down from generation to generation. It is the cultural imprint and wisdom crystallization of a nation. The exquisite traditional skills, profound ethnic thinking, and unique cultural genes contained in intangible cultural heritage are exactly what is lacking in the current period of globalization homogenization. However, intangible cultural heritage is a product of the agricultural civilization era. Although it has re-entered human vision, it seems incompatible with modern economic and social development, and its development and inheritance have encountered obstacles and difficulties. Since the rise of the cultural and creative industry, it has influenced the economic and cultural development of many countries and regions in the wave of globalization, and has a wide and complex connection with other industries in unique forms and operating methods. Culture is the cornerstone and carrier of the cultural and creative industry, and is a valuable resource that precipitates unique cultural heritage. It is necessary to fully tap cultural resources and transform them into the power source of creative industries to improve their competitiveness. Intangible cultural heritage provides cultural materials and creative sources for the cultural and creative industry. The cultural and creative industry also brings unprecedented development opportunities for intangible cultural heritage, providing it with innovative mechanisms and platforms for integration into modern society.

By combining the two, the creative value of intangible cultural heritage can be explored and transformed into creative capital, not only enhancing the competitiveness of the cultural and creative industries, but also endowing the intangible cultural heritage with new vitality. Today, people are beginning to value cultural creativity, and the value of local culture is gradually being excavated. The "Huojing" culture, which has an important historical status in the BaShu region, can serve as a new carrier for the protection and inheritance of Chengdu's non-heritage culture. While emphasizing the protection of the non-heritage industry, innovations in cultural inheritance can also be made, achieving the goal of "serving Chengdu, and Chengdu serving" local cultural and creative industries, thereby increasing opportunities for integrated development and economic growth. Against the backdrop of the world cultural industry environment, various industries are now turning their attention to the research and development of cultural and creative products to better spread and promote the influence of the "Huojing" non-heritage culture. Through data analysis and research, it has been discovered that there are almost no "Huojing" cultural and creative products, and similar cultural and creative products related to Zigong Salt Well, the Dahua Zhoutong Well, etc. are also relatively rare. The key issue is that they fail to meet the needs and demands of contemporary consumers. In the future, in order to explore new directions for the innovative inheritance of "Huojing" non-heritage culture, cultural and creative products in the BaShu culture should focus on expressing cultural and storytelling elements in their content, while incorporating modern fashion elements, increasing innovation, practicality, experience, and enjoyment in their design, and catering to the needs of the public, more effectively developing cultural and creative products that meet market demand.

3.2 The development path of the cultural and creative industry of the 'Huojing' of Bashu.

3.2.1 Concept for cultural and tourism industry development incorporating the "Huojing" symbol

The fusion of cultural heritage with creativity as its core is a wonderful way to protect intangible cultural heritage in a dynamic manner. Creativity and inheritance are not contradictory, and integrating creativity into modern life is also one of the important ways for the development of handicraft-related intangible cultural heritage. The extension of the industrial chain of the cultural and creative industry is a relatively important aspect of the protection of intangible cultural heritage. The combination of culture and creativity can not only bring about a certain amount of economic development, but more importantly, it can connect the "tradition and modernity" of local culture, promoting intangible cultural heritage to people's lives, allowing it to flourish while also being better protected. To protect and inherit the "Huojing" intangible cultural heritage of Bashu and to plan the future of the cultural and creative industry chain of ancient Shu civilization. By telling the Chinese story and ancient civilization story of Bashu's "Huojing", expressing it through international and modern digital creative means, and letting the world understand and recognize the culture of Bashu's "Huojing", we can strive to create a series of creative works that are valuable, meaningful, and marketable for the international market, ultimately achieving the cultural "going global" strategy.

The development of cultural and creative products for the "Huojing" in Bashu should be closely integrated with the local characteristic resources of the "Huojing". The focus should be on enhancing the tourist experience rather than simply increasing the quantity of cultural and creative products. Therefore, it is possible to create a concept that combines the reshaping and incubation of "Huojing" culture with heavyweight intellectual property that can be marketed globally. In terms of cultural protection, it is important to continue to explore the cultural connotations of "Huojing" and strengthen the protection of special non-legacy ancient towns, in order to present a cultural and tourism consumption scene that retains the original ecological features of these towns. Regular on-site research by experts and greater attention and support for the guardians, inheritors, and beneficiaries of intangible cultural heritage can also contribute to cultural preservation. In terms of cultural creativity, a continuous inheritance can be achieved through the use of digital cultural and creative media, such as animation, film, and short videos, as well as related products like stationery and toys, special cultural and creative museums, and theme parks. Exclusive leisure cultural industries such as "Huojing" and "Bashu Civilization" can be developed, including the creation of exclusive "escape room" dramas. Currently, cultural hot spots found in the "Huojing" ancient town include the "World's First Well of Fire", "Zhuge's Well", "Wen Jun's Hometown", and "Huojing Tea". Cultural events such as the "worship of the fire god ceremony," the Huojing town "driving day" activity, and the Huojing town forest fire warning day promotion activity have also been held. Other historical cultural hot spots that could be associated with the development of "Huojing" culture in the future include the "Most Colorful in the World - Qionglai Kilns," and "Huang Chongbao, the first female top scholar in China, hometown." Coordinated development of scenic spots along the Bashu civilization line is possible, which could promote a sustainable cultural tourism brand.

3.2.2 A Design Practice Plan for the "Huojing" Cultural and Creative Product

Below is an extraction of the inspirations from the "Huojing" culture, to plan and design an experiential tourism and educational cultural and creative product. The conceptual design of this cultural and creative product is a "Huojing Salt Making" experience toolkit. The tourists participating in the educational tour will understand the history and culture of the "Huojing" through 1 to 2 hours of cultural learning. They will personally make a handmade product using this toolkit, which can deepen their understanding of the cultural connotations of the "Huojing" through modern interpretation, thus achieving the goal of cultural heritage. The exterior design of the handmade kit adopts the pattern excavated from the Han dynasty salt well depicting the process of brine extraction and heating of Sichuan salt, using illustration to illustrate the scene on the packaging box of the toolkit. The interior of the handmade kit contains materials related to the production of salt blocks and ceramic image bricks, such as salt grains and clay. The kit comes with bamboo tubes, bamboo knives, and tracing paper etc. The instructional manual is produced in paper and includes new media scanning codes, providing details on the "Huojing" culture, salt block making, and image brick tracing processes. By organizing online and offline creative markets or other forms of business, a promotional platform is established to comprehensively showcase and expand the charming non-material cultural heritage of "Huojing," thereby enhancing its attractiveness and achievement conversion ability.

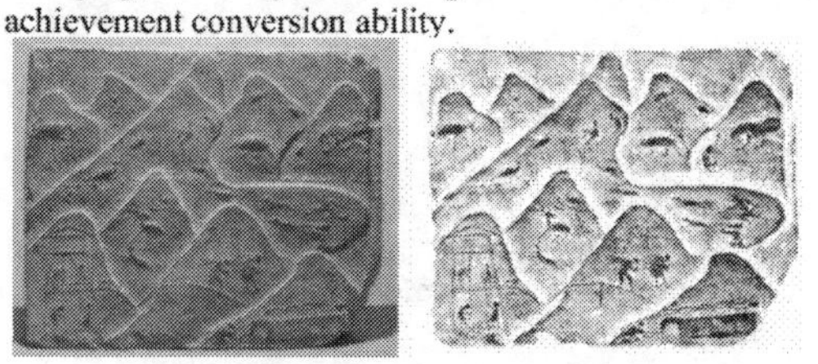

Figure 1. The pattern on Han Dynasty brick and rubbing of the image from a salt well (Huojing) in Eastern China.

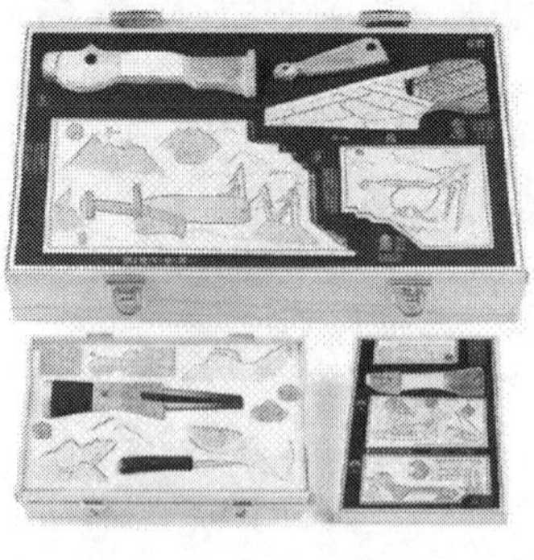

Figure 2 shows the effect of the product design package for the "Huojing salt-making" cultural and creative experience tool.

https://doi.org/10.1051/shsconf/202316702021

Figure 3 shows the promotional design for the "Huojing" cultural and creative market.

4.Conclusion

In the context of the global cultural industry, how to better promote and enhance the influence of the "Huojing" intangible cultural heritage. Creative tourism products are not only for profit, but also contain the humanistic feelings of a local area. By integrating culture into life through design, the essence of culture is reflected, providing spiritual needs for tourists and becoming a good carrier of memories. As a derivative of tourism, creative tourism products can not only promote the development of tourism, but also expand the reputation of the destination, becoming a "silent" business card.

Ultimately, industries pursue the maximization of profits. If the intangible cultural heritage is overly commercialized and excessively developed, it will also have an adverse impact on the protection of intangible cultural heritage. Therefore, a monitoring and evaluation system for the protection of intangible cultural heritage and a "cultural warning" mechanism should be established when establishing the cultural and creative industry. Related departments should take certain protective measures before the development and utilization of intangible cultural heritage, which can help to minimize damage and play a positive role in protecting and inheriting intangible cultural heritage. Intangible cultural heritage is not only a witness to history, but also a valuable cultural resource with significant value. Protecting intangible cultural heritage is not only a requirement of the country and the nation, but also a necessary requirement for the sustainable development of human society. This paper analyzes and explores the protection and inheritance modes of intangible cultural heritage in the context of cultural and creative industries using the background of the "Huojing" in Bashu. However, more people are needed to explore new models that are more suitable for each type of intangible cultural heritage.

REFERENCES

1. Huojingg Ancient Town - Baidu Baike: https://baike.baidu.com/item/Huojing Ancient Town/7811034?fr=aladdin
2. Huojingg - Baidu Baike: https://baike.baidu.com/item/Huojing/630448?fr=aladdin
3. Ban Gu: "Jiaoshi Zhi" from "Hanshu", page 1035; "Dilizhi" from "Hanshu", page 1295, published by Zhonghua Book Company in 1999. [4] Zhang Hua: "Bowuzhi", cited in "Taiping Yulan" Volume 869, published by Zhonghua Book Company in 2011.
4. Ren Naiqiang: "Huayang Guozhi Jiaobu Tuzhu", published by Shanghai Ancient Books Publishing House in 1987.
5. 6. Su Shi: "Shu Yan Shuo".

AUTHOR INDEX

Axu, Hu ... 126
Binzhe, Yang ... 1
Caiyu, Zhang ... 121
Chuan, He ... 171
Cui, Na ... 135
Dingtai, Li ... 158
Elena, Ng ... 5
Enhui, Xing ... 131
Fenglin, Xing ... 182
Guanwen, Diao ... 79, 163
Hao, He ... 5
Huangfu, Shanghua ... 45
Huhe, Qiao ... 66
Huiwen, Zhang ... 70
Huiwen, Zhao ... 79
Jiaqi, Chi ... 139, 144, 148
Jie, Liu ... 89
Jin, Lin ... 171
Jinhao, Li ... 89
Jiongfeng, Song ... 171
Kaixiang, Cui ... 49
Keke, Zhang ... 178
Lei, Liao ... 32
Ling, Li ... 135
Long, Zhang ... 84
Mengyang, Wei ... 196
Mingcen, Zhao ... 144
Mingyang, Liu ... 15
Nong, Tian Yi ... 163, 167
Peiying, Gu ... 40
Qi, Li ... 24
Shijian, Li ... 94
Shiran, Zhou ... 53
Shu, Liu ... 126
Shuang, Shi ... 171
Shuzhen, Ma ... 126
Sile, Hu ... 126
Siqi, Niu ... 139, 148
Surng, Gahb Jahng ... 84, 89
Szu-Ming, Chung ... 75
Tao, Zhao ... 94
Tingwei, Liang ... 158
Wei, Zhou ... 61
Wen, Zhao Hui ... 167
Wenchao, Wang ... 139, 148
Wenqian, Liu ... 113
Wenrui, Liang ... 40
Xianlu, Li ... 121
Xiaochun, Yang ... 108
Xiaomei, Nie ... 113
Xili, Gan ... 202
Xiujuan, Zhang ... 94
Xuan, Wang ... 121
Xueying, Su ... 10
Yang, Lu ... 36
Yi, Li ... 182
Yimeng, Zhang ... 20
Yining, Luo ... 100
Yue, Zheng ... 121
Yukang, Zhang ... 196
Yulin, He ... 28
Yunbo, Niu ... 187
Yunning, Wang ... 57
Zecheng, Li ... 84
Zhangyan, Li ... 104
Zhe, Yu ... 75
Zixuan, He ... 152